T0135238

From Signals to Image

Haim Azhari • John A. Kennedy • Noam Weiss
Lana Volokh

From Signals to Image

A Basic Course on Medical Imaging for Engineers

Haim Azhari
Department of Biomedical Engineering
Technion – Israel Institute of Technology
Haifa, Israel

John A. Kennedy
Department of Nuclear Medicine
Rambam Health Care Campus
Haifa, Israel

Noam Weiss
Department of Biomedical Engineering
Technion – Israel Institute of Technology
Haifa, Israel

Lana Volokh
Exoprother Medical
Haifa, Israel

ISBN 978-3-030-35328-5   ISBN 978-3-030-35326-1   (eBook)
https://doi.org/10.1007/978-3-030-35326-1

This Springer imprint is published by the registered company Springer Nature Switzerland AG
The registered company address is: Gewerbestrasse 11, 6330 Cham, Switzerland

Acknowledgments and Dedications

I thank G_d for helping me publish this (my third) professional textbook and for making me meet the many good and talented people, colleagues, and students who have helped me learn this fascinating field. As Rabbi Hanina states in the Talmud, (Taanit 7:70a):

> I have learned a lot from my teachers, more than that from my friends, but most of all from my students.

Many, many thanks are due to my beloved and supportive family.

I would like to dedicate my part in this book to the memory of my late brother-in-law Jack Helfgott and to the memory of my beloved father Gabriel Joseph. May they both rest in peace.

Haim Azhari

To Myriam and Sima for their patience and to the loving memory of my grandfather, Konrad Planicka, a humble prince who took pride in being a devoted custodian.

– John A. Kennedy

To my family, with endless love and gratitude

– Lana Volokh

To my mentors over the years, who introduced me to this amazing field, which allows us to see beyond the visible. To the memory of my late father, Assaf Weiss, who kindled my curiosity, even as a small child. And most importantly, to my beloved family

– Noam Weiss

Preface

Over the past three decades, we have developed and taught an introductory medical imaging course at the Technion-IIT for senior engineering students and graduate students. Within the limiting timeline of a single university course, we attempted to lay the foundations needed for working and conducting research in this fascinating field of medical imaging. The goal has been to introduce the students to the most prominent modalities in current medical imaging, present the principles of image reconstruction and convey the very basic physical, engineering, and mathematical methods that underlie each one of them or are common to all. Consequently, the topics range from the physical phenomena used for data acquisition, the constraints posed to imaging device design, and the elementary mathematics of image reconstruction. Given this broad scope, the course provides the essential basics for further self-study or for advanced courses specializing in particular modalities. No less important, was the goal of conveying an enthusiasm for working with medical scanners, since our local industry has made substantial contributions to the field.

This textbook is a summary of our introductory medical imaging course for engineers. The authors are grateful to those who have produced excellent texts on various topics in medical imaging which are referenced throughout the text for students who need a more comprehensive understanding. By introducing the principles used in medical imaging scanners, this textbook provides students with a context for new developments in the field. The chapters dealing with X-ray and X-ray CT, gamma cameras, positron emission tomography, ultrasound, and magnetic resonance imaging (MRI) are generally structured to address the physical phenomena, signal sources, data acquisition, and image formation. Some clinical examples and examples of quality control are also provided. General concepts are presented in an introductory chapter and a chapter on image reconstruction. Exemplary questions for each chapter and final exams with their solutions provide a tool for evaluating the students' ability to apply the principles, with many questions based on practical issues with device design or operation in the clinic. We believe this textbook can greatly aid both instructors and engineering students in an introductory course on medical imaging.

We are grateful to all our friends and colleagues in academia, in the clinic, and in the industry who have contributed to our course development throughout these many years and for sharing images and knowledge which eventually led to this textbook. We trust that readers will enjoy it and find it useful.

Haifa, Israel   Haim Azhari

p.s.

Although we did our best, there are probably quite a number of things to correct and improve. We would highly appreciate any comments sent to me or John by E-mail. Kindly indicate: "Medical Imaging book - Comments" in the subject. Thank you in advance.
John: j_kennedy@rambam.health.gov.il
Haim: haim@bm.technion.ac.il

Contents

1 Introduction 1
  1.1 Historical Background and Motivation 1
     1.1.1 X-Rays and CT 1
     1.1.2 Nuclear Medicine: Gamma Camera, SPECT, and PET ... 3
     1.1.3 Ultrasound 4
     1.1.4 MRI 5
     1.1.5 Other Imaging Modalities 5
  1.2 Basic Definitions of Medical Imaging 6
     1.2.1 What Is a Medical Image? 6
     1.2.2 Mapping of an Object into Image Domain 8
     1.2.3 What Are the Required Stages for Developing
           an Imaging Modality? 10
  1.3 Basic Features of an Image 12
     1.3.1 Gray Levels and Dynamic Range 12
     1.3.2 Point Spread Function and Line Spread Function 16
     1.3.3 Image Resolution 18
     1.3.4 Spatial Frequencies 23
     1.3.5 Modulation Transfer Function (MTF) 29
     1.3.6 Contrast 29
     1.3.7 Image Noise and Artifacts 33
     1.3.8 Signal-to-Noise Ratio (SNR) and Contrast-to-Noise
           Ratio (CNR) 36
     1.3.9 Detectability 37
     1.3.10 The Receiver Operator Curve (ROC) 39
  1.4 Standard Views in Radiology 42
  References 43

2 Basic Principles of Tomographic Reconstruction 45
  2.1 Introduction 45
  2.2 Part I: Basic Principles of Tomography 47
     2.2.1 Projections 47

	2.2.2	Tomographic Data Acquisition: The "Sinogram" and the Radon Transform	52
	2.2.3	Back Projection	57
	2.2.4	Algebraic Reconstruction Tomography (ART): The Projection Differences Method	58
	2.2.5	Algebraic Reconstruction Tomography (ART): The Projection Ratio Method	63
	2.2.6	The Weight Matrix	63
	2.2.7	The Tomographic Slice Theorem	66
	2.2.8	"Filtered Back Projection" (FBP)	69
	2.2.9	The Needed Number of Projections	75
	2.2.10	Spiral CT	78
	2.2.11	Fan Beam Reconstruction	78
2.3	Part II: Advanced Reconstruction Algorithms		80
	2.3.1	Difference ART Convergence	80
	2.3.2	Advanced Multiplicative ART	83
	2.3.3	Statistical Reconstruction Approaches	84
	2.3.4	Other Image Reconstruction Approaches	92
References			93
3	**X-Ray Imaging and Computed Tomography**		**95**
3.1	X-Rays: Physical Phenomena		96
	3.1.1	X-Ray Tubes	96
	3.1.2	X-Ray Generation	100
	3.1.3	X-Ray Interaction with Body Tissues	101
3.2	Projectional Radiography		106
	3.2.1	Detection and Data Acquisition	106
	3.2.2	Image Formation	110
	3.2.3	Projectional X-Ray Types and Modalities	115
	3.2.4	X-Ray Contrast Agents	119
3.3	X-Ray CT		123
	3.3.1	Basic Concept	123
	3.3.2	CT Evolution	124
	3.3.3	Detection and Signal Sources	128
	3.3.4	Image Formation	130
	3.3.5	Axial Scans and Helical Scans	134
	3.3.6	Image Manipulation and Clinical Examples	137
3.4	Dosimetry and Safety		143
	3.4.1	Ionizing Radiation and Health Risks	143
	3.4.2	The ALARA Concept	149
	3.4.3	Quality Versus Safety	150
3.5	Emerging Directions		152
	3.5.1	Dual-Energy CT (DECT)	152
	3.5.2	Multi-energy CT and Photon Counting	154
Bibliography			157

4 Nuclear Medicine: Planar and SPECT Imaging 159
  4.1 Physical Phenomena 160
    4.1.1 Physics of Radioisotopes and Radioactive Decay 160
    4.1.2 Propagation and Attenuation of Ionizing Radiation 165
    4.1.3 Radiation Safety with Open Sources 167
  4.2 Signal Sources 170
    4.2.1 Emission Versus Transmission Scanning 170
    4.2.2 Concept of a Radiopharmaceutical 171
    4.2.3 Gamma Sources 173
    4.2.4 Bremsstrahlung Sources 175
    4.2.5 Radiotracer Production 176
  4.3 Data Acquisition 179
    4.3.1 Collimation 180
    4.3.2 Scintillation Crystals 181
    4.3.3 Photomultiplier Tubes 182
    4.3.4 Positioning Circuits 184
    4.3.5 Solid-State Detectors 187
    4.3.6 SPECT Rotating Gantry Methods 189
    4.3.7 SPECT Stationary Gantry Methods 190
    4.3.8 Gating 190
    4.3.9 Hybrid SPECT/CT Systems 192
  4.4 Image Formation 193
    4.4.1 Planar Imaging 193
    4.4.2 Dynamic Imaging 194
    4.4.3 Image Digitization, Storage, and Display 195
    4.4.4 SPECT Image Reconstruction 197
    4.4.5 Attenuation Correction 198
    4.4.6 Scatter Correction 201
    4.4.7 Resolution Recovery 203
    4.4.8 Post-Processing 205
    4.4.9 Hybrid Image Fusion 206
    4.4.10 Quality Control 207
  4.5 Clinical Example 209
  4.6 Challenges and Emerging Directions 211
  References 214

5 Positron Emission Tomography (PET) 217
  5.1 Physical Phenomena 218
    5.1.1 Radioactive Decay and Positron Emission 218
    5.1.2 Particle/Anti-particle Annihilation 218
    5.1.3 Coincident Photon Pairs 220
  5.2 Signal Sources 220
    5.2.1 Emission Versus Transmission Scanning 220
    5.2.2 F-18 FDG: [^{18}F]-2-Fluoro-2-deoxy-D-glucose 221
    5.2.3 Other Positron Emission Radiotracers 221

5.3   Data Acquisition..................................... 223
        5.3.1   Physical Collimation: Septa...................... 225
        5.3.2   Electronic Collimation: Coincident Photon Detection. . . 225
        5.3.3   3D Acquisition............................... 228
        5.3.4   Time-of-Flight Acquisitions..................... 229
        5.3.5   Calibration................................. 231
        5.3.6   Noise Equivalent Count Rate.................... 232
        5.3.7   Solid-State Methods.......................... 233
        5.3.8   Hybrid PET/CT Systems....................... 234
5.4   Image Formation..................................... 235
        5.4.1   Tomographic Principle......................... 235
        5.4.2   Image Reconstruction......................... 235
        5.4.3   Attenuation Correction........................ 236
        5.4.4   Scatter Correction............................ 239
        5.4.5   Point Spread Function Modelling.................. 240
        5.4.6   Post-Processing.............................. 240
        5.4.7   Quantitation................................ 241
        5.4.8   Hybrid Image Fusion.......................... 243
        5.4.9   Quality Control.............................. 244
5.5   Clinical Example.................................... 246
5.6   Challenges and Emerging Directions...................... 247
References... 249

6   Magnetic Resonance Imaging (MRI)......................... 253
6.1   Introduction....................................... 253
6.2   The Nuclear Magnetic Resonance (NMR) Phenomenon........ 254
6.3   Associated Physical Phenomena......................... 256
        6.3.1   Magnetic Fields.............................. 256
        6.3.2   Magnetic Susceptibility: The Response of Matter
                 to Magnetic Fields........................... 258
        6.3.3   Magnetization of the Hydrogen Nucleus............ 259
        6.3.4   The Magnetic Moment and Its Response
                 to an External Magnetic Field................... 263
        6.3.5   Precession................................. 264
        6.3.6   The Rotating Reference Frame................... 266
        6.3.7   The Magnetic Field B_1........................ 267
        6.3.8   Off-Resonance and the Effective Field B_1
                 (For Advanced Reading)....................... 269
        6.3.9   The MRI Signal Source........................ 271
        6.3.10   The Spin-Lattice and Spin-Spin Relaxation Processes. . . 273
        6.3.11   The Bloch Equations.......................... 276
6.4   MRI Contrast Sources................................ 277
6.5   Spatial Mapping: The Field Gradients..................... 279
6.6   Spatial Mapping: Slice Selection........................ 282

6.7	K-Space Formulation	285
	6.7.1 Traveling in K-Space	289
6.8	Imaging Protocols and Pulse Sequences	293
6.9	Some Important Pulse Sequences (Advanced Reading)	298
	6.9.1 Inversion Recovery	298
	6.9.2 Spin Echo	302
	6.9.3 Fast or Turbo Spin Echo	305
	6.9.4 Echo Planar Imaging (EPI)	307
	6.9.5 Steady State	310
	6.9.6 Advanced Rapid Imaging Methods	312
6.10	Three-Dimensional (3D) Imaging	312
6.11	Magnetic Resonance Angiography (MRA)	313
	6.11.1 MRA Using Contrast-Enhancing Material (CEM)	314
	6.11.2 Time-of-Flight (TOF) MRA	315
	6.11.3 Phase Contrast MRA	317
References		318

7 Ultrasound Imaging 321
7.1	Introduction	321
7.2	Physical Phenomena	322
	7.2.1 Speed of Sound	324
	7.2.2 Attenuation	325
	7.2.3 The Acoustic Impedance	326
	7.2.4 Reflection Refraction and Transmission	328
	7.2.5 The Doppler Shift Effect	330
7.3	Ultrasonic Transducers and Acoustic Fields	331
	7.3.1 Piezoelectric Transducers	331
	7.3.2 Single Element Transducers	333
	7.3.3 Linear Phased Array Transducers	335
	7.3.4 Implications on Image Resolution	336
7.4	Imaging Modes and Image Formation	338
	7.4.1 The Signal Source	338
	7.4.2 The *A-Line (A-Mode)*	341
	7.4.3 The Time Gain Correction (TGC)	344
	7.4.4 The M-Mode	345
	7.4.5 The B-Scan (2D Imaging)	347
	7.4.6 Three-Dimensional (3D) and Four-Dimensional (4D) Imaging	350
	7.4.7 Through-Transmission Imaging	351
7.5	Doppler Imaging	354
	7.5.1 Single Zone Velocity Imaging	354
	7.5.2 Two-Dimensional (2D) Doppler and Color Flow Mapping (CFM)	357

7.6   Advanced Imaging Techniques   359
         7.6.1   Contrast-Enhanced Ultrasound (CEUS) Imaging   359
         7.6.2   Elastography   361
         7.6.3   Ultrafast Imaging   362
    References   363

Exemplary Questions for Chap. 1   365

Solutions to Exemplary Questions for Chap. 1   369

Exemplary Questions for Chap. 2   371

Solutions to Exemplary Questions for Chap. 2   379

Exemplary Questions for Chap. 3   387

Solutions to Exemplary Questions for Chap. 3   393

Exemplary Questions for Chap. 4   399

Solutions to Exemplary Questions for Chap. 4   403

Exemplary Questions for Chap. 5   409

Solutions to Exemplary Questions for Chap. 5   413

Exemplary Questions for Chap. 6   423

Solutions to Exemplary Questions for Chap. 6   427

Exemplary Questions for Chap. 7   433

Solutions to Exemplary Questions for Chap. 7   439

Medical Imaging: Final Exam A   441

Solutions to Exemplary Exam A   449

Medical Imaging: Final Exam B   453

Solutions to Exemplary Exam B   463

Index   469

Chapter 1
Introduction

Synopsis: In this chapter the reader is introduced to the basic concepts and useful terms applied in medical imaging.

The learning outcomes are: The reader will know what is expected from a medical image, will comprehend the issues involved in generating and assessing the quality of a medical image, and will be able to implement (conceptually) this knowledge in the development of a new imaging modality.

1.1 Historical Background and Motivation

1.1.1 X-Rays and CT

The saying "A picture is worth a thousand words" is well accepted as true in many fields of life. This statement is particularly valid in the context of medicine. For thousands of years, medicine lacked the ability to implement, for diagnosis purposes, our most efficient sense, i.e., vision, in examining the internal structure of the human body. Palpation was the most commonly implemented tool. However, palpation can hardly reveal a small fraction of the numerous existing pathologies. Hence, physicians were as helpless as "The blind gropeth in darkness" [1].

The major breakthrough occurred on November 8, 1895, when Wilhelm Conrad Röntgen found that when activating a cathode ray tube enclosed in a sealed thick black carton in a dark room, a paper plate covered on one side with barium platinocyanide placed near the tube became fluorescent, indicating a new type of ray. These invisible rays, later termed x-rays, passed through matter that was opaque to visible light rays. Soon after, grasping the potential of his discovery, he produced

© Springer Nature Switzerland AG 2020
H. Azhari et al., *From Signals to Image*,
https://doi.org/10.1007/978-3-030-35326-1_1

Fig. 1.1 The first x-ray image produced by Roentgen depicting his wife's hand and ring. (From Wikimedia)

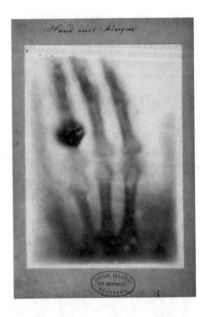

the very first medical image of his wife's hand [2], showing its internal structure. This famous roentgenogram is depicted in Fig. 1.1.

Within several months after publication of the discovery [3], roentgenograms were produced in battlefields to help locate bullets in soldiers who had been injured [4]. However, the need for using very high voltage created difficulty in its medical use. This changed in 1913, when Coolidge introduced a more practical design of high vacuum x-ray tube with a heated cathode that produced reliably high-intensity beams. Ever since, the application of x-rays in medicine has increased exponentially with improved equipment and with the more recent use of digital systems and displays. Several *billions* of planar x-ray images are currently produced annually worldwide.

A quantum leap in this area occurred on October 1, 1971, when the first computed tomography (CT) clinical image was produced by Godfrey Hounsfield and his colleagues [5]. The system which collected information via the rotation of an x-ray imaging system about a patient and applied computer reconstruction was able to produce a new kind of image. For the first time, cross-sectional images of the intact brain depicting its anatomy were available in a noninvasive manner. This event best designates the "Bagel Era," when scanners with a circular structure became dominant in the field of medical imaging. The progress and improvements continuously introduced into scanning technology have been strongly tied to the rapid advances in electronics and computing.

1.1.2   Nuclear Medicine: Gamma Camera, SPECT, and PET

Nuclear medicine was a relative latecomer as a medical specialty, for example, not having gained this designation until 1971 by the American Medical Association. Arguably, the first practical nuclear medicine procedure was developed by Saul Hertz at the Massachusetts General Hospital in the early 1940s: the treatment of Grave's disease by radioactive iodine (radioiodine) using a protocol that remains largely unchanged today [6].

The foundation of nuclear medicine was laid with the discovery of emissions from certain heavy metals by Henri Becquerel in 1896 and advanced by Pierre Curie and Marie Curie, the latter of whom coined the term "radioactivity." They recognized that some of these emissions were similar to Roentgen's x-rays and these three received the Nobel Prize in Physics in 1903 for their discoveries. George de Hevesy is credited with the radiotracer concept, now used in nuclear medicine, by using radioactivity to track small quantities of materials for plant physiology studies [7]. Technological advancements after 1930 enabled the controlled production of synthetic radioactive materials. Ernest Lawrence built the first cyclotron in 1934 at the University of California in Berkeley [8], and Enrico Fermi completed the construction of the first nuclear reactor at the University of Chicago in 1942 [9]. Ironically, while both of these technologies were rapidly advanced for the development of a weapon of mass destruction during the Manhattan Project, they both remain crucial for the production of radioactive material essential in nuclear medicine for the diagnosis and treatment of diseases. Although the use of radioactive substances to study human physiology or treat disease had been tried before, Hertz's radioiodine conception was rapidly accepted by the medical community leading to the seminal work by Samuel Seidlin in 1946 concerning the use of radioiodine in the treatment of thyroid cancer [10].

While gamma counters provided rough qualitative estimation of radioactive material within the human body, relative quantitation and mapping was first provided by images produced from Benedict Cassen's rectilinear scanner in 1951. This device made point-by-point measurements of radioactive uptake in, for example, the anterior plane of a patient lying flat on a table [11]. Hal Anger's gamma camera (1958) provided nuclear medicine clinics with a practical means of planar imaging of radioactive uptake, greatly improving image quality and reducing scan time [12]. By rotating an Anger camera about a patient, David Kuhl used the tomographic principle to image the three-dimensional distribution of radiotracers within a body in a process now known as SPECT (single photon emission computed tomography) [13]. Michael Phelps had previously used this tomographic concept on the coincidence detection of pairs of high-energy photons produced by positron emitters giving positron emission tomography (PET) images with clinically higher-quality images than SPECT for many applications [14].

Hybrid scanners, enabling the imaging of physiology combining nuclear medicine techniques and anatomy from radiological techniques like CT, are now common in clinics. This is typically accomplished using united coaxial back-to-back scanners. The anatomical information can be used to provide improved localization for the radiotracer uptake or to correct image artifacts caused by the attenuation of emission photons in SPECT or PET. In the 1990s, SPECT/CT was pioneered by Bruce Hasegawa [15], and PET/CT was developed by David Townsend [16]. These devices became commercially available around the turn of the millennium. Current developments include organ-specific cameras [17], solid-state detection, hybridization with magnetic resonance (MR-PET) imaging [18], and sophisticated computer hardware and software advances that greatly improve diagnostic image quality.

1.1.3 Ultrasound

The military arena has also contributed to the development of ultrasonic imaging. During World War I, the SONAR (Sound Navigation and Ranging) system was developed in order to detect underwater submarines. The first ultrasonic imaging system was probably introduced by Dussik [19] in 1942. He used through-transmission waves. Ironically, this system attempted to image the brain, a task which has remained a challenge to this very day for ultrasound.

Many related technical developments contributed to the progress of ultrasonic imaging in the years that followed. Some noteworthy works are that of George Ludwig, who used the pulse echo technique (known now as A-mode scanning) to examine animal tissues (his works were considered classified until 1949). John Julian Wild and John Reid were probably the first to build an ultrasonic scanner that produced real-time images (called B-scan) of the breast in 1953 [19]. Ultrasonic Doppler techniques for medical application were implemented by Shigeo Satomura and Yasuhara Nimura in 1955 for the study of cardiac valvular motion and pulsations of peripheral blood vessels [19]. George Kossoff et al. developed phased-array transducers in 1974. James J Greenleaf et al. were the first to introduce in 1974 an ultrasonic computed tomography (UCT) system [20]. The use of ultrasound for evaluating the elastic properties of tissues was suggested by Ophir et al. in 1991 [21, 22]. This has created a new diagnostic field titled "elastography." A major development was more recently established by a group including Mathias Fink and colleagues. They have implemented a planar imaging technique which can provide images with a rate of several thousand images per second [23]. The combination of ultrasonic imaging with laser has led to the development of "optoacoustic" imaging, where excitation of the tissue by a strong laser beam produces sound waves which are detectible and from which an image can be generated [24]. Ultrasonic imaging is still continuously progressing along with the general progress achieved in the fields of electronics and computer science.

1.1.4 MRI

The first measurement based on the nuclear magnetic resonance (NMR) phenomena (see Chap. 6) is attributed to Isidor Rabi in 1938, who received in 1944 the Nobel Prize in Physics for his work. His research was followed by the work of Felix Bloch and Edward Purcell (Nobel Prize in Physics 1952), whose research on NMR paved the road for the development of magnetic resonance imaging (MRI) [25].

Although Raymond V. Damadian was probably the first inventor of the MRI scanner (received a patent on 1974), he was not awarded the Nobel Prize. The prize was awarded instead to Paul Lauterbur and Peter Mansfield in 2003. This created a controversy in the scientific community and outside of it as well [26]. Regardless of this issue, the first MR image was produced by Paul Lauterbur in 1973. He used reconstruction algorithms similar to those implemented in CT (such as back projection methods and iterative reconstruction techniques (see Chap. 2)). In 1975 Richard Ernst proposed the use of phase and frequency encoding and the Fourier transform [27] which revolutionized the acquisition and reconstruction methodology of MR imaging. The first MR image in a living human was produced on July 3, 1977, by Damadian and colleagues.

The field of MRI has grown exponentially since then with continuous improvements. The progress is both in terms of hardware with scanners having higher magnetic fields, better gradients, and better coils to improve image acquisition quality and in software with pulse sequencing to generate better images and establish new contrast sources. Some notable improvements (among numerous others) are the development of echo planar imaging (EPI) by Peter Mansfield et al. [28]. This fast imaging technique enabled functional MR imaging of the living brain and provided a basis for a new branch of MRI called functional MRI (fMRI). Also, the introduction of diffusion imaging by Le Bihan et al. [29] in 1985, the first parallel imaging by Daniel Sodickson et al. [30] in 1997, and the introduction of compressed sensing to MR image reconstruction by Michael Lustig et al. [31] in 2006 are some notable contributions made to this field.

MRI is a fascinating field which is probably the most interdisciplinary imaging modality combining physics, mathematics, engineering, computer science, and of course medicine. It is considered hazardless and extremely versatile with numerous methods of distinguishing tissues with details down to the millimeter scale and with very-high-quality images. However, it is still rather expensive, slow, and cumbersome in terms of patient handling and accessibility for frequent clinical use.

1.1.5 Other Imaging Modalities

Apart from the major imaging modalities listed above, there was, and still is, a continuous research effort to expand the field. Some ideas were eventually developed into clinical tools, and many are still considered immature for practical daily use. Among these the most notable system is electrical impedance tomography (EIT), which maps the electrical conductivity of the body in a noninvasive manner by "injecting" currents and measuring tissue response. Another established modality is optical coherence tomography (OCT) which utilizes optical scatter and interferometry near-infrared light to produce images showing micron-sized features. Thermal imaging was also considered as a clinical option, and many instruments for scanning the body with infrared light were also built. New research directions on the horizon include terahertz waves, phase-contrast x-ray imaging, fluorescent imaging, Cherenkov radiation imaging, and the use of cellular phones for image reconstructions.

The use of contrast-enhancing agents is a field by itself. These agents are used in all imaging modalities. They augment the generated image signal at target organs or tissues, thus enabling functional imaging as well as anatomical imaging. Recently, the use of nanomaterials has emerged. These materials are expected to provide a "theranostic" ability, i.e., the combination of diagnosis and therapy using imaging for guidance and monitoring.

Another emerging field is "multimodal imaging" (as mentioned above), where hybrid scanners provide a combination of images which depict both the anatomy and the functionality of the imaged organ. Finally, the important role that imaging modalities play in minimally invasive surgery should be acknowledged. Imaging is currently used for guiding laparoscopic surgery, inserting needle type applicators, conducting cryogenic treatment procedures, and monitoring thermal ablation procedures which use high-intensity focused ultrasound (HIFU). Imaging plays also a crucial role in radiotherapy and other applications. *Evidently, medical imaging is one of the most exciting fields in medicine and biomedical engineering.*

1.2 Basic Definitions of Medical Imaging

1.2.1 What Is a Medical Image?

To begin our intellectual journey, consider the following set of pictures (Fig. 1.2) which contains several medical images. Study it for a while. Now, try to answer the following question: What do all these images have in common?

Hints
- Evidently, not all images depict the anatomy.
- Also, quite clearly, each imaging modality produces something different in nature.
- In addition, the quality of each image is different in terms of sharpness and details

What do all these images have in common?

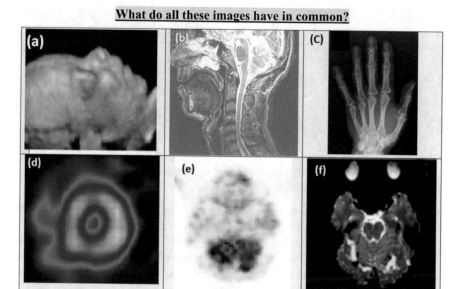

Fig. 1.2 An exemplary collection of medical images: (**a**) 3D ultrasound image. (**b**) MR image. (**c**) X-ray image. (**d**) SPECT image. (**e**) PET image. (**f**) fMR image

Answer

They all convey visual information about some property of the body

You probably have seen additional types of images, for example, x-ray CT scans, echocardiography, gamma camera images, x-ray angiography, colored Doppler ultrasound, and more. Each image is produced by a different type of a scanner which utilizes a different physical phenomenon. To be more explicit, MRI utilizes magnetic relaxation properties commonly of hydrogen nucleus; ultrasound utilizes the reflections of sound waves; CT utilizes the absorption of x-rays; gamma camera and SPECT utilize radioactive emission and absorption; PET utilizes the positron (a positively charged electron) annihilation phenomenon; Doppler imaging is based on frequency shifts of moving objects; and there are also other examples and imaging modalities.

The fact that such a variety of options coexist and are employed daily, implies that the clinical information they produce is needed and may be essential or complementary to other modalities in order to obtain reliable diagnoses. The fact that the information is also location dependent indicates that the image is in fact some kind of a map, similar to geographical maps which depict the terrain of a country using color-coded graphics. Medical images use different shades of gray and colors to convey their information. Accordingly, we define the following:

Definitions
- *Image*
  A means to depict a MAP of a measured PROPERTY
- *Medical Imaging*
  A process implemented for generating an IMAGE which
  represents a PROPERTY of the body or an organ
  for medical use

Following the above, it can be concluded that in order to produce an image, we have to deliver two basic information elements:

(i) *Visual distinction* – that is, we have to show by graphical means that one point in the image differs in its mapped property from its neighbors.
(ii) *Spatial address* – we have to indicate, using a two-dimensional (2D) or a three-dimensional (3D) coordinate system, the spatial address of each element in the picture. (The coordinate system may be rotated, scaled, or warped to accommodate the shape of the organ or pathology of interest.)

To demonstrate this idea, a schematic image is depicted in Fig. 1.3.

1.2.2 Mapping of an Object into Image Domain

As defined above, the image is but a map of some property of the scanned object. Thus, we have to know what are the more significant features of information that the scanner has to transfer from the real world to the image domain. Importantly, the

Fig. 1.3 A reliable image should indicate by differences in shades of gray or in colors that square #1 has property A and square #2 has property B which differs from A. In addition, it should indicate their relative locations and allow measuring the spatial distance d between the two

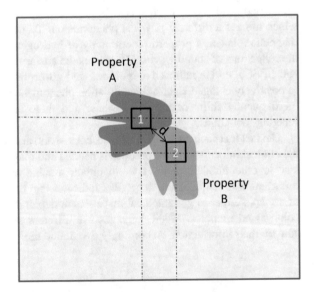

Fig. 1.4 The image on the left represents reality, while the image on the right represents the image generated by an imaginary scanner

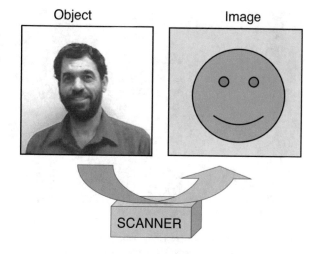

amount of information presented in the image is not as significant as the *quality* of the information displayed. Consider the example depicted in Fig. 1.4.

The imaginary scanner in this figure samples some properties of the real object (left side) and generates an image (right side) which visually depicts the processed information. Although the amount of information is substantially larger on the left image, the image on the right is sufficient to depict the mapped property. Can you tell what property is mapped in this case?

The mood of the imaged person

Although this example may seem unrealistic, it is in fact not so uncommon in medical imaging. Categorically, there are generally two types of images: (i) those that depict mainly the anatomy and (ii) those that depict mainly the physiology and are commonly referred to as "functional" images. An example of such a division is shown in Fig. 1.5.

The MR image depicted on the left side portrays a snapshot of a contracting heart. It displays a cross-sectional view of the four chambers of the heart. The heart's muscle (the myocardium) appears darker than the blood pool which has a grainy white appearance. The arrows indicate the septal (left arrow) and free walls (right arrow) of the left ventricle (LV). The two lungs appear as black regions from both sides, and a vertebra is seen at the bottom (the gray "Gaussian" shape slightly to the left); the white circle at the bottom is a cross section of the aorta. Although the anatomy is visible, we have no information on the perfusion (blood supply) to the myocardium.

On the other hand, the functional SPECT image on the right side does not show much anatomical details. All we can see is a blurry colored C-shaped object. This object is in fact a longitudinal cross section of the heart's left ventricular myocardium. The pseudo colors represent the myocardial perfusion, where hotter (red-yellow) colors depict higher perfusion and cold (blue) colors represent low

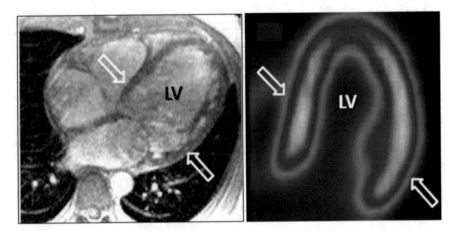

Fig. 1.5 Both images display a longitudinal cross section of the heart's left ventricle (LV). The MR image on the left depicts anatomical information of the entire heart, while the SPECT image on the right displays only functional information (perfusion) of the LV. The arrows indicate the septal (left arrow) and free walls (right arrow) of the LV. As can be noted, the heart's anatomy is not visible on the SPECT image

perfusion. Nonetheless, the information delivered by the SPECT image may be clinically much more valuable than the detailed MR image. Note, for example, the discontinuity at the apical region which is located at the upper zone in the SPECT image. This discontinuity may indicate lack of perfusion (ischemia).

1.2.3   What Are the Required Stages for Developing an Imaging Modality?

Now that you are familiar with the general concept of medical imaging, try to think (without peeping at the text below) what are the stages that are needed in order to introduce a new imaging modality. From the above, it follows that we need eventually some kind of a machine (a scanner) that will produce medical images. In order to reach this goal, the following steps have to be taken:

A. *Selecting the PROPERTY to be imaged*
   The first step should be to know what is the medical need that we wish to address. In other words, "What clinical benefit shall it produce?". Accordingly, we have to understand what may be the physiological property that best conveys the needed information.

   – For example, if the clinical need is to know what is causing a heart murmur, then mapping the blood flow directions around the heart valves may be a suitable property to choose.

B. *Choosing a PHYSICAL phenomenon which is sensitive to the said property*
   The second step would be to seek some physical effect that changes in response
   to variations in the mapped property and that can be measured by some means.
   Furthermore, it should be sensitive enough to detect changes occurring within
   the physiological range.

   - Following the above example, we may decide to use the Doppler effect in
     ultrasonic pulse echoes.

C. *Developing a MATHEMATICAL MODEL for relating the said property and the
   measured physical signals*
   This is a crucial step, because an accurate mathematical model allows one to
   quantify the relation between the mapped physiological property and the physical phenomenon. Such quantification is essential for assessing the sensitivity
   and the range of expected changes that need to be measured.

   - Following the above example, we can use the known relation relating the
     frequency shift resulting from the Doppler effect and the velocity of the
     blood.

D. *Developing a RECONSTRUCTION algorithm*
   At this point we have to generate an image from the measured physical signals
   that would display the mapped property. This is not a trivial task. Sometimes,
   multiple steps and complicated algorithms are required to transform signals
   acquired by the imaging system into a useful medical image. Moreover, the
   relation maybe implicit, or we may need to use iterative reconstruction algorithms. In other cases, there might be a constraint in the acquisition process. For
   example, certain views of the object may be obstructed by boney structures. Or
   there might be a need to acquire the information in real time yet the amount of
   needed calculations may be too large to implement on conventional hardware.

   - Relating to the above example, the need to map the Doppler shift in thousands
     of pixels in real time imposed a technical challenge that was solved (as will be
     explained in the ultrasound imaging chapter), by implementing autocorrela-
     tion and approximating the temporal phase derivatives.

E. *Building a device or system for scanning the patient*
   Eventually after the theoretical framework has been established, the ideas should
   be translated into hardware. The first thing to do is to find a suitable sensor
   (detector) that will be capable of measuring the said physical phenomenon, and
   the sensor should be sensitive enough to measure it within the relevant range.

   - In accordance with the example above, a suitable sensor may be a piezoelec-
     tric transducer that can translate the pressure produced by sound waves into
     electrical voltage. Almost equally important is designing the proper data
     collection strategy. This strategy should be compatible with the reconstruc-
     tion algorithm. The rest is engineering.

F. *Signal and image processing for improving image quality*
Apart from artifact (see definition in the following) removal and better noise reduction, some post-processing algorithms may be needed for better image representation. Ultimately, the aim of medical imaging is to provide the clinician with images that are useful for execution of a specific clinical task, for example, lesion detection, localization of tumors, changes in anatomy and texture, and quantitation of response to treatment. Therefore, image quality needs to be optimized from the perspective of that specific clinical task (as will be described in the following).

1.3   Basic Features of an Image

1.3.1   Gray Levels and Dynamic Range

The first image feature to be introduced is the visual display. Consider a scanner that maps a certain *physiological* property of the body. The *physical* effect that changes in response to variations in the mapped physiological property is usually measured by sensors that eventually convert the physical property into electric signals. Those electric signals are commonly characterized by changes in terms of voltage or current or frequency. If, for example, our scanner measures the intensity of x-rays passing through the body, then common sensing systems will transfer the varying intensity into variations in voltage. On the other hand, if blood velocity is the mapped physiological parameter, then the physical measurements will quantify the frequency shift (Doppler effect). Stemming from the above, we have to know first what is the relevant *physiological* range that we wish to map and relate it to a range of the produced physical signals. For example, the range of blood velocities in the body changes substantially from one organ to another. At the aorta, the range is between 0 and 100 [cm/sec], but the mean velocity at the retina is much smaller; it is only about 4 [cm/sec]. Evidently, the sensitivity of our measurements and the range of signal values that can be measured by the hardware should be different for each case. The range of signal values over which our scanning system can operate within tolerable limits of distortion is referred to as the system's "dynamic range."

Once the configuration of the hardware is set, we will have to choose at what range of physical levels, for example, voltages, the system will be capable of producing useful signals needed for reconstructing an image. Basically, the lowest possible visual display is commonly *black*. Thus, a threshold level should be set so that any signal below that level will appear as a black pixel in the image. Next, we will have to decide to what signal level we assign the brightest visual display which is typically *white*. This value will serve as an upper threshold. That means that any signal exceeding this threshold will have the same visual display and appear as a white dot. The concept is schematically depicted in Fig. 1.6.

For the intermediate levels, it is common to assign shades of gray. The span of levels displayed between the low threshold signal and the high threshold signal is referred to as

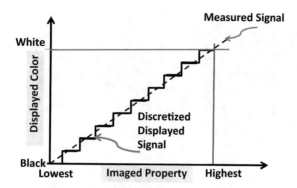

Fig. 1.6 The imaged tissue property is displayed as varying shades of gray. Black color is assigned to any measured physical signal which equal or is smaller than a defined threshold. And white color is assigned to any measured signal which equals or is higher than a defined threshold. Intermediate values are discretized and assigned gray colors

the "displayed dynamic range." The number of gray levels set within this range is commonly assigned a power of 2, i.e., $2^8 = 256$ (one byte) or $2^{16} = 65,536$ (two bytes). An exemplary grayscale which contains $2^6 = 64$ levels is depicted in Fig. 1.7. The human eye can commonly distinguish between about 50 gray levels. As can be noted in Fig. 1.7, a set of vertical dark lines seem to mark each displayed level. This is only an optical illusion! This phenomenon which is called the "Mach effect" stems from an edge enhancement mechanism of the eye. This effect can be a source of distortion in radiology.

The division into discrete levels actually bounds the resolution of the displayed mapped property. The more gray levels there are, the better is the potential ability to detect variations in the mapped property. However, in order to have a meaningful display, the selection of the displayed dynamic range and the number of gray levels should be compatible with our measurement ability. For example, if we wish to map the changes in body temperature (e.g., by using MRI), in the range of 0–64 °C, then using 64 gray levels will set the resolution to 1 °C. On the other hand, if we set it to 512 levels, the theoretical resolution should be improved to 0.125 °C. Nonetheless, this "improved" resolution would be meaningless in cases where our scanner is not sensitive enough to detect such subtle changes, if, for example, the accuracy in temperature measurement is only ±0.5 °C.

Setting the number of displayed gray levels affects the global quality of the image. Stemming from the above, the physician's ability to reach reliable clinical diagnosis depends (among other factors) on the ability to detect changes in the mapped property based on the visual information. If the display is too crude, then some of the variations will be lost. For example, two images are displayed in Fig. 1.8. The image on the left is displayed with an adequate number of gray levels, while the one on the right is displayed with a smaller and inadequate number of gray levels. As can be noted, some details are lost in this case, and the image quality is low.

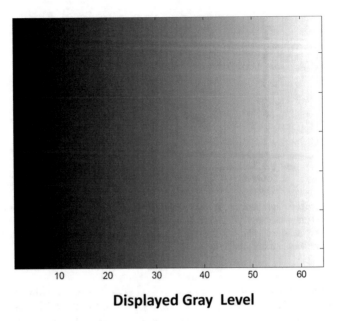

Displayed Gray Level

Fig. 1.7 A grayscale consisting of levels. The human eye can usually recognize only about 50 levels. (The black vertical lines are an illusion called the "Mach effect")

Setting the displayed dynamic range also affects the quality of the image. If the upper signal values to be displayed are too low, many regions of the image will exceed this value, and many parts of the image will be displayed as white areas. As a result, regions in the image which in fact correspond to tissues which have different values will look the same, thus misleading the person examining it. Similarly, if a too high low signal limit is set, many regions will be below it. Consequently, the image will have many dark regions and areas with different tissue properties may look the same. This is demonstrated in Fig. 1.9.

In modern display systems, color has become more prevalent. Although black and white scaling is more intuitive and offers superior delineation of fine details, sometimes colors may be preferable. This is often the case when displaying functional information atop the anatomy. In such cases, the grayscale is used for displaying structural information, while the color designates the more important mapped property. There are numerous optional color maps. Two commonly used examples which are the "Jet" and the "Hot" color maps (offered by MATLAB®) are demonstrated in Fig. 1.10 (see Fig. 5.16 for common clinical color maps). Additionally, an inverse grayscale may be used in which white pixels represent no signal and progressively darker pixels represent greater signals.

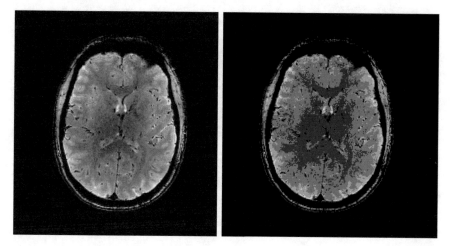

Fig. 1.8 Two displays of the same MRI image of the brain. While the image on the left was displayed with an adequate number of gray levels, the image on the right was displayed with a too small number of gray levels. As a result, fine variations in tissue properties are lost, and the quality of the visual information is reduced

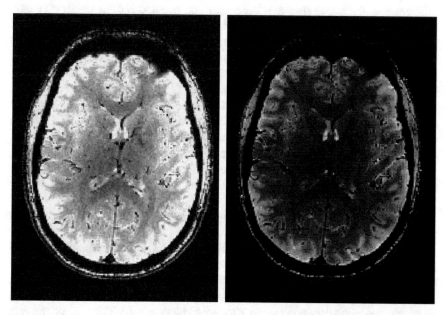

Fig. 1.9 Two displays of the same MR image with inadequate displayed dynamic range. (Left) The upper limit is too low; therefore, many regions are saturated and are depicted as white regions. (Right) The lower limit is too high; therefore, many areas are displayed as dark regions

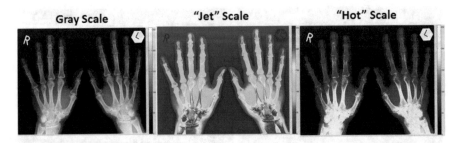

Fig. 1.10 Comparison between a grayscale display and two optional color maps. The same image is displayed in three different color maps. Note that the value of the pixels remains the same for each color scheme display

1.3.2 Point Spread Function and Line Spread Function

A crucial issue in understanding the limitations of the imaging process is to investigate how the most basic feature in an image, namely, *a single point object*, is transformed from the real space into the image domain. As explained above, the visual information conveyed by the image contains two components: (i) visual (graphic) contrast and (ii) spatial address. When scanning a single point in space and comparing it to the corresponding image produced by a specific imaging system, one may obtain a distorted image. The size and shape of the scanned point may be different than the actual size; the borders may be smeared and the object may look blurry; the gray levels may be inconsistent with the real properties of the object; and unreal image features may appear. An example is depicted in Fig. 1.11.

The imaging system response to a point source, i.e., the image of a point object generated by the scanner, is called the "point spread function" or *PSF* for short. The PSF is affected by many parameters. The main factors are typically the physical process which produces the signal, the system sensors utilized to collect the data, and the reconstruction procedure implemented to produce the image. The PSF may be independent of the position within the scanned plane (or volume). In this case it is referred to as "shift invariant PSF." Alternatively, it could be position dependent.

The mathematical relation between the PSF and the obtained image is typically described (for a shift invariant system) by a convolution process, i.e.,

$$\text{Image} = \text{object} \otimes \text{PSF} \tag{1.1}$$

where the operator $\otimes$ symbolizes convolution. Accordingly, unless the PSF is a delta function (unit impulse function), the image will always be a distorted version of the real image. Commonly, as a result, features in the object will be smeared and blurred.

In many cases the PSF may be asymmetric. Furthermore, it could be location dependent (as is the common case in ultrasonic imaging, e.g.). This may complicate processes which aim to quantify the imaging system quality. In the simple case where the PSF is approximately symmetric and shift invariant, analysis of the PSF

Fig. 1.11 A point object in the real world (left) may look substantially different in the image generated by the scanning system (right)

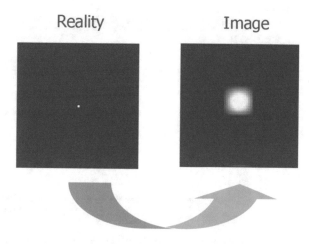

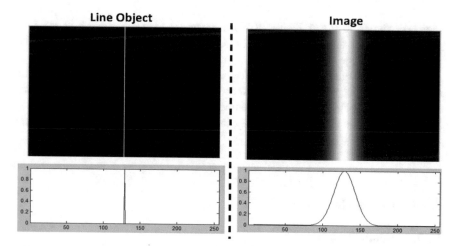

Fig. 1.12 A line object in the "real" world (Top left) with its corresponding profile (Bottom left). The LSF (Bottom right) and the simulated effect on the corresponding image (Top right)

profile may suffice to characterize the system response. In such cases, the system response to a scanned line object could also be useful. The obtained profile in such case is referred to as the *line spread function (LSF)*. A LSF with a Gaussian profile and the simulated corresponding effect on the image are depicted in Fig. 1.12. As can be noted, the line appears as a thick strip in the image with smeared gray level distribution and blurred edges.

Naturally, as the PSF or LSF become wider, the image will be more smeared and details will be blurred. This will affect the spatial resolution (see the following). In order to allow quantitative comparison between systems, the width of the LSF profile is characterized by a parameter termed "full width at half maximum" or FWHM for

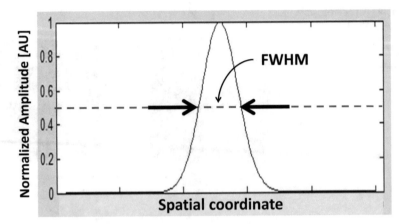

Fig. 1.13 Graphical depiction of the "full width at half maximum" (FWHM)

short. As shown in Fig. 1.13, the FWHM measures the spatial width of the LSF profile at 50% of the maximal amplitude. The smaller is the FWHM value, the sharper is the image and vice versa.

In modern imaging systems and acquisition geometries, the properties of the image vary with location. Such systems are called "shift variant." For such systems the PSF varies within the imaged volume, and the notion of "system response" function is more appropriate. Thus, mapping of point signal response of the system throughout the field of view is needed.

1.3.3   Image Resolution

1.3.3.1   Spatial Resolution

As explained above, the reconstructed image should also provide information on the spatial addresses of every mapped element within the object. However, resulting from the PSF, a point target in the scanned object is commonly transferred into a blurred spot in the image domain. Consequently, the accuracy of the spatial mapping process is reduced. Consider the example shown in Fig. 1.14. Two identical point objects are positioned close to each other at a distance d. Each of the objects is represented by a smeared spot in the image. And as the spots are larger than the actual objects, they may overlap in the image. As a result the observer cannot distinguish between the two, and may mistakenly think that there is only one object.

To demonstrate this effect with the PSF shown in Fig. 1.11, let's examine two pairs of point objects positioned side by side and scanned by an imaginary scanner. As shown in Fig. 1.15, while the upper pair in the image is separable, the lower pair is not.

Fig. 1.14 Two identical objects located close to each other may appear in the image as two overlapping spots

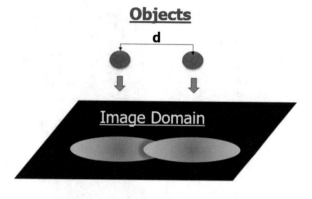

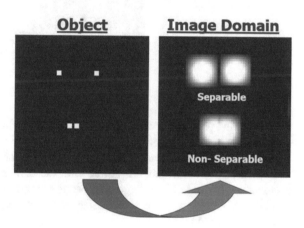

Fig. 1.15 Two pairs of point objects appear as smeared spots in the image domain. While we can distinguish in the image between the upper pair of points, the lower pair is inseparable

Accordingly, we define the system's spatial resolution as follows:

Definition

Spatial Resolution

The minimal distance at which the system provides the ability to distinguish in the image domain between two proximal point objects having similar or identical properties

The spatial resolution is thus quantified by the minimal distance between such two adjacent point targets, which still appear as a separable pair of spots in the image domain. The resolution is commonly measured by scanning special test objects which are called "resolution phantoms." These phantoms commonly have a gradually changing structure with known properties and geometry. By scanning these objects, one can quantitatively evaluate the performance of the imaging system. An example is shown in Fig. 1.16, where an MRI scan of a resolution phantom is

Fig. 1.16 An MRI scan of a resolution phantom consisting of a series of lines with circular targets having decreasing diameters and positioned closer to each other. The arrow marks the visible spatial resolution limit

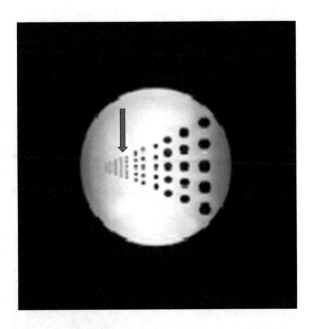

depicted. This phantom consists of a series of lines. In each line a set of circular objects with known diameters and with known gaps between them is positioned. Moving from right to left, the diameter of the circular targets is decreased. As can be noted, the individual targets are visible until the seventh column. Then, the image becomes too blurred and the targets appear as a single gray strip. In many cases, the ability to differentiate between two adjacent objects can be taken equal or slightly larger than the FWHM.

Another related parameter is the system detection ability, which refers to the smallest detectable object. However, this ability is affected by two parameters: (i) geometry, the shape and size of the smallest object that can be detected in the image, and (ii) contrast. The combination of the two for the smallest detectable object can vary for a certain range of values. That is to say, that a relatively large object with low contrast may be undetected, while a smaller object with higher contrast can be detected using the same system.

1.3.3.2 Pixels and Voxels

In the past, images were recorded by a photographic process where a plastic film coated by an emulsion containing very small silver halide crystals was used to capture the image. Exposure to light or x-rays turn, within a fraction of a second, specks within the silver halide crystals into silver specks which change the crystal color when exposed to a certain chemical, from transparent to black, yielding a "negative" picture. The finer the crystal size of the film, the better was the potential resolution that can be obtained. In current systems however, all images are digitized

Slice Thickness

Pixel Width

Pixel Height

Fig. 1.17 A digitized image consists of a mosaic arrangement of colored picture elements called "pixels" or "voxels." It should be noted that even when using a 2D image, one should recall that the picture elements (pixels) are in fact three-dimensional where the thickness is termed the "slice thickness"

and stored in the scanner's computer as a 2D, 3D, or 4D (space and time) matrices. The image is consequently displayed by using picture elements which are called "pixels" when referring to 2D images or volumetric elements called "voxels" when referring to 3D images. The concept is similar to a mosaic picture where many small colored stones are placed in a certain order to create a picture. An artistic presentation of the concept is depicted in Fig. 1.17. It should be noted however that even in the case of a 2D image, the pixels have thickness which is called the "slice thickness."

The digitized picture has many advantages. It can be conveniently stored in the computer; it can be easily transferred through the Internet or other communication lines; the image matrices can be manipulated mathematically to filter or improve the display. However, this utility is not without penalty regarding assumed homogeneity and finite spatial resolution:

Definition

Pixels and Voxels
A spatially defined volume for which we assume *homogeneity* in the imaged property. The pixel or voxel size, therefore, *sets a limit to the image spatial resolution.*

The implication is important in cases where the exact definition of the borders is important. Consider, for example, a case where a blood vessel passing through an organ of interest is imaged. As depicted in Fig. 1.18, there are some regions in the

Fig. 1.18 The disadvantage of pixel representation. (Left) The marked square actually contains about half blood vessel and half tissue. (Right) The corresponding pixel may appear as containing only tissue

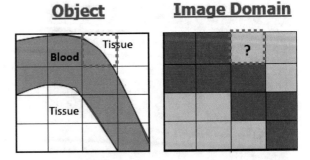

image that contain a mixture of the blood vessel and the tissue. However, each pixel must be assumed as homogenous, representing an average characteristic value represented by one assigned intensity. Thus, if we set a threshold, some pixels may appear as containing only blood, and some may appear as containing only tissue. Now consider a medical procedure where a needle has to be inserted close to the blood vessel wall without causing bleeding. Obviously, the physician has to count on the information provided by the image. Thus, while the marked pixel in the image may appear to have only tissue, it is in fact "half blood vessel and half tissue." Thus, there is a good chance that by inserting a needle into that pixel, bleeding may occur. This blended information effect where two tissue types or tissues and fluids may occupy the same voxel space is termed the "partial volume effect." This effect may lead to erroneous interpretation. Naturally, the partial volume effect is smaller for small pixels and voxels, i.e., for higher spatial resolution.

1.3.3.3   Temporal Resolution

In cases where rapid changes occur within the body, time related images may convey important clinical information. This is particularly true for cardio-vascular imaging where the heart beats at about 60 times per minute and contracts typically within 300 msec. It also applies to functional MRI where brain activity occurring within a fraction of a second is mapped, and to image guided interventional procedures where real time monitoring is essential. In such cases, the system ability to visually display temporal changes may be crucial. One should distinguish between the display frame rate which can be done in some applications off-line and the actual data acquisition rate. Accordingly, we define temporal resolution as:

> **Definition**
> *Temporal Resolution*
>   The minimal elapsed time required between the completion of the data *acquisition* process needed to reconstruct an image and the completion of data acquisition process needed to reconstruct a consecutive image

It should be noted that the time required for the image reconstruction process is important only if real-time display (frame rate of 30 Hz and higher) is obligatory.

Temporal Resolution in Physiological Time: There are certain applications where the temporal resolution of the system is too low to properly observe the physiological changes. In such cases, it is common to make use of the inherent periodicity of the body. The natural physiological "clocks" are the heart rate and the respiratory cycle. Assuming a periodic function, part of the data can be acquired relative to a signal generated by the body. The most commonly used signal is the electric cardiogram of the heart, known as the ECG. The ECG is characterized by a spike like feature relating to the onset of contraction (systole). The maximal lobe of the spike, which is called the "R wave," can be easily detected using standard signal processing. Thus, it provides a temporal landmark relative to which the physiological events can be measured. Accordingly, if we wish to image the heart, for example, at time τ after the beginning of contraction and the acquisition process of the scanner is too slow, we can divide the acquisition into N subsampling steps. Each subsample of the needed data is then collected within a small temporal window after the "R wave". The reconstruction of the full image is done upon completion of data acquisition. (Sometimes partial reconstruction is carried out in parallel to the acquisition). This mode of data acquisition is termed "cardiac gating." The concept is schematically depicted in Fig. 1.19.

The advantage of cardiac gating is that it enables one to visualize events that occur in a time scale which is much shorter than the scanner temporal resolution. For example, if we wish to see the changes in the heart during systole with a Δt millisecond temporal resolution, but our scanner needs at least 1 sec to complete the acquisition process, we can use $1000/\Delta t$ gated acquisitions. Each will subsample a different part of the information needed to reconstruct a frame at time τ_1 after the R wave. Then, shift the gate by Δt milliseconds and repeat the process. The concept is schematically depicted in Fig. 1.20. The risk in such an approach stems from the fact that the heart may change its pace occasionally. In such a case, the physiological state may slightly change and hence introduce inaccuracies and reduce the quality of the image.

1.3.4   Spatial Frequencies

Another way to quantitatively study an image is by examining its spatial spectrum properties. In a manner similar to that applied to signals in the time domain, the Fourier transform (FT) can be also applied to pictures in the spatial domain. In order to have an intuitive understanding of the spatial frequency concept, consider the following example: A person walking on a hilly terrain (see Fig. 1.21) will encounter along his path some creeks, valleys, hills, and cliffs. Accordingly, he will have sometimes to climb up and sometimes to descend down. In some cases, the changes in the terrain will be gradual, and in some cases, the changes may be more abrupt (e.g., a cliff or a creek). If we quantify the number of slope changes per given number of steps or distance, we may refer to the ratio as a representative measure of the spatial frequency. Hence, a terrain with a small number of changes per unit distance

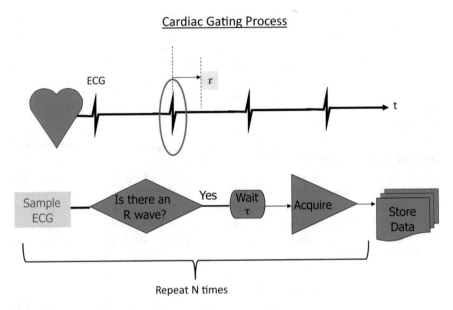

Fig. 1.19 A schematic depiction of the cardiac gating process

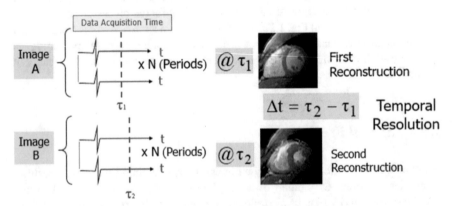

Fig. 1.20 A schematic depiction of the physiological temporal resolution concept. Using ECG gating the first frame is acquired at time τ_1 and the second frame at time τ_2 after the "R" wave. Although each frame data is collected using N subsamplings, the displayed temporal resolution will be $\Delta t = \tau_2 - \tau_1$

may be considered a region with low spatial frequencies, and a terrain with a large number of changes may be considered a region containing high spatial frequencies.

When studying the FT of an image, textural changes with notable periodicity will yield peaks in the spatial spectrum. In order to obtain another intuitive understanding of the spatial spectrum nature and its implications, consider the following picture of an auditorium (Fig. 1.22). Clearly, there are visible periodic patterns in the image. The methodical arrangement and the repeatable shape of the chairs and the stairs dominate the features of the image. Such periodic patterns transform into noticeable features in the spectrum, which are designated by bright elements in the 2D Fourier Transform of the picture. This is demonstrated in the corresponding frequency domain image of the auditorium which is shown next to its picture (Fig. 1.22).

Naturally, in the case of a two-dimensional (2D) image, we shall have to apply a two-dimensional Fourier transform (2D-FT). Using the following mathematical symbols, the spatial frequencies are marked by K_x and K_y corresponding to the horizontal and vertical directions, respectively. Importantly, spatial frequencies are directional. Hence, using vector notation, a spatial frequency may sometimes be also designated as $\overline{K} = \hat{x} \cdot K_x + \hat{y} \cdot K_y$. The corresponding physical units of spatial frequencies are [Radians/unit distace]. The spatial frequency can also be associated with a matching wavelength (λ) so that $K = 2\pi/\lambda$. Accordingly, if the image is marked as $f(x, y)$ and its corresponding 2D-FT image is marked as $F(K_x, K_y)$, the Fourier transform pair is given by (up to a constant)

$$
\begin{cases}
F\left[K_x, K_y\right] = \int\limits_{-\infty}^{\infty} \int\limits_{-\infty}^{\infty} f(x, y) \cdot e^{-jK_x \cdot x} e^{-jK_y \cdot y} dx dy \\[2mm]
f(x, y) = \int\limits_{-\infty}^{\infty} \int\limits_{-\infty}^{\infty} F\left[K_x, K_y\right] \cdot e^{+jK_x \cdot x} e^{+jK_y \cdot y} dK_x dK_y
\end{cases}
\tag{1.2}
$$

This relation is extremely important in MR imaging and very useful in CT image reconstructions and for image filtering as well. For clarity, the corresponding coordinate system notation used in this book is depicted in Fig. 1.23.

Fig. 1.21 An intuitive notion of spatial frequencies. The variations in gray level values can be considered as changes in height

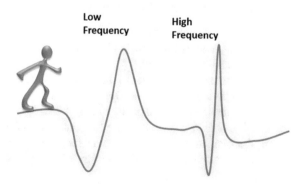

Low Frequency   High Frequency

Fig. 1.22 (Left) An image of an auditorium. Note the periodic textural patterns along the arrows. (Right) The corresponding Fourier domain (K-space) image depicted in a log scale of the intensity. Note the appearance of strong bright features

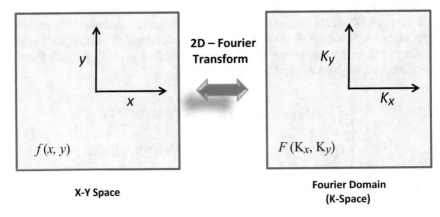

Fig. 1.23 The coordinate system notation used in this book, in image space and K-space, respectively

Although the reader is expected to be familiar with the FT in one dimension, it is worthy at this point to recall several important properties of the FT in the 2D context:

(a) First of all, the value of the central point in K-space (commonly referred to as the DC point) equals the sum of all the pixels in the image (or the average of all pixels in some notations).
(b) Secondly, the points closer to the center in K-space correspond to low frequencies, while those located far from the central point correspond to high frequencies (as depicted schematically in Fig. 1.24).
(c) Thirdly, objects in x–y space are "size antagonists" to their corresponding image in K-space, meaning that whatever looks big in x–y plane will look small in K-space and vice versa, as demonstrated in Fig. 1.25.

X-Y Space
(Image Domain)

K- Space
(Fourier Domain)

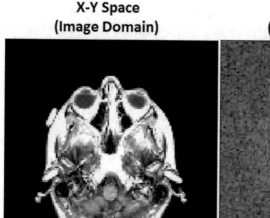 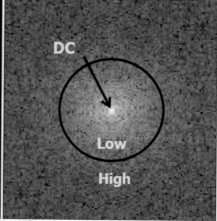

Fig. 1.24 (Left) An MR image of a head. (Right) The corresponding 2D-FT image which is referred to herein as K-space (depicted in log scale). The central point (DC) value equals the sum of all the pixels in the image. Pixels in K-space represent different spatial frequencies. Pixels close for the center correspond to low frequencies and pixels far from the center correspond to high frequencies

(d) Finally, the 2D-FT is orientation dependent, meaning, if an object is rotated by an angle θ, its corresponding 2D-FT representation will be also rotated by the same angle as shown in Fig. 1.26.

K-space manipulation is very useful in image filtering. By controlling the values of the different frequencies, the image can be sharpened or smoothed, and the image noise (defined below) can be removed. The low frequencies in K-space mainly correspond to gross textural and intensity changes in the image. The high frequencies on the other hand mainly correspond to borders and fine details in the image. Accordingly, by taking a 2D-FT of an image and removing or attenuating significantly the high frequencies and then retrieving the manipulated image by inverse 2D-FT, a "low-pass" filtered image can be obtained. And similarly by attenuating significantly the low frequencies, a "high-pass" filtered image can be obtained. This is demonstrated in Fig. 1.27.

As can be observed, the low-pass filtered image depicts the general texture of the hands with substantial blurring (note the R and L markings at the top). On the other hand, the high-pass image depicts mainly the bones' contours.

Fig. 1.25 (Top row) A
small objet in image space
will look big in K-space.
(Bottom row) A big object
in image space will look
small in K-space (depicted
in log scale)

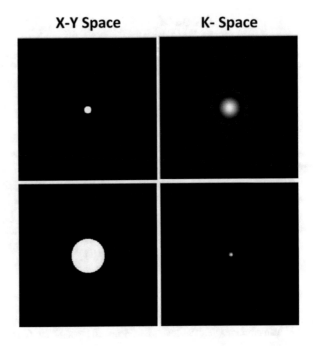

Fig. 1.26 A demonstration
of the 2D-FT sensitivity to
rotation

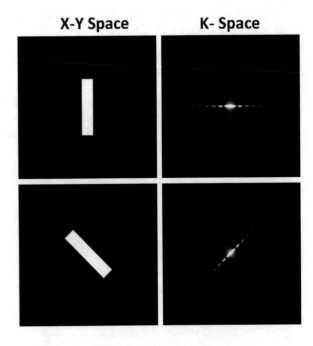

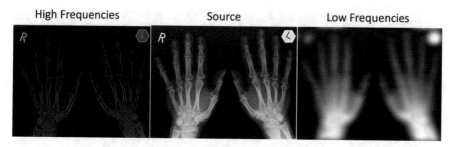

Fig. 1.27 (Middle) An x-ray image of two hands. (Left) High-pass filtered image of the same scan. (Right) Low-pass filtered image of the same scan

1.3.5   Modulation Transfer Function (MTF)

As explained above, we can control the features of an image by applying manipulations onto its K-space. However, the most dominant factor that affects the spectral properties of an image is the scanner's modulation transfer function (MTF). The MTF is the spatial frequency response of the imaging system. For various reasons, stemming from the hardware or physical phenomenon, the scanner cannot respond equally to all spatial frequencies. As a result, the actual information is distorted and effectively undergoes an undesired filtering process. For example, if the scanning system has difficulty in transferring high frequencies, the image will be blurred as demonstrated in Fig. 1.28.

The MTF can be obtained from the measured PSF by applying a Fourier transform (in 2D or 3D)

$$\text{MTF}(\overline{K}) = |F\{\text{PSF}(\overline{R})\}| \qquad (1.3)$$

where $\overline{R}$ designates the spatial location, $\overline{K}$ corresponds to the spatial frequencies, and F{ } represents the Fourier transform.

1.3.6   Contrast

In many clinical cases, the role of the medical image is to allow the physician to detect abnormalities in the body. For that aim, the physician relies mainly on three information sources: (i) the known anatomy of a healthy body, (ii) textural changes in tissues, and (iii) the differences in gray levels between adjacent areas, which is known as "contrast." Accordingly, we define the contrast in the context of medical imaging as follows:

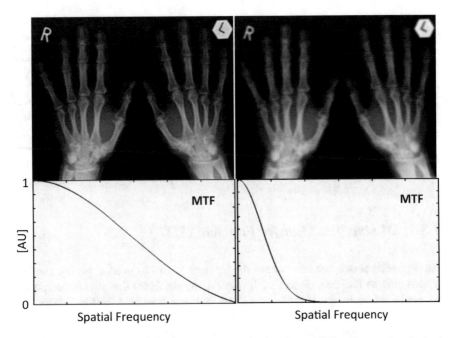

1

[AU]

0

Spatial Frequency   Spatial Frequency

Fig. 1.28 A demonstration of the modulation transfer functions (MTF) effect on the obtained image. (Left) A wide band MTF will show more details in the reconstructed image, while a narrow band (Right) will induce loss of information and induce reduction in image quality (such as blurring as shown here). MTF plots are depicted in normalized units of log amplitude ([AU])

Definition

Contrast

The mean difference in gray levels in a medical image between a region of interest (ROI) and that of the tissues that surround it

The term *mean* was applied here since it is very rare in current imaging scanners to have meaningful information from a single pixel. The use of a region of interest (ROI) is very common in medical image handling since most clinical abnormalities detectable by radiological diagnosis are localized. To present a clinical example, consider the following MR image which depicts a cross-sectional view of a brain (Fig. 1.29). As can be observed, there is a round spot (marked by the arrow) which appears in an irregular location and which has brighter gray levels than the tissue which surrounds it. The spot (which is the ROI in this case) is suspected to indicate a multiple sclerosis lesion. The higher the contrast between the suspicious spot and its surroundings healthy tissue, the easier and more reliable is the detection of abnormalities.

In order to improve the detection of abnormalities such as lesions and tumors, it is desirable to maximize the visible contrast. For that purpose, an arsenal of materials

Fig. 1.29 An MR cross-sectional image of a brain. The contrast between the suspicious ROI (marked by the arrow) and the background tissues may indicate the existence of a multiple sclerosis lesion

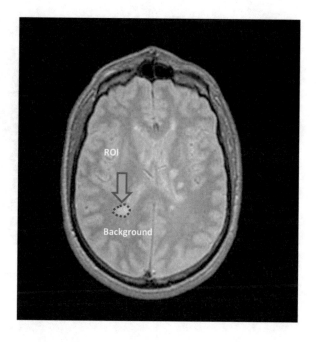

called "contrast-enhancing agents" (CEA) has been developed. These materials include (among other options) gadolinium-based compounds for MRI, iodine- or barium-based compounds for x-ray imaging, and microbubbles for ultrasound. When introduced into the body (by injection or ingestion), the CEA enhance substantially the contrast, allowing better and more reliable diagnosis.

The contrast can be enhanced either by increasing the gray levels of the ROI relative to the background or by reducing it. As reported already in 1860 (see [32]), the threshold contrast for the human eye is about 1% for a wide range of targets and conditions. This can be demonstrated by observing Fig. 1.30. Consider the three horizontal strips with their varying gray level values as a representative depiction of an ROI (the middle strip) and background (top and bottom strips). *Can you detect the column for which there is no contrast?* In order to realize how your eyes can fool you, put two pieces of a white paper with a small *vertical* gap between them. Slide slowly the pair of paper pieces and note how the contrast changes. *Can you detect it now?*

In order to quantify the contrast, several indices have been suggested. Three indices are commonly applied in vision science [32]. For a visual target on a uniform background, these indices are defined as follows:

(i) The Weber contrast is defined as

$$C = \frac{I_{\max} - I_{\min}}{I_{\text{background}}} \tag{1.4}$$

Fig. 1.30 A template demonstrating varying contrast values along the vertical direction. Note that the ROI represented by the middle strip can be either darker or brighter than the background (top and bottom strips)

where I_{max}, I_{min}, and $I_{background}$ are the maximal, minimal, and background luminance correspondingly.

(ii) The Michelson contrast is defined as

$$C = \frac{I_{max} - I_{min}}{I_{max} + I_{min}} \tag{1.5}$$

In this case the contrast is normalized and its value is bounded between 0 and 1 (i.e., its maximal value can be 1).

(iii) The root mean square (RMS) contrast is defined as

$$C = \frac{\sigma_I}{\bar{I}} \tag{1.6}$$

where $\bar{I}$ and σ_I are the mean and standard deviation luminance correspondingly.

In the context of medical imaging, it is suggested herein to use a variation of the Weber contrast, i.e.:

$$C = \left| \frac{\bar{I}_{ROI} - \bar{I}_{background}}{\bar{I}_{background}} \right| \tag{1.7}$$

where $\bar{I}_{ROI}$ and $\bar{I}_{bacground}$ are the mean gray level values in the region of interest (ROI) and the background, respectively. The absolute value is applied since the ROI can be less intense than the background, as, for example, when using iron oxide nanoparticles CEA in MRI.

1.3.7   Image Noise and Artifacts

Two "bad" features are mainly responsible for reducing image quality: (i) image noise and (ii) artifacts. Both of them are actually manifestations of the same thing: *erroneous visual information*. They induce into the image pixels or patterns of pixels which do not correspond to the real object and can mislead the observer of the image. While image noise is commonly random in nature and stems from deficiencies in signal quality or inclusion of unrelated signals, artifacts are commonly ordered in nature and present nonexistent objects or image features. Accordingly, in this textbook (other sources may have different definitions), the following definitions will be applied:

Definitions
- *Image Noise:* Stochastic variation in visual information
- *Image Artifacts:* Structured or non-random erroneous visual information

In order to study and quantify image noise and artifacts, we must establish first a "ground truth" image (also commonly referred to as "gold standard"). For that purpose, it is common to build a special object called a "phantom," which is defined as follows:

Definition
Phantom

   A physical or virtual object with precisely known geometry and physical properties, which can be used to assess the performance of the scanning system

The phantom may be made of any material that can be scanned by our system, be it plastic, glass, wood, metal, etc. A very simple phantom can be prepared simply by filling a test tube with water or other liquid solution. More complicated phantoms may attempt to mimic realistically an organ and may have dynamic parts, such as circulating fluids for imitating blood flow imaging or moving parts for simulating cardiac or respiratory motion. For assessing reconstruction algorithms and for computer simulations, it is common to use a numerical phantom. The most famous numerical phantom is the "Shepp-Logan" brain phantom introduced in 1974 by Larry A. Shepp and Benjamin F. Logan, from Bell Laboratories [33]. The numerical phantom is comprised of a set of ellipses which have varying "intensities" and are arranged in a manner that somewhat resembles a CT cross section of a brain. The phantom generated by using a MATLAB® code is depicted in Fig. 1.31.

The "Shepp-Logan" phantom   "Noisy Image"

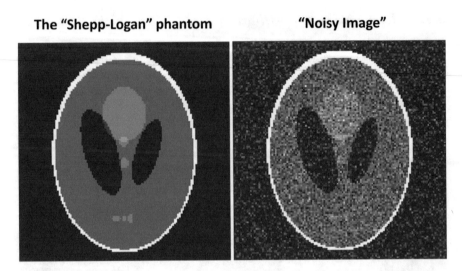

Fig. 1.31 (Left) A "clean" image of the Shepp-Logan numerical phantom. (Right) The same image contaminated by noise

As test objects for imaging, effective phantoms should have two competing characteristics: (i) Phantoms should have a known, measurable, configuration against which the image can be compared. (ii) Phantom image acquisition should emulate relevant imaging conditions, including the effects of anatomy. The first drives the phantom design to be simple, and the second drives the phantom design to be more complicated. The Shepp-Logan phantom illustrates these competing principles: on the one hand the contrasts and layout of the visible structures emulate CT contrast of brain anatomy and pathology, while on the other hand all the features are simple ellipses to facilitate mathematical analysis.

Noise and its removal are an extensively studied topic in the fields of signal and image processing and are beyond the scope of this textbook. Nonetheless, it is worthy to mention here some characteristic noise models. The first and commonly used is a *Gaussian noise* model. This model has a spectral probability density function (PDF), which is given by [34]

$$P(g) = \sqrt{\frac{1}{2\pi\sigma^2}} \cdot e^{-\frac{(g-\mu)^2}{2\sigma^2}} \tag{1.8}$$

where g is the gray level value, μ is the mean noise value (which in many cases is equal to 0), and σ is the distribution standard deviation.

In many cases the noise model is described as "white noise." This refers to a random signal which is assumed to have equal intensity for all the frequencies. In other words, its PDF has a constant power for the entire spectrum [34].

Commonly, the effect of the noise is modeled as an additive process, i.e., when added to an image, it will randomly alter the gray level of each pixel, i.e.,

$$\tilde{I} = I + n \tag{1.9}$$

where $\tilde{I}$ is the corrupted image, I is the source input image, and n is a noise image with a certain mean and preset variance.

The amount of changes inserted by the white noise is in many cases characterized by a Gaussian distribution. Thus, the model becomes "white Gaussian noise." (It should be noted that the term *white* is borrowed from optics where frequencies are associated with colors and an even mixture of all the colors yields white light. It has nothing to do with the display or color of a medical image.)

Another important noise model in medical imaging is the "speckle noise." This type of noise is associated with ultrasonic imaging. It is multiplicative in nature and follows the gamma distribution. The corresponding effect can be modeled by random value multiplications with pixel values of the image and is given by [35]

$$\tilde{I} = I + n \cdot I \tag{1.10}$$

where $\tilde{I}$ is the corrupted image, I is the source input image, and n is a noise image with zero mean and preset variance.

Another notable noise model is Poisson noise, which is also known as photon noise. This model may be applicable to gamma ray-based imaging in which the distribution is characterized by the Poisson distribution:

$$P(t) = \frac{\lambda^t \cdot e^{-\lambda}}{t!} \tag{1.11}$$

where t is an integer (i.e., $t = 1, 2, 3, 4\ldots$) and λ is the mean and also the variance.

There are naturally other noise models, and the most suitable one should be applied for each scanning system after studying its various sources. As stated above, typically, noise is random in nature and merely corrupts the image without depicting particular patterns. Artifacts, on the other hand, are characterized by structured obstructions or structured artificial enhancements in the image. Their source may be some systematic flaw in the data acquisition process or image reconstruction. Most common in the context of medical imaging are artifacts caused by metallic objects. These may occur, for example, in x-ray CT imaging of organs containing orthopedic or dental implants. But they may also be caused by patient motion during the scan (mainly in MRI) and even insufficient data acquisition. Some exemplary artifacts are depicted in Fig. 1.32.

Reconstruction Artifact   MRI Artifact   CT Artifact

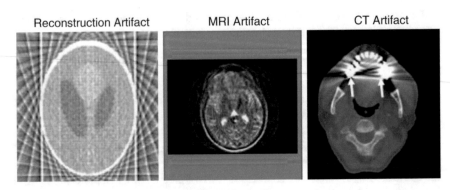

Fig. 1.32 (Left) Streaking artifacts caused by image reconstruction from too few projections. (Middle) MRI motion artifact which occurred due to patient motion. (Image provided by GE Healthcare). (Right) X-ray CT image depicting metal artifact indicated by the arrows

1.3.8 Signal-to-Noise Ratio (SNR) and Contrast-to-Noise Ratio (CNR)

In order to assess the quality of the scanner and its produced images in quantitative terms, several indices can be used. The most commonly used one is the signal-to-noise ratio (SNR). High SNR indicates that the image is "cleaner" and superior to a similar image with a low SNR. This index is applied in many fields that contain signal or image processing. Furthermore, there are several alternative definitions for the SNR (e.g., see a review in [36]). The basic definition is as follows:

$$
\begin{aligned}
\text{SNR} &= \frac{\text{Signal Power}}{\text{Noise Power}} = \frac{E\{\text{Sig} \times \text{Sig}^*\}}{E\{\text{Noise} \times \text{Noise}^*\}} \\
\text{SNR}_{\text{dB}} &= 20 \cdot \log_{10} \left\{ \frac{E\{|\text{Sig}|\}}{E\{|\text{Noise}|\}} \right\}
\end{aligned}
\tag{1.12}
$$

where E designates the expected value operator and the $*$ sign here designates complex conjugation. The bottom row provides the SNR value in decibel (dB) terms. Note that the number 20 in the multiplication stems from the fact the power is proportional to the square of the amplitude and decibels are typically calculated as: $10 \times \log_{10}$ (measured power/reference power).

The practical problem associated with this definition is how to know what the actual net signal power is and what is the net noise power. Obviously, when using numerical phantoms such as the Shepp-Logan, the precise values can be determined by dividing the sum of squares of all the pixels in the "clean" image by the sum of squares for the net noise image. However, under realistic circumstances, the best we can do in many cases is to scan a phantom and analyze its image features. Accordingly, if relevant, one can scan a homogenous object (a simple phantom) and use the reciprocal value of the coefficient of variance as an alternative estimate of the SNR:

$$\text{SNR} = \frac{|\bar{I}_{\text{object}}|}{\sigma_{\text{object}}} \tag{1.13}$$

where $\bar{I}_{\text{object}}$ is the mean gray level value of the object in the image and σ_{object} is the corresponding standard deviation of the gray levels for the object.

The contrast-to-noise ratio (CNR) indicates a metric for the visibility of an object of interest. The higher the CNR value, the better are the chances that an abnormality would be visually detected. However, the quantification of the CNR is not unique and has several alternative definitions. For example, [36] lists five different definitions used in fMRI. One modified definition suggested herein is

$$\text{CNR} = \frac{|\bar{I}_{\text{ROI}} - \bar{I}_{\text{background}}|}{\sigma_{\text{Noise}}}$$

$$\text{CNR}_{\text{dB}} = 20 \cdot \log_{10}\left\{\frac{|\bar{I}_{\text{ROI}} - \bar{I}_{\text{background}}|}{\sigma_{\text{Noise}}}\right\} \tag{1.14}$$

where σ_{Noise} designates the noise standard deviation. In practical terms, this value can be estimated by measuring the variance in a region within the image which evidently do not contain any object, such as air or the variance in a region near the ROI that is approximately homogenous. Another, useful estimate of the CNR is given by

$$\text{CNR} = \frac{|\bar{I}_{\text{ROI}} - \bar{I}_{\text{background}}|}{\sqrt{\sigma_{\text{ROI}}^2 + \sigma_{\text{background}}^2}} \tag{1.15}$$

where $\bar{I}_{\text{ROI}}, \bar{I}_{\text{background}}$ are the mean gray levels in the region of interest and the background, respectively, and $\sigma_{\text{ROI}}, \sigma_{\text{background}}$ are the standard deviations within the ROI and the background, respectively.

A key point to remember is that the metrics of image quality reviewed here and elsewhere are useful in the clinic insofar as they can indicate what constitutes a diagnostically relevant image. What engineers or scientists see as the "better" image is not necessarily true, medically speaking. Beyond image quality, each medical test carries its own burdens: economic, ergonomic, and associated risks. For practical purposes, the clinicians are the arbiters of what medical image is effective for finding pathology.

1.3.9   Detectability

Detectability describes the ability of an imaging system to render a feature of interest, such as a lesion, in a manner that it can be discerned by the reader. Because human perception is a part of the detection process, detectability is, to some degree, subjective. Regardless, rules of thumb have been developed that can provide guidance as to the limits of detectability in an imaging system. Detectability can be expressed as the conditions under which a feature of interest is visually evident. These conditions typically include the feature size, its contrast to the background, and the level of noise in the image. Other parameters might be included, such as the luminance or the intensity of the image. Medical imaging usually employs displays with dynamically adjustable display brightness, so luminance in itself is less of an issue. However, the dynamic range and the spatial resolution of the displays, in addition to the room lighting conditions under which the image display is viewed, are all parameters that can affect detectability, even though they are not part of the image acquisition device. For the purpose of this discussion, spatial resolution, contrast, and noise are considered with respect to their effects on detectability, but the topic is much broader.

A large feature with high contrast compared to its surroundings in a noiseless image will be far easier to detect than a small feature, with a signal of similar magnitude to the background, in a noisy image. The detectability of the former is greater than the latter. Considering a small lesion, the system must have sufficiently good spatial resolution so that its signal is not so blurred as to become indistinguishable from adjacent features or background. A lesion may have a distinguishable signal when its intensity is higher than that of the background. An x-ray CT of a lung tumor is an example: a tissue mass of the lesion is surrounded by the much less dense material of the lung. Conversely, the lesion signal might be lower than that of the background. For example, healthy myocardial tissue will become well perfused with a radiotracer in a nuclear medicine heart scan giving a bright image. An ischemic region will appear as an absence of radiotracer, with lower intensity on the image, indicating a lack of blood flow. Taking contrast (C) as the difference between the lesion signal (S) and the background signal (B) normalized to the background signal, then $C = (S - B)/B$, similar to Eq. 1.7. The first case has positive contrast and the second case has negative contrast. If the magnitude of the lesion contrast is greater than the minimal contrast that can be represented by the system, the lesion is detectable.

Although dynamic range may be a limiting factor to this threshold contrast, more commonly image noise is a major factor. All the imaging methods described above have image noise. For x-ray and nuclear medicine systems, the signal is generated by photon counting devices which follow Poisson statistics in the recording of events used to make the image. Clearly, if the level of image noise is about the same magnitude as the lesion contrast, the lesion may not be detectable since its signal is washed out or lost in the noise (σ). Traditionally, a contrast-to-noise ratio (R) was defined as $R = C/\sigma$, and a requirement of $R > 1$ was used to indicate detectability: i.e., the contrast exceeds the noise. However, this criterion was found to be insufficient because, among other considerations, it did not exclude the possibility of false detection errors: perceiving a lesion in the noise when, in truth, none exists. Fundamental studies on feature detectability established more practical conditions such as the Rose criteria [36]. For photon counting systems, if n is the number of photons per square centimeter (photon count density), and d is the lesion size, then

$$nd^2C^2 \geq k^2 \tag{1.16}$$

where k ranges from 3 to 5 depending on imaging conditions. For example, k might be smaller for clinical medical images with fewer pixels. For photon counting systems, the noise decreases with increased photon count density. Accordingly, if lesion size or contrast magnitude becomes smaller, the photon count density must increase in order to reduce the noise and maintain lesion detectability.

Figure 1.33 shows nine simulations of a nuclear medicine image with a 64 × 64 matrix and a nominal pixel size of 1 mm. The contrast of the four lesions in each is 2, and their diameters are 2, 3, 4, and 5 mm. For the upper left-hand image, the photon count density is 300 counts/cm^2, and the spatial resolution is 5 mm FWHM. Each column of images has the same photon count density, and each row of images has the same spatial resolution. The images in the two columns to the right of the upper left-hand image increase the photon count density by factors of 4 and 16, respectively. The images of the two rows below the upper left-hand image improve the spatial resolution by factors of 2 and 4, respectively. Taking $k = 5$ and applying Eq. 1.16 for the first column of images yield $300d^2 2^2 \geq 5^2$ or $d \geq 1.4$ mm. The lower left-hand image shows the case where the resolution (5/4 = 1.25 mm FWHM) is approximately the pixel size (1 mm), and we can see that the small lesions are at the limits of detectability, consistent with the calculation from Eq. 1.16. The image at the lower right with the best spatial resolution and the highest photon count density has the best detectability and all the lesions are clearly visible.

Systems should be designed to optimize detectability for clinical tasks. Given a system design, greater detectability usually comes at a cost. Increasing acquisition time can be used to improve the spatial resolution of an MR scan or the count density of a nuclear medicine scan, potentially improving detectability. Increasing the tube current for x-ray or x-ray CT imaging improves the photon count density, and likely the detectability, but increases the patient radiation exposure. Good physical design and engineering principles can help optimize detectability.

1.3.10   The Receiver Operator Curve (ROC)

Receiver operator curves are a method of representing how effective a test is; in this case, a medical imaging test. When evaluating medical tests, they are compared against a "gold standard" which is accepted as the true result. For example, an x-ray image indicating a metastatic lesion might be compared to the laboratory results of a biopsy of the lesion. If the histological results of the biopsy show that the lesion was indeed metastatic, then the "gold standard" of the biopsy confirms the "positive" x-ray image as a "true positive": pathology has been confirmed. However, if the biopsy indicates normal tissue, the "positive" x-ray image is a "false positive" in that the image is positive for pathology but it is false: no pathology exists. Conversely,

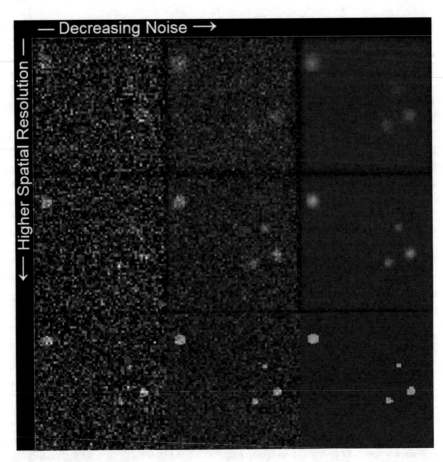

Fig. 1.33 Simulated nuclear medicine images. Four simulated lesions that are not clearly evident in the upper left-hand image that has the poorest spatial resolution and the highest noise levels due to low photon count density. Images in each column have the same level of noise, and in each row they have the same spatial resolution. The image at the bottom right has the lowest noise and highest spatial resolution showing good detectability of all four lesions. Consistent with the Rose criteria, in the image at the lower left, the smaller lesions are at the limits of detectability

perhaps the x-ray is clean and therefore "negative" for pathology. A biopsy of the region of interest may confirm normal tissue, indicating that this image is a "true negative": both the image and gold standard are negative for pathology. However, if the image is negative and the biopsy is positive, this imaging test is a "false negative," indicating no disease when there is disease. This is merely an example. Gold standards are established by consent within the medical community and are typically the most accurate tests. Gold standards, like surgical confirmation, might be too difficult, expensive, or risky to apply to a large population, so other tests like medical imaging are often applied first.

The medical question as to how successful a test is at detection is answered by a concept called sensitivity [37]. This is not the same "sensitivity" that physicists and engineers use when describing the efficiency of detectors in a medical image scanner. Here, sensitivity is the fraction of positive images that are truly positive. A companion measure is how successful the test is at determining that there is no pathology. This is the fraction of negative images that are true negatives and is called specificity. The need for both sensitivity and specificity becomes obvious if we propose, for example, an imaging test for lung cancer that is positive if the patient has lungs. Surely, this will "catch" all the pathology in that all the true positives will be called positive: our sensitivity is 100%. However, no one is excluded so there are no negatives detected, so our specificity is 0%. The test is not specific: it doesn't exclude patients without pathology. Such a "test" is, of course, useless. Ideally, both the sensitivity (p) and specificity (s) are 100%, indicating that all positive images are truly positive for pathology and all the negative images are truly negative for pathology.

Finally, very few tests give a yes or no answer. Reading an equivocal x-ray as positive for pathology will likely have a certain level of confidence associated with the radiologist reading it. In some studies, the radiologist rates his or her confidence on a numerical scale, 0 to 10, for example. A score of 0 in our example above would indicate definitely no metastatic lesion as per the x-ray image, whereas a score of 10 is a definite "yes." Perhaps the intensity or size of a lesion on an image can be measured quantitatively and some cutoff point chosen for indicating a positive image. A receiver operator curve (ROC) plots the sensitivity (p) versus 1 minus the specificity ($1 - s$). The upper right-hand corner of such a plot is the case mentioned above where the sensitivity is 100% and the specificity is 0%, and the lower left-hand corner, the origin, is the opposite case. A diagonal line between these corners represents "guessing": i.e., a test that gives random results.

The ROC is obtained by plotting p vs. ($1 - s$) for various levels of our scale. First p and ($1 - s$) are plotted for the most confident reading (e.g., a score of 10), then the next most confident reading (a score of 9 and above), etc. Ideally, we have scores of only 10s and 0s, and we're completely accurate so the curve is 1.0 throughout and the area under the curve (AUC) is 1.0. More likely, the curve is less than ideal (see Fig. 1.34) and is below 1.0 throughout and the area under the curve is less than 1.0. Conversely, if the test is always exactly wrong, the area under the curve is 0. The area under the curve is the probability that given two images, one truly positive and one truly negative, the radiologist will correctly identify the positive image. The area under a 45° diagonal line mentioned above is 50%: indicating "guessing." These curves can be used to compare different tests in that a higher AUC indicates a superior test, all other things being equal. Thus, generating ROCs can help answer questions of "detectability": a reader could be presented with images known to contain lesions or not, in order to compare lesion detectability under different image quality conditions, or between two different scanners. With more information about the patient population, they can also be used to help decide on an appropriate cutoff criterion for indicating a "positive" image, whether that be a confidence level or some other metric.

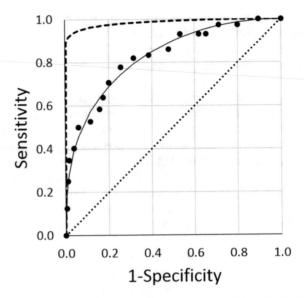

Fig. 1.34 Receiver operator curve (ROC) sensitivity vs. 1 − specificity. Simulated sensitivity and specificity data are used here to build an ROC curve (dots ●) by varying levels of confidence in reading a medical image as "positive" for pathology. A fitted curve (solid line —) gives an area under the curve (AUC) of about 0.85. The diagonal (dotted line ·····) represents a test no better than "guessing" with an AUC of 0.5. A curve representing an extremely good test (dashed line − − −) has an AUC of approximately 1

1.4   Standard Views in Radiology

Finally, it should be mentioned that three standard views are used in clinical radiology. These views are schematically depicted in Fig. 1.35. They are defined as follows:

- *Coronal view* – corresponds to images depicting planar cross sections which are parallel to the scanner's bed.
- *Axial view* – corresponds to images depicting planar cross sections which are orthogonal to the scanner's bed. Its orientation relative to the patient is from side to side (i.e., left to right).
- *Sagittal view* – corresponds to images depicting planar cross sections which are perpendicular to the two other planes. It is orthogonal to the scanner's bed but its orientation relative to the patient is from top to bottom.

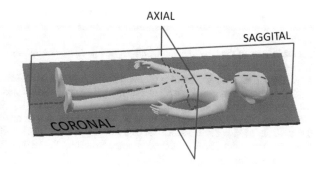

Fig. 1.35 The three standard views used in radiology

References

1. The Bible, Deuteronomy, Ch. 28; 29.
2. http://www.nobelprize.org/nobel_prizes/physics/laureates/1901/rontgen-bio.html.
3. Röntgen W. Ueber eine neue Art von Strahlen. Vorläufige Mitteilung. In: Aus den Sitzungsberichten der Würzburger Physik.-medic. Gesellschaft Würzburg; 1895. p. 137–47.
4. Reed AB. The history of radiation use in medicine. J Vasc Surg. 2011;53:3S–5S.
5. Hounsfield GN. Computerized transverse axial scanning (tomography): part 1. Description of system. Br J Radiol. 1973;46:1016–22.
6. Hertz S, Roberts A. Radioactive iodine in the study of thyroid physiology, VII: the use of radioactive iodine therapy in hyperthyroidism. JAMA. 1946;131:81–6.
7. de Hevesy G. The absorption and translocation of lead by plants: a contribution to the application of the method of radioactive indicators in the investigation of the change of substance in plants. Biochem J. 1923;17:439–45.
8. International Atomic Energy Agency. Cyclotron produced radionuclides: physical characteristics and production methods, Technical Reports Series No. 468. Vienna: IAEA; 2009.
9. Fischer D. History of the International Atomic Energy Agency: the first forty years. IAEA-STI-PUB-1032. Vienna: IAEA; 1997.
10. Seidlin SM, Marinelli LD, Oshry E. Radioactive iodine therapy effect on functioning metastases of adenocarcinoma of the thyroid. JAMA. 1946;132:838–47.
11. Cassen B, Curtis L, Reed C, Libby R. Instrumentation for ^{131}I use in medical studies. Nucleonics. 1951;9:46–50.
12. Anger HO. Scintillation camera. Rev Sci Instrum. 1958;29:27–33.
13. Kuhl DE, Edwards RQ, Ricci AR, et al. The Mark IV system for radionuclide computed tomography of the brain. Radiology. 1976;121:405–13.
14. Phelps ME, Hoffman EJ, Mullani NA, Ter Pogossian MM. Application of annihilation coincidence detection of transaxial reconstruction tomography. J Nucl Med. 1975;16:210–5.
15. Hasegawa BH, Gingold EL, Reillly SM, Liew SC, Cann CE. Description of a simultaneous emission-transmission CT system. Proc SPIE. 1990;1231:50–60.
16. Townsend DW, Beyer T, Kinahan PE, Brun T, Roddy R, Nutt R, et al. The SMART scanner: a combined PET/CT tomograph for clinical oncology. In: Conference record of the 1998 IEEE nuclear science symposium, vol. 2; 1998. p. 1170–4.
17. Ben-Haim S, Kennedy J, Keidar Z. Novel cadmium zinc telluride devices for myocardial perfusion imaging—technological aspects and clinical applications. Semin Nucl Med. 2016;46(4):273–85.

18. Catana C, Wu Y, Judenhofer MS, Qi J, Pichler BJ, Cherry SR. Simultaneous acquisition of multislice PET and MR images: initial results with a MR compatible PET scanner. J Nucl Med. 2006;47:1968–76.

19. Woo J. A short history of the development of ultrasound in obstetrics and gynecology. http://www.ob-ultrasound.net/history1.html. Accessed 12 Sept 2011.

20. Greenleaf JF, Johnson SA, Lee SL, Herman GT, Wood EH. Algebraic reconstruction of spatial distributions of acoustic absorption within tissue from their two-dimensional acoustic projections. In: Acoustical holography, vol. 5. New York: Plenum Press; 1974. p. 966–72.

21. Ophir J, Céspedes I, Ponnekanti H, Yazdi Y, Li X. Elastography: a quantitative method for imaging the elasticity of biological tissues. Ultrason Imaging. 1991;13(2):111–34.

22. Ophir J, Alam SK, Garra BS, Kallel F, Konofagou EE, Krouskop T, Merritt CRB, Riggetti R, Souchon R, Srinivasan S, Varghese T. Elastography: imaging the elastic properties of soft tissues with ultrasound. J Med Ultrason. 2002;29:155–71.

23. Bercoff J. In: Minin O, editor. Ultrafast ultrasound imaging - medical applications: InTech; 2011. p. 1–24. https://doi.org/10.5772/19729. Available from: https://www.intechopen.com/books/ultrasound-imaging-medical-applications/ultrafast-ultrasound-imaging.

24. Yao J, Wang LV. Photoacoustic tomography: fundamentals, advances and prospects. Contrast Media Mol Imaging. 2011;6(5):332–45.

25. https://nationalmaglab.org/education/magnet-academy/history-of-electricity-magnetism/pioneers/isidor-isaac-rabi. Accessed July 2017.

26. Kauffman G. Nobel prize for MRI imaging denied to Raymond V. Damadian a decade ago. Chem Educ. 2014;19:73–90.

27. Aue WP, Bartholdi E, Ernst RR. Two-dimensional spectroscopy. Application to nuclear magnetic resonance. J Chem Phys. 1976;64:2229. https://doi.org/10.1063/1.432450.

28. Stehling MK, Turner R, Mansfield P. Echo-planar imaging: magnetic resonance imaging in a fraction of a second. Science. 1991;254(5028):43–50. http://www.jstor.org/stable/2879537.

29. Le Bihan D, Breton E, et al. MR Imaging of intravoxel incoherent motions: application to diffusion and perfusion in neurologic disorders. Annual Meeting of the RSNA, Chicago, 1985.

30. Sodickson DK, Manning WJ. Simultaneous acquisition of spatial harmonics (SMASH): fast imaging with radiofrequency coil arrays. Magn Reson Med. 1997;38(4):591–603.

31. Lustig M, Donoho D, Pauly JM. Sparse MRI: the application of compressed sensing for rapid MR imaging. Magn Reson Med. 2007;58:1182–95. https://doi.org/10.1002/mrm.21391.

32. Pelli DG, Bex P. Measuring contrast sensitivity. Vis Res. 2013 September 20;90:10–4. https://doi.org/10.1016/j.visres.2013.04.015.

33. Shepp LA, Logan BF. The Fourier reconstruction of a head section. IEEE Trans Nucl Sci. 1974;21(3):21–43.

34. Boyat AK, Joshi BK. A review paper: noise models in digital image processing. Signal Image Process Int J (SIPIJ). 2015;6(2):64–75.

35. Welvaert M, Rosseel Y. On the definition of signal-to-noise ratio and contrast-to-noise ratio for fMRI data. PLoS One. 2013;8(11):e77089. https://doi.org/10.1371/journal.pone.0077089.

36. Rose A. Vision: human and electronic. New York: Plenum Press; 1973.

37. Zhou X-H, Obuchowski NA, McClish DK. Statistical methods in diagnostic medicine. 2nd ed. Hoboken: Wiley; 2011.

Chapter 2
Basic Principles of Tomographic Reconstruction

Synopsis: In this chapter the reader is introduced to the basic principles and tools of tomographic reconstruction. The chapter is divided into two sections: Part I provides the basics of computed tomography. Part II describes more advanced descriptions and methods.

The learning outcomes are: The reader will understand what is the relation between information collected in the form of projections and the imaged objects, will comprehend the different approaches that can be utilized for data collection and image reconstruction, and will be able to apply this knowledge in order to implement basic algorithms for image reconstruction and assess their performance.

2.1   Introduction

The word tomography is derived from ancient Greek: τόμος (tomos), which means "slice, section" and γράφω (graphō), which means "to write". A slice is a thin cross section taken through the object as demonstrated in Fig. 2.1.

Tomographic reconstruction is the process of generation of cross-sectional images of the body (and sometimes also 3D images) from multitude of external views. It is a key image formation element and is essential for realization of the clinical potential of major imaging modalities. The unique strength of tomographic imaging is its ability to depict the internal structures within the body and visualize tissues and physiological processes inside it in a noninvasive manner.

© Springer Nature Switzerland AG 2020
H. Azhari et al., *From Signals to Image*,
https://doi.org/10.1007/978-3-030-35326-1_2

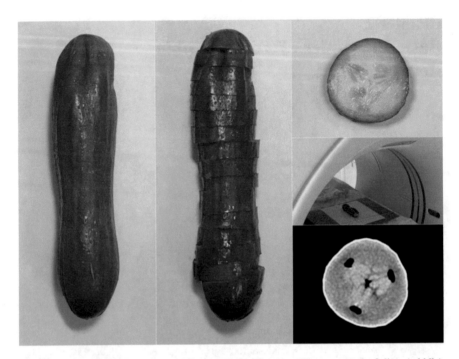

Fig. 2.1 The tomographic concept. A 3D object (left). The same object as a stack of slices (middle), consisting of many individual cross-sectional slices (upper right). A CT scan (middle right) provides a reconstructed CT slice (lower right) showing the internal structure of an individual slice

Both theory and practice of tomographic reconstruction are closely linked with multiple areas of basic and applied science. Among them are linear algebra, statistics, physics, optimization theory, signal processing, and more.

While basic image reconstruction algorithms date back (mostly) to the mid-twentieth century, significant developments are constantly being made. Below are just a few milestones in the history of medical image reconstruction:

- The Radon transform and its inverse transform which was introduced by Johann Radon (1917) [1].
- Algebraic reconstruction by Kaczmarz (1937) [2].
- Successive substitution vs direct Fourier methods (1956) [3].
- Iterative method for emission tomography (1963) [4].
- X-ray CT imaging (1972) [5].
- Algebraic reconstruction tomography (ART) (1970) [6].
- Weighted least squares for 3D-SPECT (1972) [7].
- Richardson/Lucy iteration for image restoration (1972, 1974) [8, 9].
- Proposals to use Poisson likelihood for emission and transmission tomography (1976, 1977) [10, 11].

- Expectation-maximization (EM) algorithms for Poisson model by Shepp and Vardi (1982) [12].
- Approximate filtered back-projection (FBP)-style algorithm for circular CT (1984) [13].
- Bayesian Poisson emission reconstruction (1985) [14].
- Ordered subsets EM (OSEM) algorithm by Hudson and Larkin (1994) [15].
- First exact FBP-like reconstruction for cone beam tomography (2002) [16].

Contemporary reconstruction solutions are tailored to the imaging modality, acquisition system, acquisition protocol (i.e., geometry and dose), and the target clinical application (meaning the information of clinical importance to be delivered by the image).

Among the key development drivers of medical tomographic reconstruction is the constant introduction of fast, low-cost, and miniature computers allowing practical implementation of more realistic, albeit complicated, solutions. Another driver is the development of novel imaging systems and data acquisition schemes, requiring development of dedicated algorithms. The third factor is the development of novel imaging applications. The need to reduce the radiation dose required to obtain clinically useful images gave rise to research on noise properties of reconstructed images and the development of novel applied approaches. Machine learning and artificial intelligence tools allow for further refinement and optimization of algorithms.

Theory and practice of tomographic reconstruction for medical applications continue to be an area of active research.

2.2 Part I: Basic Principles of Tomography

2.2.1 Projections

Projections are the fundamental building blocks of the data used in the tomographic process. Conceptually, a projection can be considered as a partially transparent shadow cast by the studied object onto the detectors when some kind of energy, commonly in the form of radiation, passes through it. The basic idea is schematically depicted in Fig. 2.2. Here, for example, a vertical ray passing through the object will change its properties according to the relevant properties of the tissues it encounters along the interaction path. A detector positioned on the other side will quantify one of the modified properties (e.g., amplitude, energy, arrival time, etc.). When scanning an object, the data is collected along a predefined trajectory by many such rays. For example, in this case, the values collected along the horizontal line constitute the projection onto the x-axis.

More generally speaking, data collected in tomographic modalities are divided into *transmission* and *emission* modes. In transmission tomography, the source of energy that is used to generate the signal is located *outside* the object being imaged.

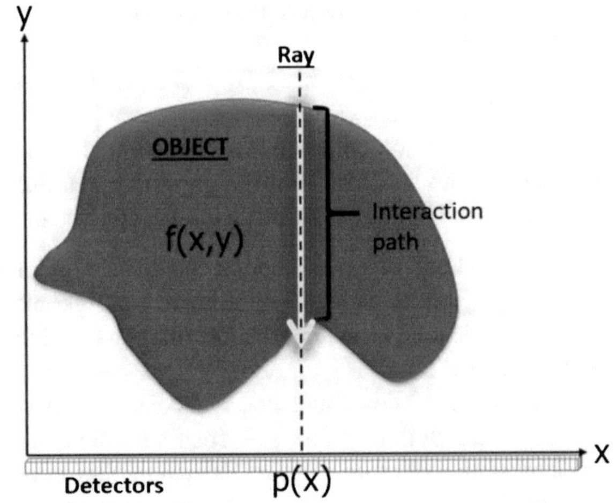

Fig. 2.2 A ray passing through an object may change its properties. The detected values of the ray's properties on the other side (the horizontal line in this case) constitute the projection

Fig. 2.3 Schematic depiction of transmission and emission tomography modes. (Top) projection data formation in x-ray CT; (middle) projection data formation in SPECT; (bottom) projection data formation in PET

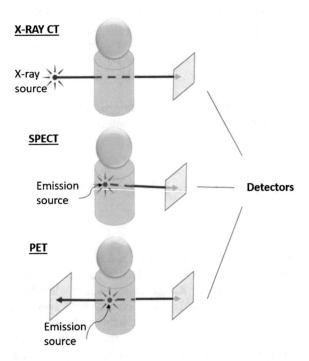

In emission tomography, the *object itself* is the source of the signal. X-ray CT (computed tomography) and ultrasound are examples of transmission tomography, while SPECT (single photon emission CT) and PET (positron emission tomography) represent emission tomography. These types of data collection modes are schematically depicted in Fig. 2.3. Also, historically, some of the first MRI acquisitions used a projection imaging technique that could be modeled as an emission process.

Fig. 2.4 A projection
through the object onto a
detector array positioned at
angle θ. The line integrals
along the multiple detection
paths result in the
corresponding detected
projection elements

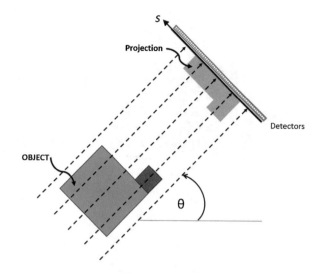

Most imaging systems acquire data in the form of projections from multiple
viewing angles. A projection element constitutes an overlay, or, in mathematical
terms, line integral, of all information contributing signals along a detection path
L through the object f to the corresponding detector element s at angle θ (see
Fig. 2.4).

$$p(s, \theta) = \int_{-\infty}^{+\infty} f(x, y)\,dl \tag{2.1}$$

where $p(s, \theta)$ is the corresponding projection value. The infinite borders of the
integral are just set for mathematical convenience in further derivations (e.g., to
comply with the Fourier transform, etc.). In practice, naturally, the integral is applied
only along the relevant distance L within the object.

The line integral transform $f(x, y) \rightarrow p(s, \theta)$ is known as the x-ray transform and is
given in its simplest discrete form by the sum

$$p(s, \theta) = \sum_{i,j \in L} f(i, j) \tag{2.2}$$

where $f(i, j)$ is the discrete presentation of the image and the indices $\{i, j\}$ designate
the relevant pixel addresses. It should be noted that the projection operation results in
the reduction of one dimension, i.e., $3D \rightarrow 2D \rightarrow 1D$. In our example, a projection of
a two-dimensional object results in a one-dimensional vector. Consequently, as a
result of the projection operation, important information may be lost.

This is demonstrated in the example shown in Fig. 2.5 for a 2D object represented
by a 5×5 matrix of pixels. The object has two regions of different size and signal

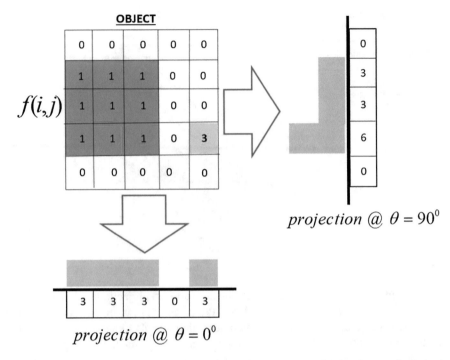

Fig. 2.5 Two one-dimensional projections of the same two-dimensional object obtained along orthogonal view directions yield two different forms of information about the object

intensity ("1" and "3") which are being captured by some hypothetical imaging system. Observing the two projections acquired along the two orthogonal detector positions, it can be noted that in the first detector position (projecting to the right), it is not possible to recognize the presence of the two separated regions. It is also not possible to localize the origin of the high-intensity signal detected along the fourth row. On the other hand, in the second detector position (projecting downward), we can deduce from the projection that there are two separate regions in the object. However, it is not possible to distinguish between the low-intensity region extending deep into the field of view and the small, high-intensity region that may represent a pathologic lesion.

In a similar manner, a projection of a 3D object is a 2D image. Examples of diagnostically useful 2D projections are an x-ray radiograph like a "chest x-ray" or planar imaging in nuclear medicine like a "whole-body bone scan." These 2D diagnostic images do not require image reconstruction per se and do not provide the 3D tomographic data discussed above.

Another quality related to projections is the associated change in contrast. For example, lesion detection is one of the frequent tasks addressed by emission tomography. Consider an example of an image comprising of 5 × 5 pixels with a "lesion" (designated by a gray level of 5) present within a uniform background (with gray levels of 1) as illustrated in Fig. 2.6. The "lesion" is clearly visible within the

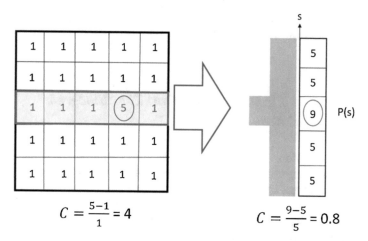

$$C = \frac{5-1}{1} = 4$$

$$C = \frac{9-5}{5} = 0.8$$

Fig. 2.6 A representative image comprising of 5 × 5 pixels with a simulated lesion (Left). The corresponding in-slice contrast estimator equals 4. (Right) The corresponding projection depicts a much smaller contrast. As can be noted the contrast estimator C is reduced from 4 to 0.8

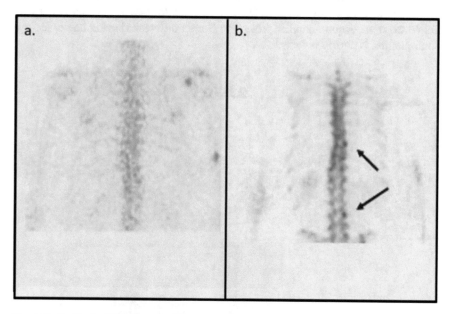

Fig. 2.7 (Left) A SPECT projection. The spine lesions are not detectable. (Right) The reconstructed coronal slice through the spine. Now the lesions are clearly visible

in-slice contrast (quantified by the number C). Contrary to that, the contrast is substantially reduced in the projection. This is also demonstrated in the clinical example shown in Fig. 2.7.

To conclude, it can be realized that in medical imaging, projection images alone may often possess low or even misleading diagnostic merit.

2.2.2   Tomographic Data Acquisition: The "Sinogram" and the Radon Transform

Despite their limitations, as stated above, projections are the main building blocks of tomographic imaging and often the only type of data available. Thus, the question is: How can we use them in a more beneficial manner?

The answer is simple, though so not simple to accomplish (as will be discussed in the following sections): *We can combine and use information provided by multiple projections acquired from different directions!*

To demonstrate this concept, consider the situation depicted schematically in Fig. 2.8. In this case a projection image of two cylinders (representing, e.g., two blood vessels) is obtained. As can be observed, it is impossible to determine from this image which blood vessel is in front of the other (or maybe they both intersect). However, if we acquire another projection from a different viewing angle, the true configuration is clearly revealed.

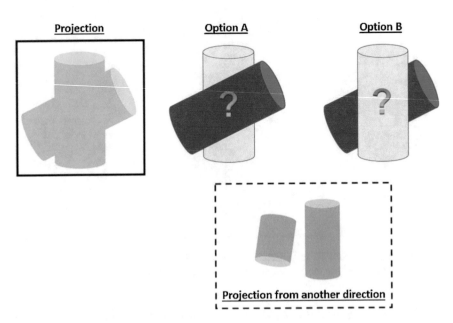

Fig. 2.8 A schematic demonstration for the need to acquire projection data from different viewing angles. (Left) A projection image of the two cylinders is ambiguous. We cannot determine whether Option A or Option B is correct. (Bottom) A projection acquired from a different viewing angle reveals the true configuration

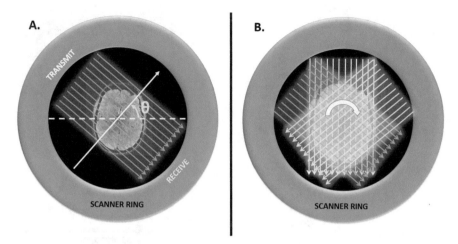

Fig. 2.9 (**a**, Left) A ring-shaped x-ray CT tomographic scanner has transmission and reception components both of which rotate together around the scanned object. Each projection is acquired at a predefined angle θ. (**b**, Right) A set of projections is collected around the object in order to generate a "sinogram"

It follows from the above that in order to reconstruct an image, we must acquire projection data from various viewing angles. Although generally speaking we can acquire information using any arbitrary configuration; handling this information is easier if the data is acquired in a systematic manner. The simplest configuration is to use a ring and acquire the projections at equiangular positions as schematically depicted in Fig. 2.9.

The physical property of each beam is detected by the corresponding detector. The detector transforms the physical property into a signal (commonly electrical), which is converted using an analog to digital card (ADC) into a number. The array of numbers collected from all the detectors represents the projection. The data collected in this manner is stored in the scanner's computer and arranged in a matrix called a "sinogram." Each row in the sinogram corresponds to a specific projection angle. Each column designates a specific detector address along the array of detectors used for acquisition. The process is schematically depicted in Fig. 2.10. An exemplary sinogram of an ellipse is depicted in Fig. 2.11. It is worth noting that the projections for $p(\theta)$ and $p(\theta + 180°)$ should be identical but flipped in the through-transmission mode.

It should be noted at this point that the procedure described above relates to a discrete form which is convenient for practical computerized processing. Nonetheless, the original transformation into projection data in a continuous form was originally developed mathematically by Johann Radon in 1917. Hence, the procedure is called the "Radon transform." To define the "Radon transform" in mathematical terms, let us first consider a single ray which is part of the projection view of the target object taken at angle θ. Let $\{x, y\}$ be the coordinates of reference and

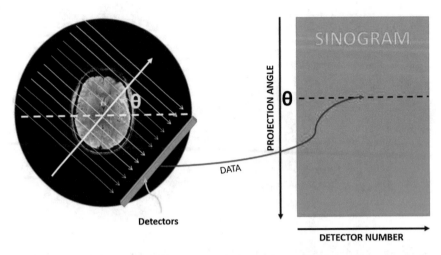

Fig. 2.10 (Left) Schematic depiction of the physical procedure applied for acquiring a projection. (Right) Each projection datum is digitized and stored in a matrix called "sinogram," where each row corresponds to a certain projection angle and each column to a specific detector

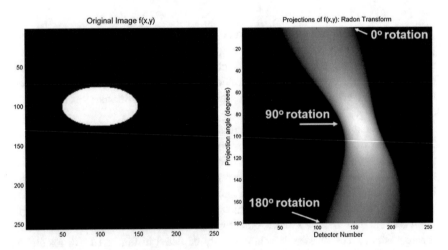

Fig. 2.11 (Left) An elliptic object in image domain. (Right) Its corresponding Radon transform (sinogram)

$\{x', y'\}$ be the rotated coordinate system at angle θ, where the ray is perpendicular to coordinate x', as shown schematically in Fig. 2.12.

Any point $Q(x, y)$ located on that ray is positioned at distance t from the coordinate y'. It can be easily shown that it is subjected to the following relation:

$$t = x \cdot \cos(\theta) + y \cdot \sin(\theta) \tag{2.3}$$

where t is practically the address of the detector that will sense that ray.

Fig. 2.12 A ray used for
projection, acquired at
viewing angle θ, is located
at distance t from the rotated
coordinate y'. Point $Q(x, y)$
is located on that ray

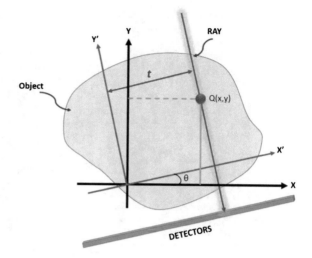

Next let us recall Dirac's delta function, which can be thought of heuristically as

$$
\begin{cases}
\delta(t) = \begin{cases} \infty & t = 0 \\ 0 & t \neq 0 \end{cases} \\
\text{s.t.} \\
\int\limits_{-\infty}^{\infty} \delta(t)dt = 1
\end{cases}
\tag{2.4}
$$

As can be noted, the delta function is non-zero only if its argument is zero. It can also be shown that

$$
\int\limits_{-\infty}^{\infty} \delta(\tau - t) \cdot f(t) \cdot dt = f(\tau)
\tag{2.5}
$$

Thus, the projection obtained from this ray is given by

$$
p(\theta, t) = \int\limits_{-\infty}^{\infty} \int\limits_{-\infty}^{\infty} f(x, y)\delta(t - x\cos\theta - y\sin\theta)dxdy
\tag{2.6}
$$

Although seemingly cumbersome, this analytical description is useful for the "filtered back projection" image reconstruction as will be shown in the following. The sinogram is not the only useful method of organizing projection data for reconstruction, but it is convenient for scanners which acquire data using a cylindrical geometry.

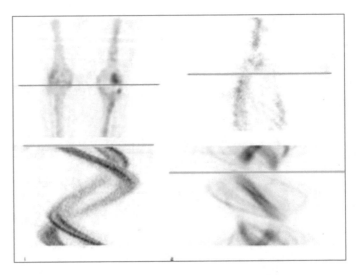

Fig. 2.13 Examples of two clinical SPECT projections (top) and their corresponding sinogram images (bottom). Left: an anterior knee projection. Features of high signal on the knees trace out a sine wave in the sinogram. Right: a chest SPECT projection acquired from the right side of the patient. The green line on the projection image represents the location of the slice for which the sinogram was calculated. The green line on the sinogram image designates the location of the projection view in the sinogram space

Sinogram
A sinogram (also known as the Radon transform of an object) is a format used for depicting the collected projection data, whereby (in the 2D case) each line corresponds to the projection angle and each column to one detector.

Two exemplary sinograms obtained from clinical SPECT scans are illustrated in Fig. 2.13.

Historical Note: Radon and X-ray Transforms
The Radon transform, introduced in 1917, establishes the relationship between an n-dimensional object and all $(n-1)$-dimensional "hyperplanes" passing through the object. In case of a two-dimensional image, the hyperplanes reduce to lines passing through the object, or line integrals.

Radon Transform Definition: For an n-dimensional object, the Radon transform is the collection of all $k = n - 1$ dimensional "hyperplanes" that pass through (or intersect) the object. For $n = 2$, $k = 1$ so the "hyperplanes" are

(continued)

just line integrals. For $n = 3$, $k = 2$ so the "hyperplanes" are now the 2D planes intersecting the object, not the same as line integrals (which is what our data are).

X-ray Transform Definition: For an n-dimensional object, the X-ray transform is the collection of all $k = 1$ dimensional "hyperplanes" (i.e. line integrals).

In 2D, as in a tomographic slice, the Radon and X-ray transforms are the same, but not in 3D.

2.2.3 Back Projection

Now the *BIG* question is: *How can we go back from the measured projections information (the sinogram) and reconstruct a tomographic image?*

There are more than several options to answer this question as will be outlined in the following sections. In fact, there are books and numerous articles dealing with this "inverse" problem. Before we continue, it is worth explaining why not use a simple straightforward approach. Conceivably, since every projection point actually provides a linear equation (see Eq. 2.2), it follows that if we have enough projection rays, i.e., at least equal to the number of the unknown pixels, then we can write a set of mutually dependent equations and solve algebraically for the value of each pixel. However, that approach is impractical (at least currently). Consider, for example, that we have an image consisting of 512×512 pixels. That means that we have to solve simultaneously a set of 262,144 mutually dependent equations. This is computationally a heavy task, not to mention the complications which may rise from measurement noise and overdetermined or underdetermined situations. Thus, we seek alternative and computationally less demanding approaches.

For the sake of clarity, we shall start with the simplest method termed: "back projection." The idea is very simple, recalling the fact that every projection point is actually the sum of all the pixel values along the beam which was used to generate it. We count the number of participating pixels and assign to each of those pixels an equal value. That is to say that if the projection value for beam β is p_β, then each participating pixel is assigned a value of p_β/m, where m is the number of pixels contributing to beam β. The result is a line of pixels tracking the beam's trajectory. This is demonstrated graphically in Fig. 2.14 for two beams passing through a 5×5 pixel image.

To explain how this method can be applied for reconstructing an image, consider the following exemplary image of only 5×5 pixels and four of its projections as shown in Fig. 2.15.

Applying the back projection process yields the four matrices shown in Fig. 2.16 (on the left). The reconstructed image is obtained by summation of these four matrices. It should be pointed out that this is the simplest back projection implementation algorithm. More sophisticated algorithms can account accurately for

Back Projecting

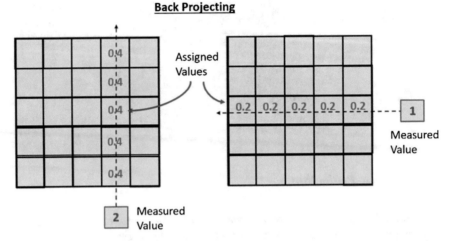

Fig. 2.14 A simple demonstration of the back projection operation. (Left) The measured vertical projection value was 2, and there were five pixels contributing to the beam. Thus, the assigned pixels value is 2/5 = 0.4. (Right) Similarly for this horizontal projection, the assigned value is 1/5 = 0.2

diagonal distances and for the beam width – as will be discussed in the following. As can be noted, the reconstructed image is not similar to the source image. Nonetheless, the four non-zero pixels in the source image had the highest gray levels in the reconstructed image. Adding more projections to the process may further improve the quality of the reconstructed image. For example, acquiring 180 projections (i.e., every 1°), for the ellipse shown in Fig. 2.11, and applying the back projection algorithm yield the reconstructions shown in Fig. 2.17.

The main advantage of the back projection algorithm is that it is very simple to implement. It can be further improved by accounting for the precise geometric configuration, i.e., normalizing to the actual projection line length, to the number of beams passing through each pixel and accounting for the actual beam width, i.e., taking into account the relative contribution of each pixel to each beam. The disadvantages are that the images are commonly blurred with diminished contrast as shown in Fig. 2.17 and artifacts and false contrast may appear as shown, for example, in Fig. 2.18. So how can this be corrected?

2.2.4   Algebraic Reconstruction Tomography (ART): The Projection Differences Method

In order to improve the reconstruction process, let us first examine the results obtained from the back projection algorithm for the 5 × 5 pixel image shown in Fig. 2.16. Comparing it to the source image, it is quite evident that the two differ substantially.

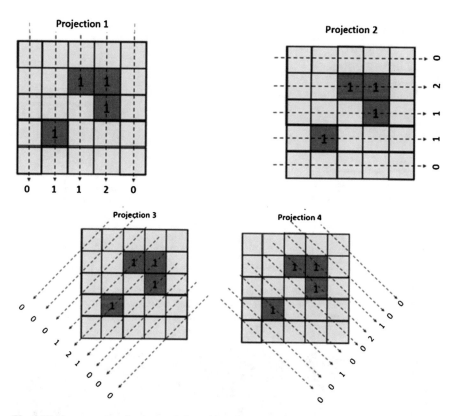

Fig. 2.15 An exemplary image consisting of 5 × 5 pixels and four of its projections. Red pixels have a value of 1 and the gray ones have a value of 0

However, one should recall that the only information available to us are the measured projections. We do not know what are the actual pixel values and how reliable is our reconstruction at this point. Nevertheless, we can obtain new insights by applying "forward projection" to the reconstructed image. Taking the reconstructed image, we can synthetically generate projections along the same directions of the measured ones. Then, comparing the measured projections to the calculated projections, we can obtain a clear indication that the reconstruction is currently wrong. This is a valuable quantitative source of information as shown in Fig. 2.19.

Now that we have both the measured projection for each beam p_β and the corresponding computed forward projection $\widetilde{p}_\beta$, we can calculate the discrepancy E_β for that beam

$$E_\beta = p_\beta - \widetilde{p}_\beta \tag{2.7}$$

If E_β is positive, that means that the reconstruction underestimated the pixel values. Thus, we have to increase their values. On the other hand, if E_β is negative,

Back Projection 1

0	0.2	0.2	0.4	0
0	0.2	0.2	0.4	0
0	0.2	0.2	0.4	0
0	0.2	0.2	0.4	0
0	0.2	0.2	0.4	0

Back Projection 2

0	0	0	0	0
0.4	0.4	0.4	0.4	0.4
0.2	0.2	0.2	0.2	0.2
0.2	0.2	0.2	0.2	0.2
0	0	0	0	0

Reconstruction

0.00	0.70	0.50	0.65	0.40
0.40	0.60	1.35	1.50	0.65
0.50	0.65	0.80	1.35	0.50
0.45	1.10	0.65	0.60	0.70
0.40	0.45	0.50	0.40	0.00

Back Projection 3

0	0	0	0.25	0.4
0	0	0.25	0.4	0.25
0	0.25	0.4	0.25	0
0.25	0.4	0.25	0	0
0.4	0.25	0	0	0

Back Projection 4

0	0.5	0.3	0	0
0	0	0.5	0.3	0
0.3	0	0	0.5	0.3
0	0.3	0	0	0.5
0	0	0.3	0	0

Fig. 2.16 Back projecting each "measured" projection onto its corresponding line of pixels ("beam") yields the four matrices shown on the left. Their summation yields the reconstructed image shown on the right. The highlighted zones correspond to the actual object

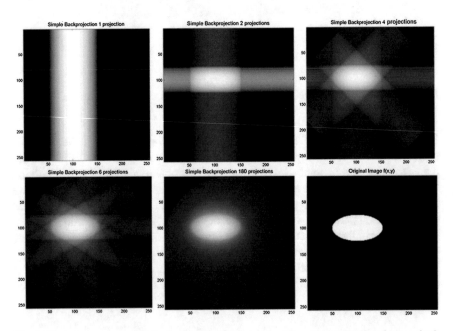

Fig. 2.17 A demonstration of the back projection reconstruction process. Starting from a single projection (top left) until using 180 projections (middle bottom). The source image is also depicted for comparison (bottom right)

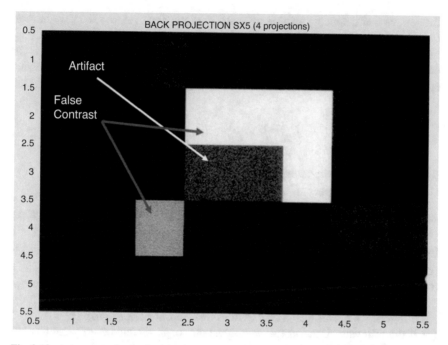

Fig. 2.18 A demonstrative back projection reconstruction of the 5 × 5 image depicted above. Note the artifact in the center and the false contrast between the square at the bottom left and the rotated L-shaped object at the top right side

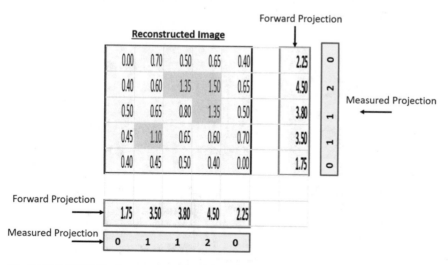

Fig. 2.19 (Middle) The image reconstructed from the four projections using the back projection algorithm. (Bottom) The corresponding vertical projection of the reconstructed image and the measured one. (Right) The corresponding horizontal projection of the reconstructed image and the measured one. Note the large discrepancy between the pairs

that means that the reconstruction overestimated the pixel values. Thus, we have to decrease their values accordingly. The idea of the projection differences reconstruction method is simply to back project across the m pixels each discrepancy E_β for each beam back onto the current reconstructed image. Thus, if the gray level value for a certain pixel after h iterations is $g_{i,j}^{(h)}$, then its corrected value $g_{i,j}^{(h+1)}$ would be

$$g_{i,j}^{(h+1)} = g_{i,j}^{(h)} + \frac{p_\beta - \widetilde{p}_\beta^{(h)}}{m} \tag{2.8}$$

where $\widetilde{p}_\beta^{(h)}$ is the corresponding forward projection value after h iterations.

In order to reconstruct an image, the following steps are implemented:

(a) Start with an initial guess of the image. If you do not have any a priori knowledge, then the first guess can be simply a matrix filled with zeros.
(b) For each relevant beam, calculate the forward projection $\widetilde{p}_\beta^{(0)}$.
(c) Using Eq. 2.8 correct the pixel values for each beam and for all the beams.
(d) Compare quantitatively the newly reconstructed image to the previous reconstruction (e.g., by using the sum of squares of all the pixel discrepancies). If the difference is higher than a certain threshold, repeat steps (b)–(d).
(e) If the difference is lower than the threshold, or if the number of iterations has exceeded a certain limit, then stop!

In order to better understand the method, let us view a simple example of a 2 × 2 pixel image, as shown in Fig. 2.20. In this case the reconstruction converges to the right image upon completion of one iteration.

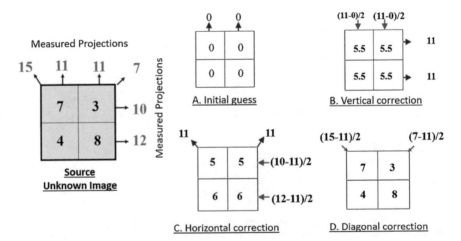

Fig. 2.20 ART using the projection difference method. (Left) The source image is unknown, but its projections are "measured." (**a**) The initial guess is a zero matrix. (**b**) Correcting for the vertical direction using Eq. 2.4. (**c**, **d**) correcting for the horizontal and diagonal directions, respectively

It should be emphasized that the method presented here is the simplest approach for implementation and is given without proof of convergence. More advanced methods will be presented in the following.

2.2.5 Algebraic Reconstruction Tomography (ART): The Projection Ratio Method

The ART projection differences method is simple to implement, but the number of required computations and iterations may be large and time-consuming. In certain applications, such as nuclear medicine (PET and SPECT), the image is rather sparse. That is to say that only small regions may contain important clinical information. In such cases a different approach may be implemented. Rather than taking the projection differences, the projection ratios are used. Using mathematical terms, Eq. 2.8 is replaced by

$$g_{i,j}^{(h+1)} = g_{i,j}^{(h)} \cdot \frac{p_\beta}{\widetilde{p}_\beta^{(h)}} \tag{2.9}$$

One of this method's advantages is that zero-valued projections null their corresponding trajectories. Thus, the reconstructed image is "sparsified" rapidly and the computation burden may be reduced. Naturally, division by zeros and small numbers is a pitfall to be avoided. In order to demonstrate how this method is implemented, let us use again the same 2×2 pixel object. The results are given in Fig. 2.21. Although in this case one iteration was not enough to converge, the pixel values are close to the source image values.

2.2.6 The Weight Matrix

The ART algorithms rely upon accurate estimation of the forward projection values $\widetilde{p}_\beta$ for each beam. The more accurate is the model used for computing $\widetilde{p}_\beta$, the more reliable is the correction applied to the reconstructed image. In the back projection examples given above, we have used the simplest approach defined by Eq. 2.2. However, in reality each beam used for generating a projection has a certain width which should be accounted for. Assuming for simplicity a 2D configuration, it can be easily observed that for beams which are not parallel to the axes, some of the pixels may contribute more and some less than the others as shown in Fig. 2.22.

As can be observed from Fig. 2.22, pixels located outside the beam path do not contribute at all to the resulting projection. But more than that, it can be observed that even pixels located within the beam path do not contribute equally to that

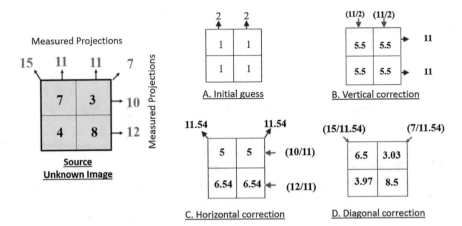

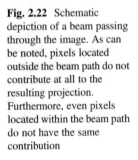

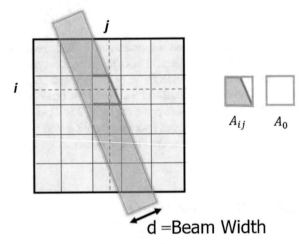

Fig. 2.21 ART using the projection ratio method. (Left) The source image which is unknown, but its projections are "measured." (**a**) The initial guess is a matrix of "1" (to avoid division by zero). (**b**) Correcting for the vertical direction using Eq. 2.5. (**c, d**) Correcting for the horizontal and diagonal directions, respectively

Fig. 2.22 Schematic depiction of a beam passing through the image. As can be noted, pixels located outside the beam path do not contribute at all to the resulting projection. Furthermore, even pixels located within the beam path do not have the same contribution

corresponding projection. Thus, Eq. 2.2 should be modified in order to improve the estimation of the forward projection:

$$p_\beta = \sum_{i,j} W_{i,j}^\beta f(i,j) \qquad (2.10)$$

where $f(i,j)$ is the image value for pixel i, j and $W_{i,j}^\beta$ is the weight corresponding to beam β assigned to each pixel in the image. It is important to note that the summation is done here over the entire image. Thus, the weights for each beam path β can be represented by a matrix which dimensions are equal to the dimensions of the image (typically $n \times n$). Stemming from the fact that most pixels do not contribute anything to that specific projection, most of the weights will be assigned a zero value, and the weight matrix will be sparse.

As for setting the optimal weights for each of the pixels and for each of the beams; this is an issue which requires an accurate physical model of the acquisition system. There are several suggested options. For example, in the 2D model depicted in Fig. 2.22, we can calculate the area covered in each pixel by the beam and normalize it to the pixel area, i.e.:

$$W^{\beta}_{i,j} = \frac{A^{\beta}_{i,j}}{A_0} \tag{2.11}$$

where $A^{\beta}_{i,j}$ is the pixel area covered by the beam β and A_0 is the area of a single pixel.

Seemingly, the calculation of the weight matrices is a simple task. However, it is a time-consuming and memory-consuming procedure. Consider, for example, an image comprising 256×256 pixels ($n \times n$ pixels), scanned by $m = 402$ angular projections (the choice of this number which equals $n\pi/2$ will be explained later), and each projection was sampled by 360 detectors (the length, l, of the diagonal size: $l = n\sqrt{2}$). Thus, it follows that we need to calculate and store about $n^2 \cdot m \cdot l = 9,484,369,920$ values. Although feasible, especially since most of the values equal zero, it is inconvenient and impractical for straightforward real-time computation.

Historically, alternative options which attempt to minimize the computational and storage burden were suggested. The simplest of all is to create a binary weighing matrix were each pixel within the beam is assigned "1" and all the rest are assigned "0." Naturally, this reduces the accuracy but provides faster and more practical algorithms. With some care in their construction, some practical computer algorithms store only the non-zero values of $W^{\beta}_{i,j}$ and avoid most of the null-multiplications implied by Eq. 2.10. With advances in clinical computing power, more sophisticated representations of the weight matrix have come to be routinely used to more accurately represent geometrical and physical effects (e.g., photon attenuation and collimator response in SPECT [17]). These weights are often referred to as the system matrix.

Finally, as will be shown in the following sections, it is more convenient to work with vectors and matrices. Thus, Eq. 2.10 can be rewritten as follows. Consider an image containing $N = n \times n$ pixels. Let us represent the image by a vector $\bar{f}$ of length N. This can be achieved by stacking the pixel rows one after the other in a raster mode. Now let $\bar{p}$ represent the vector of all the obtained projections (size $M = m \cdot l$). Then, Eq. 2.10 becomes

$$\bar{p} = \overline{\overline{W}} \cdot \bar{f} \tag{2.12}$$

where $\overline{\overline{W}}$ is the weight matrix of size $M \times N$ and will be indexed in this form as $W_{\beta i}$: this is the probability that a signal from the pixel i of image $\bar{f}$ will contribute along

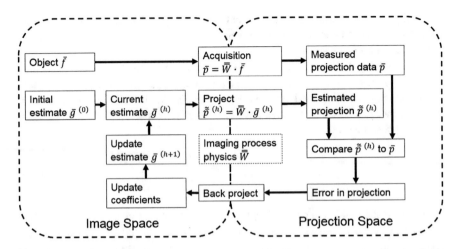

Fig. 2.23 Generic iterative image reconstruction scheme. Ideally, the image estimate $g(h)$ should converge to the "true" image of the object f

beam path to detector β in the projection $\bar{p}$. A generalized, iterative reconstruction scheme is shown in Fig. 2.23, using the notation described above. Ideally, the image estimate $g^{(h)}$ should converge to the "true" image of the object f. Convergence is commonly defined in terms of root mean square error (RMSE): where the discrepancy d between the measured and estimated projections at each iteration, h, is given by:

$$d^{(h)} = \sqrt{\frac{1}{M} \sum_{\beta} \left(p_{\beta} - \sum_{i} W_{\beta i} \cdot g_i^{(h)} \right)^2} \qquad (2.13)$$

and where β indexes each beam path and i indexes each image pixel. Practically, the number of iterations used in clinical imaging for such reconstructions is determined heuristically with feedback from radiologists or nuclear medicine physicians. Details are given below in Sect. 2.3 on advanced reconstruction algorithms.

2.2.7 The Tomographic Slice Theorem

Although ART methods are popular, they require iterative computations. (At this point we will skip the proof and terms of their convergence.) If the number of projections is sufficient (as will be determined later), an analytical robust algorithm for fast image reconstruction termed "filtered back projection" (FBP) is available [18]. In order to understand how it works and derive its mathematical basis, we have to introduce first the "tomographic slice theorem" which is defined as follows:

The Tomographic Slice Theorem (2D Case)
The 1D Fourier transform of a projection line of an object acquired at angle θ is equal to a line inclined at the same angle in the 2D Fourier transform of the object (K-space), passing through its origin.

The slice theorem is also depicted graphically in Fig. 2.24 for clarity.
In order to prove the theorem, we shall start with the Radon transform of Eq. 2.6:

$$p(\theta, t) = \int\limits_{-\infty}^{\infty} \int\limits_{-\infty}^{\infty} f(x, y)\delta(t - x\cos\theta - y\sin\theta)dxdy \qquad (2.6)$$

where $f(x, y)$ is the object image, δ is Dirac's delta function, and the integration is performed from $-\infty$ to ∞. For a projection at angle $\theta = 0$, this simply becomes (see Eq. 2.5)

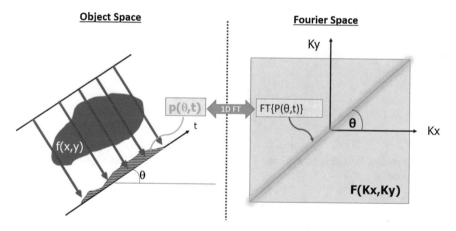

Fig. 2.24 (Left) The acquired projection of the object $p(\theta, t)$ undergoes a 1D Fourier transform. (Right) The result is equal to a line inclined at the same angle in the 2D Fourier transform of the image (K-space)

$$p(\theta = 0, t) = \iint f(x, y) \cdot \delta(t - x) \cdot dxdy = \int\limits_{-\infty}^{\infty} \left[\int\limits_{-\infty}^{\infty} f(x, y) \cdot dy \right] \cdot \delta(t - x) \cdot dx$$

$$= \int\limits_{-\infty}^{\infty} f(x, y) \cdot dy$$

$$(2.14)$$

Taking the 1D Fourier transform, (FT) of this projection yields

$$FT[p(\theta = 0, t)] = \int\limits_{-\infty}^{\infty} [p(\theta = 0, t)] \cdot e^{-jK_x x} \cdot dx \qquad (2.15)$$

$$FT[p(\theta = 0, t)] = \int\limits_{-\infty}^{\infty} \left[\int\limits_{-\infty}^{\infty} f(x, y) \cdot dy \right] \cdot e^{-jK_x x} \cdot dx$$

Rearranging the terms yields

$$FT[p(\theta = 0, t)] = \int\limits_{-\infty}^{\infty} \int\limits_{-\infty}^{\infty} f(x, y) e^{-jK_x x} \cdot 1 \cdot dxdy \qquad (2.16)$$

$$FT[p(\theta = 0, t)] = \int\limits_{-\infty}^{\infty} \int\limits_{-\infty}^{\infty} f(x, y) e^{-jK_x x} \cdot e^{-j0y} \cdot dxdy$$

Comparing this equation to the 2D Fourier transform of the object where $FT[f(x, y)] = F(K_x, K_y)$

$$F[K_x, K_y] = \int\limits_{-\infty}^{\infty} \int\limits_{-\infty}^{\infty} f(x, y) e^{-jK_x x} \cdot e^{-jK_y y} \cdot dxdy \qquad (2.17)$$

It is clear to note that

$$FT[p(\theta = 0, t)] = F[K_x, K_y = 0] \qquad (2.18)$$

Recalling that $K_y = 0$ corresponds to the line passing through the K-space origin at angle $\theta = 0$ (actually the K_x-axis) provides a proof of the slice theorem for that angle. In order to extend the proof to any other projection angle, we recall that the axis selection is arbitrary and the Fourier transform is rotation related, i.e.:

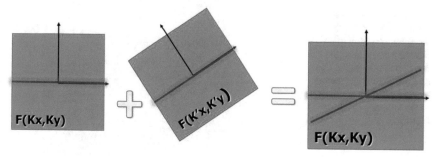

Fig. 2.25 According to the proof of the slice theorem outlined above, we can obtain the horizontal line in any reference frame orientation. These lines can be combined into a single matrix in the image K-space

$$f(\theta + \theta_0, t) \Leftrightarrow F(\theta + \theta_0, K) \tag{2.19}$$

where both the image and its 2D Fourier transform are presented in a polar coordinate system. That implies that we can rotate the reference frame as we desire and obtain the corresponding "horizontal" line in K-space. Then, we can re-rotate the reference frame to the original axes and obtain the inclined line according to the slice theorem, as shown schematically in Fig. 2.25.

The advantage of the slice theorem is that by applying it to the sinogram, line by line, we can collect many radial lines in K-space and fill it with information. Then interpolate into a Cartesian grid and apply the inverse Fourier transform to obtain the image. Or alternatively apply the non-uniform Fourier transform (NUFT), as shown schematically in Fig. 2.26.

2.2.8 "Filtered Back Projection" (FBP)

After proving the slice theorem, we now have a new image reconstruction tool that creates a direct analytical connection between our measurements (the Radon transform) and the Fourier transform of the image. All we have to do now is fill in the Fourier domain (K-space) of the image and perform an inverse transform to obtain the reconstructed image. However, as explained above, the K-space filling pattern is in the form of an asterisk. Thus, in order to reconstruct the image, we need to apply an interpolation to a Cartesian grid in the frequency domain. This is highly undesirable, since any change even in a single pixel in K-space may affect the entire image (see Chap. 6). Thus, we seek a robust reconstruction method that overcomes the limitations set by the asterisk pattern data acquisition. Let us start by rewriting the inverse Fourier transform of the image

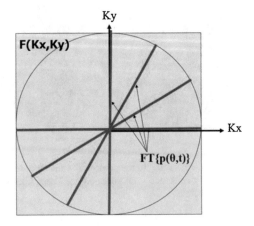

$$f(x,y) = \int\limits_{-\infty}^{\infty} \int\limits_{-\infty}^{\infty} F(K_x, K_y) e^{+jK_x x} \cdot e^{+jK_y y} \cdot dK_x dK_y \qquad (2.20)$$

where $f(x, y)$ and $F(K_x, K_y)$ are the image and its 2D Fourier transform (FT), respectively. For simplicity, the scaling factor (typically $1/2\pi$, given the non-unitary angular frequency form of Eq. 2.17) has been dropped. Now, let us switch into polar representation of the Fourier domain by applying the following relations:

$$\begin{cases} k = \sqrt{k_x^2 + k_y^2} \\ \phi = \tan^{-1}\left[\dfrac{k_y}{k_x}\right] \end{cases}$$
$$\begin{cases} k_x = k \cdot \cos\phi \\ k_y = k \cdot \sin\phi \end{cases} \qquad (2.21)$$
$$\int F(k_x, k_y) dk_x dk_y = \int F(k, \phi) \cdot J \cdot dk d\phi$$

where J is the Jacobian needed to switch between the Cartesian and polar coordinate systems and is defined by

$$J = \begin{vmatrix} \dfrac{\partial K_x}{\partial K} & \dfrac{\partial K_y}{\partial K} \\ \dfrac{\partial K_x}{\partial \varphi} & \dfrac{\partial K_y}{\partial \varphi} \end{vmatrix} = K\cos^2\varphi + K\sin^2\varphi = K \qquad (2.22)$$

It is important to note that $K \geq 0$ in this formulation. Thus, the FT can be rewritten as (Please note that K and k are the same. K is used here for emphasizing terms),

$$\int_{-\infty}^{\infty} \int_{-\infty}^{\infty} F(k_x, k_y) dk_x dk_y = \int_{K=0}^{\infty} \int_{\phi=0}^{2\pi} F(k, \phi) \cdot K \cdot dk d\phi \qquad (2.23)$$

Or if written explicitly,

$$f(x, y) = \int_{-\infty}^{\infty} \int_{-\infty}^{\infty} F(K_x, K_y) e^{+jK_x X} e^{+jK_y Y} dK_x dK_y$$

$$= \int_{K=0}^{\infty} \int_{\phi=0}^{2\pi} F(k, \phi) \cdot e^{+jK[x \cdot \cos(\phi) + y \cdot \sin(\phi)]} \cdot K \cdot dk d\phi \qquad (2.24)$$

Next let us split the integral into two "half planes," i.e., for the upper half plane, φ is an element of $\{0 \rightarrow \pi\}$, and for the bottom half plane, φ is an element of $\{\pi \rightarrow 2\pi\}$:

$$f(x, y) = \int_{K=0}^{\infty} \int_{\phi=0}^{\pi} F(k, \phi) \cdot e^{+jK[x \cdot \cos(\phi) + y \cdot \sin(\phi)]} \cdot K \cdot dk d\phi$$

$$+ \int_{K=0}^{\infty} \int_{\phi=0}^{\pi} F(k, \phi + \pi) \cdot e^{+jK[x \cdot \cos(\phi + \pi) + y \cdot \sin(\phi + \pi)]} \cdot K \cdot dk d\phi \qquad (2.25)$$

However, as recalled,

$$\begin{cases} \cos(\phi + \pi) = -\cos(\phi) \\ \sin(\phi + \pi) = -\sin(\phi) \end{cases} \qquad (2.26)$$

And also,

$$e^{+jK[x \cdot \cos(\phi + \pi) + y \cdot \sin(\phi + \pi)]} = e^{-jK[x \cdot \cos(\phi) + y \cdot \sin(\phi)]} \qquad (2.27)$$

In addition, let us define the following relation for the bottom "half plane":

$$F(k, \phi + \pi) = F(-k, \phi) \qquad (2.28)$$

That is to say that a negative k value applies for that lower "half plane." Thus, Eq. 2.25 can be rewritten as

$$f(x,y) = \int\limits_{K=-\infty}^{+\infty} \int\limits_{\phi=0}^{\pi} F(k,\phi) \cdot e^{+jK[x \cdot \cos(\phi) + y \cdot \sin(\phi)]} \cdot |K| \cdot dk d\phi \qquad (2.29)$$

Importantly, the Jacobian is taken in its absolute value $|K|$ and the integration over k is applied to the range $\{-\infty \rightarrow \infty\}$. Recalling the relation $t = [x \cdot \cos(\phi) + y \cdot \sin(\phi)]$, this equation can be written as

$$f(x,y) = \int\limits_{K=-\infty}^{+\infty} \int\limits_{\phi=0}^{\pi} F(k,\phi) \cdot e^{+jK \cdot t(x,y,\phi)} \cdot |K| \cdot dk d\phi \qquad (2.30)$$

Or after rearranging the terms,

$$f(x,y) = \int\limits_{\phi=0}^{\pi} \left[\int\limits_{K=-\infty}^{+\infty} F(k,\phi) \cdot e^{+jK \cdot t(x,y,\phi)} \cdot |K| \cdot dk \right] \cdot d\phi \qquad (2.31)$$

Now let us study the term in the square brackets. As recalled the slice theorem (ST) states that $FT\{p(\phi,t)\} = F(K,\phi)$. Thus, by reverting the ST, i.e., by applying the inverse FT to both sides, yields

$$FT^{-1}[F(K,\varphi)] = p(\varphi,t) \qquad (2.32)$$

Or after writing it explicitly,

$$\int\limits_{-\infty}^{\infty} F(K,\phi)e^{+jK \cdot t}dK = p(\phi,t) \qquad (2.33)$$

This expression is almost identical to the square brackets in Eq. 2.31 except for the multiplication of the expression $|K|$. As recalled from signal processing theory, such a product of a signal in the frequency domain is actually a filter applied to the signal in the ordinary space. So if we define $F(k,\varphi) \cdot |K| = \tilde{F}(k,\varphi)$ where the wave sign (tilde) indicates that our signal has been filtered, the expression becomes

$$\int\limits_{-\infty}^{\infty} \tilde{F}[k,\varphi]e^{+jK \cdot t}dK = \tilde{p}(t,\varphi) \qquad (2.34)$$

This means that the expression in the square brackets in Eq. 2.32 above is actually a filtered projection of the image. Therefore, the relation between the reconstructed image and the projections can be written as

$$f(x,y) = \int_0^\pi \tilde{p}(t,\varphi)d\varphi \tag{2.35}$$

This important equation implies that by integrating the *filtered* projections over all the projection angles, the target image can be reconstructed. This is a robust analytical expression for image reconstruction. Stemming from the fact that the projections are filtered first, it was titled the filtered back projection (FBP) method.

Furthermore, this method can also be implemented in the object domain. Studying the definition of the filtered projections

$$\tilde{p}(t,\varphi) = \int_{-\infty}^{\infty} |K| \cdot F(K,\varphi)e^{+jK\cdot t}dK = \mathrm{FT}^{-1}\{|K| \cdot F(K,\varphi)\} \tag{2.36}$$

And recalling that multiplication in the frequency domain is equivalent to a convolution in the object domain, the following can be rewritten

$$|K| \cdot F(K,\varphi) = \mathrm{FT}\{g \otimes p\} \tag{2.37}$$

Therefore,

$$\tilde{p}(t,\varphi) = \mathrm{FT}^{-1}\{\mathrm{FT}\{g \otimes p\}\} = g \otimes p \tag{2.38}$$

where the corresponding filter is given by

$$g = \mathrm{FT}^{-1}\{|K|\} \tag{2.39}$$

The above filter is called the "K-filter." It is given explicitly for a band limited case $\{-K_{max} \rightarrow K_{max}\}$ by:

$$g(t) = \int_{-K_{max}}^{K_{max}} |K|e^{+j\cdot 2\pi \cdot K \cdot t}dK \tag{2.40}$$

$$g(t) = \frac{K_{max}}{\pi \cdot t} \sin(2\pi \cdot K_{max} \cdot t) - \frac{\sin^2(\pi \cdot K_{max} \cdot t)}{\pi^2 \cdot t^2}$$

Note that K is defined in this equation as inverse distance or cycles per distance, giving units such as cm^{-1}.

To summarize, the FBP algorithm consists of the following steps:

1. Acquire a sinogram.
2. Filter each projection (i.e., each line) in the sinogram by convolving it with the kernel g as defined above.

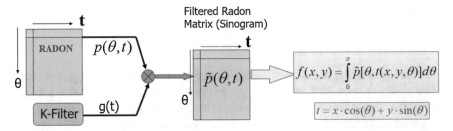

Fig. 2.27 Block diagram of the FBP algorithm. The sinogram is first filtered line by line by convolution with the K-filter. The resulting filtered sinogram is integrated for image reconstruction

3. For each pixel in the image $f(x, y)$, calculate $t = [x \cdot \cos(\phi) + y \cdot \sin(\phi)]$ for every projection angle ϕ.
4. Integrate the values of the filtered projection at location t over all the projection angles.
5. Repeat steps 3 and 4 for all the pixels.

For clarity the procedure is also depicted graphically in Figs. 2.27 and 2.28. Alternatively, step 2 of the FBP algorithm could be replaced by taking the 1D Fourier transform of each sinogram line with respect to detector location t, and multiplying this result by a ramp filter $|K|$, and then applying a 1D inverse Fourier transform to get the filtered sinogram result. Demonstrative images depicting the filtered vs. nonfiltered back projection reconstructions are shown in Fig. 2.29.

FBP reconstruction's limitations are evident when reconstructing noisy projection data. The truncated ramp filter $|K|$, with $|K| \leq K_{max}$, reduces the "blurriness" of unfiltered back projection (Fig. 2.29), but by its very nature, it retains the high spatial frequency components that are most likely to contain the noise data from photon counting systems such as CT, SPECT, and PET. One method to reduce this drawback is to "roll off" the $|K|$ filter before it reaches K_{max}, for example, by using a Hann filter. If $G(K)$ is the K-filter in the frequency domain, then instead of $G(K)$ being $|K|$, with $|K| \leq K_{max}$, the Hann filter is defined as

$$G(K) = \frac{|K|}{2}\left(1 + \cos\frac{\pi K}{K_{max}}\right) \tag{2.41}$$

Here, at low frequencies, Hann filter matches the $|K|$ filter, but it rolls off to zero at the highest frequency, K_{max}. It should be noted however that with the advent of sufficient computing power in the clinic, iterative construction techniques have enabled reconstruction with a superior suppression of image noise.

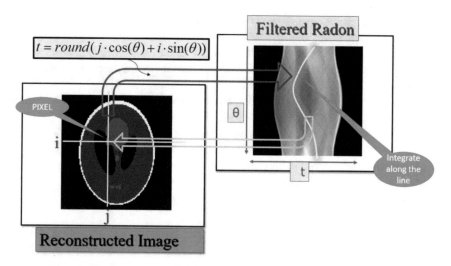

Fig. 2.28 Graphic depiction of the integration step in the FBP. For each pixel and for each projection angle, the address t is calculated (a rounded number is used in the discrete form). The corresponding line in the filtered sinogram is then integrated as shown

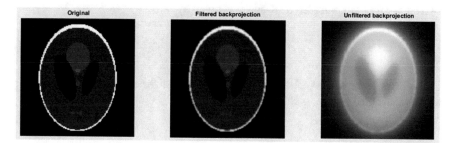

Fig. 2.29 (Left) A Shepp-Logan digital phantom image. (Middle) FBP reconstruction (Right) Nonfiltered back-projection reconstruction. Note the high quality of the FBP reconstruction as compared to the blurred low-resolution and low-contrast image obtained by the nonfiltered reconstruction

2.2.9   The Needed Number of Projections

As explained above, the FBP actually utilizes the slice theorem. This implies that the sampling in K-space has an asterisk pattern as depicted schematically in Fig. 2.26. Clearly, the more projections acquired, the denser is the filling of the corresponding K-space. The question is: how many projections are needed in order to obtain a reliable reconstruction?

But first let us observe what happens if we use too few projections. This is demonstrated in Fig. 2.30, where a 256×256 phantom image is reconstructed by only 36 projections (by acquiring a projection at every $5°$). As can be noted the reconstructed image is covered with "streak" artifacts. These are more clearly visible in the background of the object.

So how can this be avoided? In order to answer the question, observe the sampling pattern depicted schematically for two adjacent lines in K-space as shown in Fig. 2.31.

The circle within which the data is sampled in the frequency domain has a radius which is determined by the maximal spatial frequency that can be used. If the pixel size is Δ (in x-ray CT, e.g., this size is determined mainly by the detectors spatial configuration), then the spatial sampling frequency is given by $2\pi/\Delta$.

According to the Nyquist-Shannon sampling theorem, the maximal spatial frequency that can be used K_{max} without inducing aliasing is half of the sampling frequency and is thus given by:

$$K_{max} = \frac{\pi}{\Delta} \tag{2.42}$$

That is, the highest spatial frequency that can be represented across an image is one cycle (2π) every two pixels (2Δ). This is intuitively obvious if one considers that at least two pixels are needed to represent the "crest" and "trough" of a spatial "wave." If N points are sampled in each projection, then the *radial* distance between two adjacent points sampled in K-space is given by

$$\Delta K_R = \frac{2K_{max}}{N} \tag{2.43}$$

If the number of projections acquired is H, then the angle $\Delta\theta$ between two adjacent sampled lines is given by

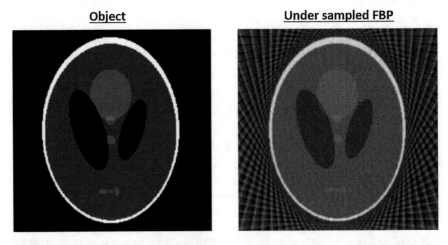

Fig. 2.30 (Left) The Shepp-Logan phantom image. (Right) FBP reconstruction obtained by using 36 projections. Note the streak artifact

Fig. 2.31 The sampling pattern in K-space is an asterisk within a circle of radius K_{max}. The radial distance between two sampled points is ΔK_R and the tangential distance is ΔK_θ. The angle between two adjacent lines (which is the angle between two adjacent projections) is $\Delta\theta$

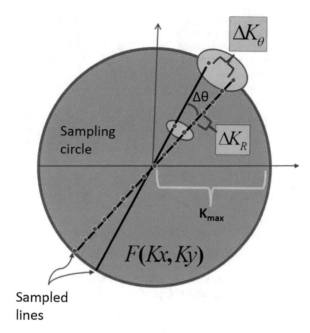

Sampled lines

$$\Delta\theta = \frac{\pi}{H} \qquad (2.44)$$

Therefore, the *tangential* distance ΔK_θ between two adjacent sampled points is given by

$$\Delta K_\theta = K \cdot \frac{\pi}{H} \qquad (2.45)$$

As can be noted, the distance increases for larger values of K (higher frequencies). The worst case is at (naturally) K_{max}:

$$\Delta K_\theta = K_{max} \cdot \frac{\pi}{H} \qquad (2.46)$$

In order to avoid artifacts, that distance should not be larger than ΔK_R, i.e.:

$$\Delta K_\theta \leq \Delta K_R$$
$$\Rightarrow K_{max} \cdot \frac{\pi}{H} \leq \frac{2K_{max}}{N} \qquad (2.47)$$

which yields the following ratio:

Fig. 2.32 In spiral CT, the patient bed is moving while the projections are acquired with a rotating gantry. The relative acquisition trajectory is spiral

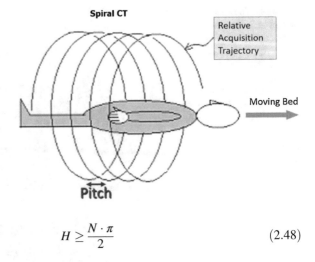

$$H \geq \frac{N \cdot \pi}{2} \tag{2.48}$$

Thus, the number of acquired projections H should be at least about 1.57 times larger than the number of pixels N sampled in each projection.

2.2.10 Spiral CT

Modern x-ray CT imaging utilizes an acquisition technique titled "spiral" or "helical CT." The idea is quite simple. The bed carrying the patient is moved continuously during the CT acquisition. And since the projections are acquired also while the scanner's gantry rotates around the bed, the resulting acquisition trajectory is of a spiral (helix) geometry as shown schematically in Fig. 2.32.

The result is a stack of many Radon transforms, where most of the information is missing. The missing data is commonly filled by interpolation as shown schematically in Fig. 2.33. The advantage is that this provides three-dimensional information which can be used for 3D rendering of the scanned object. Also, the smooth bed motion that comes from the continuous motion required enables faster scans compared to a stepped bed motion acquisition (so-called "axial" CT).

2.2.11 Fan Beam Reconstruction

The Radon transform discussed above presumes that the line projections are obtained from parallel rays. Indeed, initially, practical reconstruction algorithms presupposed sinograms acquired in geometries whereby the ray paths were parallel. However, it is possible to organize the projection data on more arbitrary geometries as can be imagined by a further look at Eq. 2.10 and Fig. 2.22: each voxel is assigned

Spiral CT

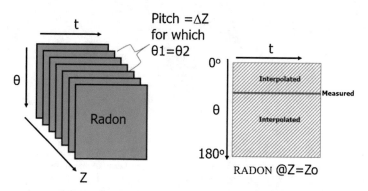

Fig. 2.33 By implementing a spiral CT scan, a stack of many parallel Radon transforms is obtained, where most of the needed information is completed by interpolation

a weight representing the probability of its contribution to a particular detector element. Although presented in parallel projection geometry, it is not limited to this. As long as the weighting matrix (system matrix) accurately represents the contribution of each voxel to the projection data along each ray path, Eq. 2.10 would hold true for arbitrary geometries. Of course, the ray paths need to be designed in a manner that would sample the entire object and at a number of angles sufficient to supply data for a tomographic reconstruction. Regardless, the traditional filtered back projection algorithm requires parallel ray projections and many other alternative reconstruction algorithms, even those using system matrices, continued to require data as standard sinograms until well past the turn of the millennium. Consequently, there has been motivation to reorganize data from non-standard geometries into sinograms representing individual tomographic slices. In the previous section, it was described how spiral CT data can be interpolated to represent projection data in a single plane.

An additional challenge is presented by more current CT acquisition designs in which data are acquired using a fan beam. X-rays from a point source in the x-ray tube fan out through the object being imaged into detectors arrayed as an arc. This data set of divergent attenuation measurements is then rearranged to produce parallel projections. This process is known as parallel re-binning and is shown schematically in Fig. 2.34. The principle is to group together data from ray paths which pass through the object at the same angle. After parallel re-binning, traditional image reconstruction methods such as filtered back projection can be used on the data.

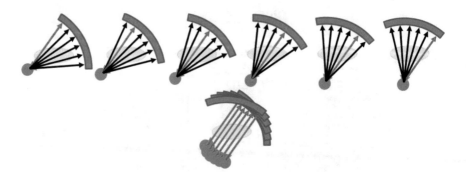

Fig. 2.34 In modern CT scanner design, data are acquired using a fan beam. X-rays (arrows) from a point source in the x-ray tube (gray) fan out through the object being imaged (yellow) into a detector array (blue). It is possible to produce parallel projections from a set of divergent attenuation measurements. As the x-ray tube and detector rotate about the object (top row), rays at the same angle (green) can be reorganized as a set of parallel rays (bottom). This process is known as parallel re-binning

2.3   Part II: Advanced Reconstruction Algorithms

2.3.1   Difference ART Convergence

As shown above in Eq. 2.12, the basic imaging problem can be written as a set of linear equations:

$$p_\beta = \sum_{i=1}^{N} W_{\beta i} f_i, \quad \beta = 1, 2, \cdots, M \tag{2.49}$$

one for each projection data element, β. Here there are M equations (e.g., M values in a sinogram) and N pixels in the "true" image of the object, $\bar{f}$. In the formal algebraic reconstruction approach, the projection measurements are viewed as constraints. The goal is to find an image estimate $\bar{g}$ that satisfies all these constraints, that is, fits the measured data $\bar{p}$. In this way we are looking to obtain an object estimate that could have produced the measured projection data.

Additional constraints imposed by our knowledge about the object and the property being imaged may be added to the system of linear equations. For example, in emission tomography, knowledge that the reconstructed image shall not contain negative values is easy to incorporate into reconstruction scheme by adding a set of constraints:

$$f_i \geq 0, \quad i = \cdots = 1, 2, \cdots, N \tag{2.50}$$

A useful approach for handling inconsistent (e.g., because of noise) or over determined problems is to look for a vector $\widetilde{p}$ which minimizes an error function. The most common choice of error function is Euclidean norm of the difference, $\left\|\widetilde{p} - \overline{p}\right\|^2$.

Kaczmarz's approach (circa 1938) [2] is one of the earliest tomographic reconstruction approaches. The method has a simple and intuitive basis. It solves the system of equations (Eq. 2.49) one by one, each time looking for the closest solution satisfying the next constraint.

Each linear equation in the system of equations (Eq. 2.49) defines a hyperplane in the image space. All of the equations need to be satisfied simultaneously for the image $\overline{f}$ to be a solution to the problem, so $\overline{f}$ lies on the intersection of all the hyperplanes, in the ideal case. Our estimated solution $\overline{g}$ lies on, or at least close to, the intersection of all the hyperplanes.

The approach is easiest to explain geometrically. Let us consider a two-dimensional case of only two pixels and two equations:

$$W_{11}f_1 + W_{12}f_2 = p_1$$
$$W_{21}f_1 + W_{22}f_2 = p_2 \tag{2.51}$$

In the notation of our estimate $\overline{g}$, this case would be written

$$W_{11}g_1 + W_{12}g_2 = \widetilde{p}_1$$
$$W_{21}g_1 + W_{22}g_2 = \widetilde{p}_2 \tag{2.52}$$

Equations 2.51 and 2.52 are equivalent and $\overline{g} = \overline{f}$ if the Euclidean norm of the difference $\left\|\widetilde{p} - \overline{p}\right\|^2$ is 0. An image containing two pixels can be presented as a point in a plane and a hyperplane in two-dimensional space is a line (Fig. 2.35). The solution to our system lies on the intersection of the two lines. The first step is to choose an initial estimate – an arbitrary point in plane $\overline{g}^{(0)}$.

The next step is to find a point closest to it on line $W_{11}f_1 + W_{12}f_2 = p_1$ by orthogonally projecting $\overline{g}^{(0)}$ onto that line. This operation results in a new estimate $\overline{g}^{(1)}$. Continue by projecting onto the second line to obtain $\overline{g}^{(2)}$. The process produces a series of update guesses $\overline{g}^{(0)}, \overline{g}^{(1)}, \overline{g}^{(2)}, \cdots \overline{g}^{(h)}$, converging to the solution of the system of equations such that $\overline{g}^{(h)} \cong \overline{f}$.

Now we move on to express the process algebraically. All we need is the expression of an orthogonal projection operation. The orthogonal projection of a vector $\overline{g}^{(\beta-1)}$ onto a hyperplane $\overline{W}_\beta \cdot \overline{g} = p_\beta$ can be shown to be given by [18]

$$\overline{g}^{(\beta)} = \overline{g}^{(\beta-1)} + \overline{W}_\beta \frac{\left(p_\beta - \overline{W}_\beta \cdot \overline{g}^{(\beta-1)}\right)}{\overline{W}_\beta \cdot \overline{W}_\beta} \tag{2.53}$$

or, written explicitly,

Fig. 2.35 Schematic of
Kaczmarz's procedure in
two-dimensional space

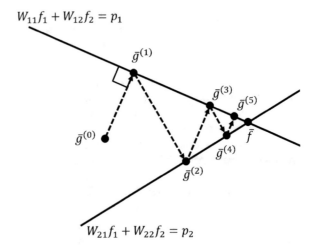

$$g_i^{(\beta)} = g_i^{(\beta-1)} + W_{\beta i} \frac{\left(p_\beta - \sum_j W_{\beta j} \cdot g_j^{(\beta-1)}\right)}{\sum_j \left(W_{\beta j}\right)^2} \qquad (2.54)$$

where $\bar{g}^{(\beta)}$ is the image estimate after β updates (i.e., one update for each projection).
Here, the summation of the vector scaler product is indexed as j ranging from 1 to N.

In order to find a solution, an initial estimate $\bar{g}^{(0)}$ is sequentially projected
(in linear algebra sense) onto hyperplanes defined by each of the constraints. An
iteration of the algorithm is completed when all the M equations have been
addressed. For the next iteration, $\bar{g}^{(M)}$ is projected onto the first hyperplane and the
whole process is repeated.

The initial estimate is usually selected to be uniform: $g_i^{(0)} = C, (i = 1, 2, \cdots, N)$,
such that C is a constant. Common choices of initial estimate are a uniform zero
image or a rough estimate of mean voxel value. Alternatively, a FBP reconstructed
image can be used as an initial estimate.

The convergence rate of ART is largely dependent on the degree of orthogonality
of the successive lines (hyperplanes), such that when the equations are nearly
orthogonal, the convergence is faster than when the equations are far from orthogonal. Also, a solution exists only if the system of equations is consistent. In our
two-dimensional example, there would be no solution if the two lines were parallel.
When there are three equations and two variables, the system is overdetermined, and
an exact solution barring a perfect intersection of all three may not necessarily exist.
Sometimes, in this case, convergence to an approximate solution can be forced by
relaxation. At every projection operation, the step size is decreased to not reach the
corresponding constraint (line or hyperplane). Relaxation is achieved by adding a
parameter α, $0 \le \alpha \le 1$ leading to only partial use of the calculated update factor:

$$g_i^{(\beta)} = g_i^{(\beta-1)} + \alpha W_{\beta i} \cfrac{\left(p_\beta - \sum\limits_j W_{\beta j} \cdot g_j^{(\beta-1)}\right)}{\sum\limits_j \left(W_{\beta j}\right)^2} \qquad (2.55)$$

The introduction of a relaxation factor speeds up convergence and increases an ART algorithm's robustness to noise. Such a procedure can be implemented to lead to an approximate least squares solution to the problem. In real-life situations, the system is often inconsistent due to noise in measured projections. It can be shown that, when a solution exists, the algorithm will converge to a solution closest to the initial estimate $\bar{g}^{(0)}$ in the sense of least squares [19]. The Kaczmarz solution can be shown to have a minimal Euclidean norm. In addition to a highly variable convergence rate, another drawback of Kaczmarz's method is that, at each update, only a single measurement is considered. A complete iteration of the algorithm requires a number of updates equal to the number of projection data elements. This makes the method practical only for low-dimensional, slice-by-slice reconstructions. Finally, Eq. 2.54 alone may update pixels to negative values. A requirement of non-negativity in the solution may only be ensured through additional constraints, such as checking and updating pixels only to non-negative values.

2.3.2 Advanced Multiplicative ART

The projection ratio method of Eq. 2.9 can be reformulated to include the system matrix (weight matrix). This multiplicative ART (MART) uses relative error and a multiplicative update. Using the notation of the previous section

$$g_i^{(\beta)} = g_i^{(\beta-1)} \cfrac{p_\beta}{\sum\limits_j W_{\beta j} \cdot g_j^{(\beta-1)}} \qquad (2.56)$$

where $\bar{g}$ is the estimate of the "true" object image $\bar{f}$, with index i ranging from 1 to $N = n^2$ for an $n \times n$ image matrix. The beam path index β ranges from 1 to M projection data points (e.g., K angular projections of a row of L detectors: $M = K \times L$). Here, the summation of the vector scaler product is indexed as j ranging from 1 to N. It's worth reiterating that these methods are not limited to the cylindrical geometry implied by a sinogram since the system (weight) matrix $W_{\beta i}$ describes the physical (including geometrical) relationship between i^{th} pixel the "true" image f_i and the projection into the βth detector p_β. A significant advantage of MART is a built-in non-negativity solution. It also allows to constrain the reconstruction by setting to zero pixels of the initial estimate that are known to be outside of the reconstructed object (outside of the "support" of $\bar{f}$). Due to the multiplicative update, all pixels set to zero in the initial estimate will retain a zero value throughout the further iterations. On the negative side, the performance of MART depends on the quality of the initial estimate.

2.3.3   Statistical Reconstruction Approaches

So far we ignored the statistical nature of tomographic data acquisition and the fact that the data acquired under realistic conditions (mainly, limited acquisition time) are noisy. All we did was to try and solve the set of linear equations. Statistical reconstruction methods explicitly incorporate a model of the statistical properties of the imaging process in addition to the geometrical model of the imaging system. These random components include noise in the projection data arriving into the detector, object variability, and noise introduced by the imaging system itself. A statistical approach to image reconstruction recognizes the variability in the object and data. Here, the detected projection data (a specific realization of the photon counting process) are viewed as the expected value of the projection $E(p)$ of the imaged object. All methods belonging to the family of statistical reconstruction approaches treat image reconstruction as a statistical estimation problem. The task of such an algorithm is: given the data $\bar{p}$, find an object estimate $\widehat{\bar{g}}$ that gives the "best fit" to the available data. The quality of the fit is determined based on a statistical estimation criterion.

2.3.3.1   Weighted Least Squares

Least squares (LS) is another widely used approach to statistical estimation. Under the LS approach, we aim at finding an image which would produce projection data closest to the observed projections in the sense of least squares (Euclidean distance).

$$\widehat{\bar{g}} = \operatorname{argmin}_{\bar{g}} \left[\left\| \bar{p} - \overline{\overline{W}} \cdot \bar{g} \right\|^2 \right] = \operatorname{argmin}_{\bar{g}} \left[\sum_{\beta} \left(p_\beta - \sum_j W_{\beta j} \cdot g_j \right)^2 \right] \quad (2.57)$$

Statistical properties of measured data can be incorporated into the LS framework by adding weights in such a manner that projection data elements which are expected to have higher variance have lower contribution to the objective function. This results in the weighted LS (WLS) solution:

$$\widehat{\bar{g}} = \operatorname{argmin}_{\bar{g}} \left[\left(\bar{p} - \overline{\overline{W}} \cdot \bar{g} \right) \overline{\overline{D}} \left(\bar{p} - \overline{\overline{W}} \cdot \bar{g} \right) \right]$$

$$= \operatorname{argmin}_{\bar{g}} \left[\sum_{\beta} D_\beta \left(p_\beta - \sum_j W_{\beta j} \cdot g_j \right)^2 \right] \quad (2.58)$$

where $\overline{\overline{D}}$ is a diagonal matrix with elements $D_\beta \cong \frac{1}{\mathrm{var}(p_\beta)}$. The weights may be constant and based on measured projections or, alternatively, updated at each iteration based on the projection of the current image estimate $\overline{\overline{W}} \cdot \widehat{\overline{g}}^{(h)}$.

In order to solve the LS or WLS problem, classical optimization techniques are used to realize an additive update scheme

$$\widehat{\overline{g}}^{(h+1)} = \widehat{\overline{g}}^{(h)} + t \Delta \widehat{\overline{g}}^{(h)} \tag{2.59}$$

where t is the step size to be taken along the update direction $\Delta \widehat{\overline{g}}^{(h)}$.

The optimal step size at each iteration is the one to minimize the quadratic form, which is the argument in Eq. 2.58:

$$\frac{\partial \left[\left(\overline{p} - \overline{\overline{W}} \cdot \overline{g} \right) \overline{\overline{D}} \left(\overline{p} - \overline{\overline{W}} \cdot \overline{g} \right) \right]}{\partial t} \Bigg|_{\widehat{\overline{g}}^{(h)} + t \Delta \widehat{\overline{g}}^{(h)}} = 0 \tag{2.60}$$

Solving for t it gives us

$$t = \frac{\Delta \widehat{\overline{g}}^{(h)'} \overline{\overline{W}}' \overline{\overline{D}} \left(\overline{p} - \overline{\overline{W}} \cdot \widehat{\overline{g}}^{(h)} \right)}{\Delta \widehat{\overline{g}}^{(h)'} \overline{\overline{W}}' \overline{\overline{DW}} \cdot \Delta \widehat{\overline{g}}^{(h)}} \tag{2.61}$$

A number of specific approaches to direction selection have been proposed and widely explored. With coordinate descent, a simple way to select the next step direction is to update one image element at a time. Then, the vector $\Delta \widehat{\overline{g}}^{(h)}$ consists of zeroes in all locations except for the one corresponding to the image element currently being updated. In this case, when updating voxel g_j, only the corresponding system matrix elements $W_{\beta j}$ will be involved. Alternatively, a standard gradient descent approach (steepest decent) may be used to determine the update direction for the whole image at once: choose to advance the solution in the direction which, given the current image estimate, leads to the fastest decrease of the objective function (the argument in Eq. 2.58). The initial step direction is computed through back projection. Finally, with the conjugate gradient method, in order to increase efficiency over the steepest descent method, we can select step directions at subsequent iterations to be conjugate to one another. The intuition under this approach is similar to efficient hyperplanes arrangement described in the algebraic reconstruction section.

The non-negativity of the steepest decent or conjugate gradient solutions is not guaranteed. Enforcing non-negativity by clipping resulting negative pixel values to zero is useful in maintaining non-negativity, but interferes with convergence dynamics. LS-based optimization is equivalent to maximum likelihood (ML) criterion (discussed below) under a Gaussian noise data model. The Gaussian model is adequate when applied to high-count data in photon counting systems. For this reason, LS and WLS approaches are more frequently used in transmission (e.g., x-ray CT) than in emission tomography (e.g., SPECT or PET). The LS solution is equivalent to ML under the assumption that all projection data elements have equal variance.

2.3.3.2 Maximum Likelihood Expectation Maximization (MLEM)

Maximum likelihood (ML) expectation maximization (EM) is one of the main approaches implemented in tomographic reconstruction, most widely used in emission tomography. Under the ML model, it is assumed that the probability distribution for the measured data $\bar{p}$ is determined by some parameter vector $\bar{g}$ (e.g., the estimated radiotracer distribution in an emission scan). The probability P to observe $\bar{p}$ given $\bar{g}$ is called the likelihood function $L(\bar{p}|\bar{g}) = P(\bar{p}|\bar{g})$. The ML framework provides us with a rule for finding an image which would be, in the maximum likelihood sense, the best estimate of the imaged object: choose the reconstructed image as an estimate of the image object for which the measured projection data would have had the highest likelihood:

$$\widehat{\bar{g}} = \mathrm{argmax}_{\bar{g}}[L(\bar{p}|\bar{g})] \tag{2.62}$$

Among all possible objects, our solution will be the one which is most likely to produce the measured data. In other words, we are looking to maximize statistical consistency with the measurements (projections).

MLEM algorithm approaches the solution of 2.62 iteratively. Each MLEM iteration consists of two steps. In the first step, the E-step, the likelihood function of the current estimate is calculated. The second step, the M-step, produces a new image estimate that maximizes the likelihood.

The maximum likelihood (ML) is one of the most widely used statistical estimation criteria. ML estimators are asymptotically unbiased, meaning that the expected value of the estimate approaches the "true" image $(E\left(\widehat{\bar{g}}\right) \to f)$. ML methods are valuable when the data are inherently noisy.

The MLEM reconstruction algorithm can be derived as follows. In order to effectively use ML and EM mathematical tools, we first define a "complete" data space. We define a random vector $\bar{c}_\beta$, whose elements $c_{\beta j}$ correspond to the number of photons emitted from voxel j and detected in projection element β. Note that our measured data vector $\bar{p}$ is determined in terms of complete data $p_\beta = \sum_j c_{\beta j}$.

In case of emission tomography, as discussed earlier, total emission from a voxel element j is described by a Poisson distribution with mean λ_j. It is assumed that the distributions are stationary throughout the acquisition. The contribution of object element j with mean λ_j to detector element β is assumed to be fully determined by the system matrix $\overline{\overline{W}}$.

The number of detected photons emitted toward each of the detector elements is a collection of independent Poisson distributed random variables ($c_{\beta j}, \beta = 1, 2, \cdots, M;$

$j = 1, 2, \cdots, N$). The total number of photons detected in projection element β is thus also a Poisson distributed random variable. Importantly, individual projection measurements are independent. In the process of reconstruction, we will be looking for an image $\bar{g}$ representing expected values of the image vector $\bar{f}$ that over many realizations (acquisitions) would give the mean values $\bar{\lambda}$. Based on Poisson counting statistics, the equation for complete data likelihood is

$$L(\bar{c}|\bar{f}) = \prod_\beta \prod_j \frac{\left[(W_{\beta j}\lambda_j)^{p_\beta} e^{-(W_{\beta j}\lambda_j)}\right]}{[p_\beta!]} \qquad (2.63)$$

Shepp and Vardi [12] showed in 1982 that this model more accurately portrays the physics of the emission process than models at the basis of other existing reconstruction algorithms. Their MLEM algorithm was derived from maximizing the log-likelihood, and they showed that by maximizing the log-likelihood, each MLEM step increases the likelihood to convergence. The iteration from step h to $h + 1$ is given by

$$g_i^{(h+1)} = \frac{g_i^{(h)}}{\sum_{\beta=1}^{M} W_{\beta i}} \sum_{\beta=1}^{M} \frac{W_{\beta i} p_\beta}{\sum_{j=1}^{N} W_{\beta j} \cdot g_j^{(h)}} \qquad (2.64)$$

where $\bar{g}$ is the estimate of the "true" image $\bar{f}$, with index i ranging from 1 to N, and the beam path index β ranges from 1 to M projection data points, for the projection data $\bar{p}$. The system matrix $W_{\beta i}$ models the probability that an emission from pixel f_i is detected in projection p_β. Here, the summation of the vector scaler product is indexed as j ranging from 1 to N.

The MLEM update formula has a very simple and intuitive structure and, actually, is quite similar to the update equation of MART (Eq. 2.56). Each iteration represents one forward projection and one back projection. At each iteration, all image elements are updated simultaneously based on all projection measurements.

MLEM algorithm behavior is characterized by the following properties:

- *Monotonic convergence.* Due to concavity of the log-likelihood, after each iteration, likelihood is increased. A series of updates per Eq. 2.64 will converge to the maximum likelihood image.
- *Positivity.* Due to the structure of the update Eq. 2.64 and non-negativity of the initial estimate, the reconstructed image preserves positivity after any number of updates. In addition, image elements that have been set to zero in the initial estimate $\bar{g}^{(0)}$ will keep zero values throughout the updates. The latter allows to preset image elements to zero based on the knowledge of the object location within the reconstructed space for transmission tomography or based on the knowledge of tracer distribution within the reconstructed space for emission tomography.

Fig. 2.36 Typical
convergence of an MLEM
algorithm. The upper four
images show a progressively
more accurate
reconstruction of a Shepp-
Logan digital phantom as
more iterations are
performed, whereas the
graph tracks the
log-likelihood of an EM
reconstruction

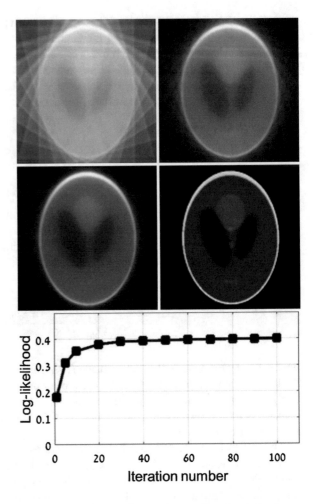

- *Count preservation.* Starting from the first update and at each subsequent iteration, the total expected number of counts in the reconstructed image stays
  constant.
- *Non-linearity.* Maximum likelihood estimator is non-linear. As a result, convergence speed differs across the reconstruction space. Slower convergence is
  observed in "cold" regions, leading to positive bias (count overestimation).

One of the main limitations of the MLEM algorithm is that the convergence speed
is slow, and decreases from iteration to iteration. A typical convergence rate is
illustrated in Fig. 2.36.

2.3.3.3 Ordered Subset Expectation Maximization (OSEM)

The convergence of MLEM was greatly increased by using the concept of ordered subsets, and gave rise to practical iterative reconstruction in the clinic [15]. The idea behind the ordered subset method is to divide the projection data into mutually exclusive subsets and derive each update of the image estimate from just a subset of acquired data. Ordered subset expectation-maximization (OSEM) reconstruction has been especially important with SPECT and PET image reconstructions which have historically suffered from image noise due to low photon counting statistics. With EM and ART methods, in order to speed up convergence, it is beneficial to rearrange acquired projection data so that projection views acquired at significantly different angles from one another are processed sequentially. This may be illustrated as follows: let data be acquired by rotating a detector around the patient into projection views at eight different angles $p(\theta_1, t)$, $p(\theta_2, t)$,..., $p(\theta_8, t)$, as shown in Fig. 2.37. Using projections sequentially does not introduce new information quickly. The projections at the first two angles $p(\theta_1, t)$ and $p(\theta_2, t)$ are similar views whereas $p(\theta_1, t)$ and $p(\theta_4, t)$ have quite different views. Reordering the sequence of the projection data can introduce new information to the iterations, sooner.

A more efficient update sequence will be based on the following sequence: $p(\theta_1, t)$, $p(\theta_4, t)$, $p(\theta_7, t)$, $p(\theta_2, t)$, $p(\theta_5, t)$, $p(\theta_8, t)$, $p(\theta_3, t)$, $p(\theta_6, t)$. The algorithm first processes all projection elements of view 1, then view 4 and so on. Such a rearrangement has been shown to speed up convergence. The drawback of such an approach is that reconstruction cannot be performed simultaneously with the acquisition process, in that all the projection views must be available for rearrangement.

The use of ordered subsets transforms the MLEM update equation (Eq. 2.64) into:

Fig. 2.37 Ordered subset schematic. In iterative reconstruction, using projections sequentially in the order of typical acquisition does not introduce new information quickly. Projection $p(\theta_1, t)$ and $p(\theta_2, t)$ are similar views whereas $p(\theta_1, t)$ and $p(\theta_4, t)$ have quite different views. Reordering the sequence of the projection data can speed convergence of iterative reconstruction

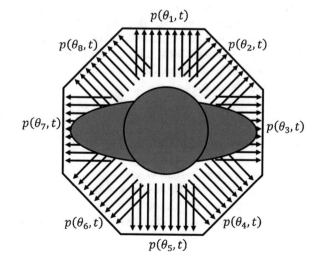

$$g_i^{(h+1)} = \frac{g_i^{(h)}}{\sum_{\beta \in S_b} W_{\beta i}} \sum_{\beta \in S_b} \frac{W_{\beta i} p_\beta}{\sum_{j=1}^N W_{\beta j} \cdot g_j^{(h)}} \qquad (2.65)$$

where h is update number, and S_b, is a subset of the projection data. In the case of a sinogram with K rows (with L detector locations per row), it can be divided into B subsets, such that index b ranges from 1 to B. There are $R = K/B$ rows per subset, so B would be chosen such that it is a factor of K and R is an integer. A complete sequence of updates makes use, in turn, of all of the subsets and comprises a single iteration of the algorithm, so that all M beam paths are used. For a sinogram: $M = B \times R \times L$. Therefore, for one complete iteration of OSEM, all projection data are used, but the image estimate is updated as many times as there are subsets (B).

It can be easily seen that if there is just one subset containing all the projection data, Eq. 2.65 is equivalent to the MLEM update in Eq. 2.64. If each subset contains just a single projection bin, we obtain the familiar MART in Eq. 2.56. For mathematical reasons, to avoid division by zero in Eq. 2.65, each projection subset shall contain data originating from every image element with non-zero probability.

The number and arrangement of subsets are important for the performance of the algorithm. Convergence of OSEM relative to MLEM is typically accelerated by a factor equal to the number of subsets. Subsets are generally selected so that each subset represents the projection data uniformly and fully. A natural and practical approach is using equally spaced projection views as subsets. In our acquisition model from Fig. 2.37, a possible subset selection would be 2 subsets: $S_1 = \{p(\theta_1, t),$ $p(\theta_4, t), p(\theta_7, t), p(\theta_2, t)\}$ and $S_2 = \{p(\theta_5, t), p(\theta_8, t), p(\theta_3, t), p(\theta_6, t)\}$, which will speed up convergence approximately by a factor of two. A rule of thumb suggests that subsets should be balanced between them in terms of total projection counts.

As opposed to MLEM, there is no proof of absolute convergence for OSEM. It is important to note that noise amplification is accelerated as well, compared to MLEM. Numerous works have shown empirically that OSEM has convergence dynamics similar to MLEM, with large image features reconstructed first and fine details converging later. The number of subsets is an additional algorithm parameter, which requires optimization to match acquisition parameters, data statistics and the clinical task at hand. Regardless, OSEM is successfully used in multiple medical imaging applications.

2.3.3.4 Bayesian Approach: Maximum A Posteriori (MAP)

In previous sections, imaged objects were reconstructed based on measured data and known data properties. The only deviation from this principle has been in addition of non-negativity constraints, which was made based on our knowledge of possible values of the captured object property. In practice, there is a lot more we may know about the object.

The Bayesian approach to image reconstruction provides a practical framework to utilize our prior knowledge about the imaged object. Among most commonly used assumptions are those of smoothness. The reason for this is the general tendency of iterative reconstructions to amplify noise with iterations, producing images that are increasingly noisy, leading to totally "salt and pepper" appearance where noise dominates object features. The smoothness assumption is reasonable because tissues and organs tend to have similar properties throughout them, and high frequency components in reality represent transitions between organs or regions possessing different properties, such as the difference between the density of bone compared to that of soft tissue or the difference in tumor-targeting tracer uptake in surrounding healthy tissue as opposed to the tumor itself.

Under Bayesian approach, we are looking for an image estimate $\widehat{\overline{g}}$ that has the highest probability given the measured data $\overline{p}$: $P(\overline{g}|\overline{p})$. This leads us to maximum a posteriori (MAP) criterion:

$$\widehat{\overline{g}} = \text{argmax}_{\overline{g}}[P(\overline{g}|\overline{p})] \qquad (2.66)$$

Making use of Bayes' law

$$P(\overline{g}|\overline{p}) = \frac{P(\overline{p}|\overline{g})P(\overline{g})}{P(\overline{p})} \qquad (2.67)$$

and omitting $P(\overline{p})$ as quantity independent from $\overline{g}$, Eq. 2.66 becomes

$$\widehat{\overline{g}} = \text{argmax}_{\overline{g}}[P(\overline{p}|\overline{g})P(\overline{g})] \qquad (2.68)$$

Taking the logarithm, we obtain

$$\widehat{\overline{g}} = \text{argmax}_{\overline{g}}[\ln P(\overline{p}|\overline{g}) + \ln P(\overline{g})] \qquad (2.69)$$

Note that equation contains log-likelihood $\ln P(\overline{p}|\overline{g})$, which we have already mentioned as an optimization criterion, and in addition, there is the logarithm of the prior probability term $\ln P(\overline{g})$. So, instead of just maximizing consistence with the measured data, we balance it with our prior knowledge regarding the "true" image $\overline{f}$, since $\overline{g}$ iterates toward $\overline{f}$. If we have no prior expectations regarding the object, that is, prior $P(\overline{g})$ is uniform, the maximum a posteriori criterion is equivalent to maximum likelihood. The prior term $\ln P(\overline{g})$ is often called *penalty function*, as it penalizes the selection of "unreasonable" solutions.

Typical MAP-EM approaches [20, 21] results in the following update equation:

$$g_i^{(h+1)} = \frac{g_i^{(h)}}{\sum_{\beta=1}^{M} W_{\beta i} + \alpha \frac{\partial U(f)}{\partial g}} \sum_{\beta=1}^{M} \frac{W_{\beta i} P_\beta}{\sum_{j=1}^{N} W_{\beta j} \cdot g_j^{(h)}} \qquad (2.70)$$

The combination of likelihood and penalty terms alters the shape of the goal function and may lead to faster convergence. Under this approach, the selection of prior function U becomes critical for the outcome of the reconstruction process, as does the scalar weight α. It can be constructed as a series of potential functions between voxels that are members of a local neighborhood or "clique." For example, U can be designed to impose local homogeneity in the reconstructed image by penalizing results that give sharp variations in local pixels. Alternatively, under a nonlocal approach, similar image values might be ensured within cliques defined by some similar property other than their location within the image space. These may be statistical properties of the neighborhood, location on an organ edge or within an organ (based on our knowledge of actual organ location or based on an auxiliary image acquired using another imaging modality like CT), or other knowledge.

There is no general recommendation on selection of the prior term, potential function, clique, or the relative weight given to the prior knowledge term. Practical application of these methods requires careful optimization of all the algorithm components and careful evaluation of the clinical usefulness of the resulting reconstructed images. Parameter sets are usually tailored to a specific system, acquisition protocol, and clinical task at hand. If so, they can provide sharper images resulting from more EM iterations while greatly suppressing image noise.

2.3.4 Other Image Reconstruction Approaches

Many valuable image reconstruction techniques are beyond the scope of this text. The above section on advanced reconstruction techniques serves only as an introduction to advanced algorithms in an era in which solid-state PET, solid-state SPECT, and 128-slice CT scanners are beginning to be routinely introduced to clinics. This section is not comprehensive. For example, the row action maximum likelihood algorithm (RAMLA) can improve the convergence of EM algorithms [22] and has been successful in commercial clinical devices, but has not been detailed above. Compressed sensing reconstruction for sparse data acquisition [23, 24] has created a tremendous impact in the field of MRI, introducing a new paradigm in rapid image acquisition from subsampled data [25]. Another emerging approach is the implementation of artificial intelligence (AI)-based reconstruction [26, 27] which has been proven effective under research conditions. Most CT vendors have implemented reconstruction techniques that combine the advantages of statistical reconstruction methods with the traditional CT image "texture" that results from FBP. This has enabled the production of diagnostic quality CT images with a substantially reduced radiation dose exposure for the patient. Automated methods of correctly handling motion of the patient and/or organs using so-called 4D reconstruction methods are being introduced. Much of the drive for further reconstruction algorithm development will continue to come from the need for reduced scan times, reduced radiation exposure, and improved diagnostic image quality.

References

1. Radon J. Über die Bestimmung von Funktionen durch ihre Integralwerte längs gewisser Mannigfaltigkeiten. Berichte über die Verhandlungen der Königlich-Sächsischen Akademie der Wissenschaften zu Leipzig, Mathematisch-Physische Klasse [Reports on the proceedings of the Royal Saxonian Academy of Sciences at Leipzig, Mathematical and Physical Section], Leipzig: Teubner. 1917;69:262–77.
2. Kaczmarz S. Angenaherte auflosung von systemen linearer gleichungen. Bull. Acad. Pol. Sci. Lett. A. 1937;6-8A:355–7.
3. Bracewell RN. Strip integration in radio astronomy. Aust J Phys. 1956;9:198–217.
4. Kuhl DE, Edwards RQ. Image separation radioisotope scanning. Radiology. 1963;80:653–62.
5. Hounsfield G. A method of apparatus for examination of a body by radiation such as x-ray or gamma radiation. 1972. US Patent 1283915. British patent 1283915, London.
6. Gordon R, Bender R, Herman GT. Algebraic reconstruction techniques (ART) for three-dimensional electron microscopy and x-ray photography. J Theor Biol. 1970;29:471–81.
7. Goitein M. Three-dimensional density reconstruction from a series of two-dimensional projections. Nucl Instr Meth. 1972;101(3):509–18.
8. Richardson WH. Bayesian-based iterative method of image restoration. J Opt Soc Am. 1972;62(1):55–9.
9. Lucy L. An iterative technique for the rectification of observed distributions. Astron J. 1974;79(6):745–54.
10. Rockmore AJ, Macovski A. A maximum likelihood approach to emission image reconstruction from projections. IEEE Trans Nuc Sci. 1976;23:1428–32.
11. Rockmore AJ, Macovski A. A maximum likelihood approach to transmission image reconstruction from projections. IEEE Trans Nuc Sci. 1977;24(3):1929–35.
12. Shepp LA, Vardi Y. Maximum likelihood reconstruction for emission tomography. IEEE Trans Med Imaging. 1982;MI-1:113–22.
13. Feldkamp LA, Davis LC, Kress JW. Practical cone-beam algorithm. J Opt Soc Am. 1984; A6:612–9.
14. Geman S, McClure DE. Bayesian image analysis: an application to single photon emission tomography. In: Proceedings of Statistics and Compensation Section of the American Statistical Association; 1985. p. 12–8.
15. Hudson HM, Larkin RS. Accelerated image reconstruction using ordered subsets of projection data. IEEE Trans Med Imaging. 1994;MI-13:601–9.
16. Katsevich A. Theoretically exact filtered backprojection-type inversion algorithm for spiral CT. SIAM J Appl Math. 2002;62(6):2012–26.
17. Seo Y, Wong KH, Sun M, Franc BL, Hawkins RA, Hasegawa BH. Correction of photon attenuation and collimator response for a body-contouring SPECT/CT imaging system. JOURNAL OF NUCLEAR MEDICINE. 2005;46(5):868–77.
18. Kak AC, Slaney M. Principles of Computerized Tomographic Imaging: IEEE Press, New York; 1988.
19. Tanabe K. Projection method for solving a singular system of linear equations and its applications. Num Math. 1971;17:203–14.
20. Hebert TJ, Leahy RM. A generalized EM algorithm for 3-D Bayesian reconstruction from Poisson data using Gibbs priors. IEEE Trans Med Imaging. 1989;MI-8:194–202.
21. Lalush DS, Tsui MW. A general Gibbs prior for maximum a posteriori reconstruction in SPET. Phys Med Biol. 1993;38:729–41.
22. Browne J, de Pierro AB. A row-action alternative to the EM algorithm for maximizing likelihood in emission tomography. IEEE Trans Med Imaging. 1996;15:687–99.
23. Candes EJ, Romberg J, Tao T. Robust uncertainty principles: exact signal reconstruction from highly incomplete frequency information. IEEE Trans Inf Theory. 2006;52(2):489–509.
24. Candes EJ, Tao T. Near-optimal signal recovery from random projections: universal encoding strategies? IEEE Trans Inf Theory. 2006;52(12):5406–25.

25. Lustig M, Donoho D, Pauly JM. Sparse MRI: the application of compressed sensing for rapid MR imaging. Magn Reson Med. 2007;58(6):1182–95.
26. Wang Y, et al. 3D conditional generative adversarial networks for high-quality PET image estimation at low dose. NeuroImage. 2018;174(1):550–62.
27. Bo Z, Liu JZ, Cauley SF, Rosen BR, Rosen MS. Image reconstruction by domain-transform manifold learning. Nature. 2018;555:487–92.

Chapter 3
X-Ray Imaging and Computed Tomography

Synopsis: In this chapter, the reader is introduced to the basic physics of x-rays and their implementation in planar and computed tomography (CT) imaging.

The learning outcomes are: The reader will understand how x-rays are produced, how they are detected, and their interaction with matter. They will review the different approaches utilized for data acquisition, and finally, they will learn how x-ray imaging can be implemented in different clinical settings.

X-ray imaging is based on the fact that when irradiating the body with an x-ray beam, different tissues will interact differently with the x-ray photons. This phenomenon is attributed to differences in the electron density of each tissue type. When irradiating a patient with x-rays, some of the radiation is absorbed or attenuated by the body itself, and some of the radiation passes through; this is the remnant radiation. When placing an x-ray-sensitive film or digital detectors behind the patient's body, to absorb the remaining photons, a map of the attenuation is formed. This is a two-dimensional image or projection of the inner structure of the scanned body. *The different attenuation properties of the different body tissues are the source of the x-ray image contrast. It is what gives the ability to differentiate one tissue from another.*

There are several groups of scan types:

- Radiography – this is a single image taken at a specific direction and orientation, which shows the internal anatomy of the scanned body. Usually it requires a very small amount of radiation.
- Fluoroscopy – this method is used for continuous x-ray imaging, where the images are presented live to the physician on a monitor, just like a movie. This technique is most frequently used for examining the flow of contrast agent through blood vessels. It might require a relative high dosage of radiation depending on the procedure time and complexity.

© Springer Nature Switzerland AG 2020
H. Azhari et al., *From Signals to Image*,
https://doi.org/10.1007/978-3-030-35326-1_3

- Computed tomography – this method involves many different successive projections, from different angles, that are taken during the continuous rotation of the x-ray tube and detectors around the patient. A reconstruction algorithm processes the data to generate a volumetric representation of the body, which can be divided into different cross sections, i.e., image slices. It can show static anatomical structure, as well as dynamic functioning, such as the beating heart. It requires a much higher x-ray dosage than conventional radiography since it combines many different projections.

Medical x-rays are used for diagnosis, screening, and intervention. They are utilized in a large variety of clinical usages such as mammography, chest x-ray, cardiac catheterization, full-body CT scan, and more.

3.1 X-Rays: Physical Phenomena

Like light, x-rays are part of the electromagnetic spectrum waves but they are too energetic to be visible to the naked eye. Spanning between the ultraviolet and γ-ray (gamma-ray) sections of the spectrum, x-ray photons are sufficiently energetic to be ionizing. In general, ionizing radiation can break inter-atomic bonds in matter, including human tissue. Whereas visible light photons have energies around 1.6–3.1 eV (electronvolts), medical x-rays typically have energies ranging from 10 to 150 keV (kilo electronvolts). That is, x-ray photons in medical imaging are about 10,000–100,000 times more energetic than visible light photons.

3.1.1 X-Ray Tubes

Medical x-rays are usually produced by x-ray tubes (e.g., a Coolidge x-ray tube). Housed within a vacuum tube (Fig. 3.1), electrons are accelerated from a cathode (i.e., the electron source) toward an anode (i.e., a metal target). Once the accelerated electrons hit the target, they rapidly decelerate, and a part of their kinetic energy is then converted into x-rays by means of physical interactions.

The emission of the electrons from the cathode is based on thermionic emission: heating a metal filament. The filament temperature is controlled by an electric current. An increase of the filament current elevates its temperature, such that a hotter filament emits more electrons.

The x-rays are generated by accelerating electrons to hit a metal anode and release part of their energy as energetic photons.

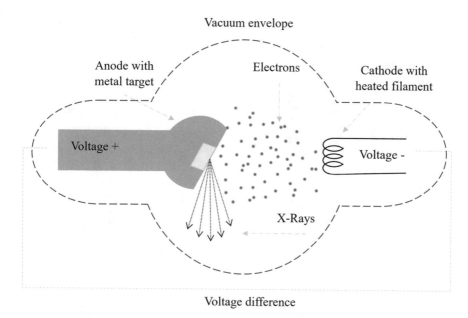

Fig. 3.1 X-ray tube illustration. The cathode emits electrons that then bombard the metal target of the anode, releasing energy as heat and as x-rays

The required temperature for an effective thermal emission can reach a couple of thousands of degrees Celsius. Usually a tungsten-based filament is used, in order to reach the necessary temperature, since it has a very high melting point. Once the metal becomes hot enough, it starts to emit electrons until a point where an electron cloud starts to form around the filament (space charge). The emitted electrons are then accelerated by applying an electrostatic electric field via a voltage difference between the source (cathode) and the target (anode).

Since the number of accelerated electrons toward the cathode is proportional to the number of x-ray photons (for a given electrostatic potential), the user can control the beam intensity of output x-rays by changing the electric current of the tube. The tube current is actually controlled by the filament current. For such thermionic emission, even a slight change in filament current has a substantial effect on the emitted electrons. Therefore, a small change in filament current can result in a significant change in the tube current and the total x-ray output. It is customary to use a feedback loop on the filament temperature and the current to control the number of emitted electrons to make sure it remains steady.

In addition, an increase of the voltage difference between the anode and cathode increases the kinetic energy of the electrons. Hence, it increases the maximal energy of the resulting x-rays.

The target (anode) can be stationary or rotating. A rotating target has the benefit of better heat distribution, which enables higher x-ray output intensities without thermal damage to the target. This is a significant aspect, since only about 1% of the

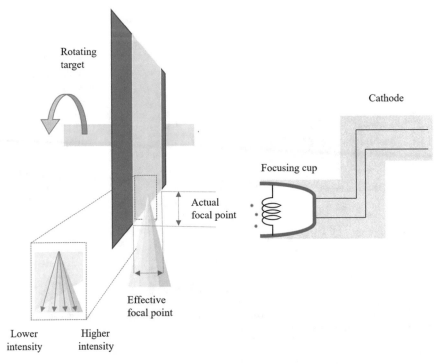

Fig. 3.2 Illustration of a rotating target. The electrons are directed with the help of the focusing cup, limiting the size of the actual focal spot. The angle of the target sets the size of the effective focal spot. The heel effect (lower left-hand image) is a result of the differences in photon path lengths within the target. Longer path lengths decrease the x-ray intensity

electrons' kinetic energy is converted into x-rays, while the rest of that energy is converted into heat.

A focusing cup located around the filament prevents the dispersion of electrons from the filament (Fig. 3.2). The cup assists in shaping the accelerated electron stream into a beam and directing it toward a small area on the target. This is called the actual focal spot.

The width of the electron beam, combined with the angle of the target area, dictates the size of the effective focal spot. The effective focal spot has a significant impact on the final image resolution. Smaller and larger focal spots yield higher and lower spatial resolution, respectively.

The electron beam that hits the target generates x-ray photons. Since the target is angled, with respect to the electron beam direction (i.e., an axis that stretches between the anode and the cathode), the x-rays are not distributed uniformly along that axes. The x-ray photons closer to the anode side have a longer path to travel "inside" the target (penetration depth), and they are more likely to be attenuated and absorbed by the angled heel compared to the photons closer to the cathode (see Fig. 3.2 inset). Consequently, the radiation intensity is greater on the cathode side of the tube. This effect can cause variations in the image, especially in planar scans. One way to maintain even image quality when acquiring a planar image is to position the more attenuating part of the body toward the side of the tube with the cathode.

In the x-ray beam path exiting the x-ray tube, a set of specially designed materials with different widths and shapes serve as physical filters to shape the x-ray beam and beam spectrum. One of the main purposes of these filters is to absorb the "soft" photons having relatively low energy. These photons are the most likely to get fully attenuated within the body of a patient, without reaching the detectors. They are undesirable as they will not contribute to the image quality but they increase the radiation dose and thereby increase the assigned health risk due to ionizing radiation. The removal of "soft" photons increases the average beam energy and is called "beam hardening." The total beam intensity (e.g., photons per second) is decreased by filtering. In addition, a compensation filter (sometimes called wedge or bowtie filter) might be used, depending on the clinical needs. These filters have an uneven shape that ultimately reshapes the beam spatially to even out the radiographic density when scanning uneven tissues (Fig. 3.3). They can be designed to optimize the imaging of relatively small regions such as the spine. Such filters absorb most of the beam intensity that spreads to the side of the beam and limit the zone of high radiation, consequently setting a boundary for the size of the scanned objects.

Lastly a collimator is used to limit the resulting x-rays into the direction toward the desired scanned location. It reduces the number of x-ray photons that might be scattered into the detectors from other unwanted locations. Such photons can cause image degradation and reduce the overall image quality while increasing the

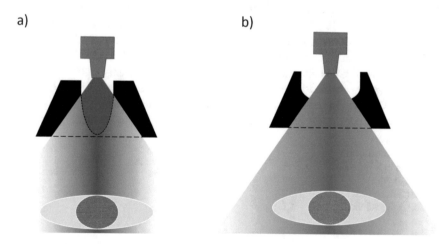

Fig. 3.3 Illustration of narrow and wide wedges (**a** and **b**, respectively) used as x-ray beam (blue) filters (black). Even though the tube's x-ray output is the same, after the wedge shaping, the beam on the left is more intense in the middle and is narrower compared to the wider beam on the right. The wider wedge on the left is more suitable for cases when the scanned object is large, such as the shoulders. On the other hand, the narrow wedge is suitable for cases when the scanned object is rather small, such as the heart

radiation dose. In addition, it serves as a shutter for the x-ray beam. When the tube begins to generate x-rays, it takes time for the filament to get hot enough for the desired current and x-ray flux to be achieved. This is called "rise time." During this time the collimator is closed, to reduce the x-ray exposure. Once the entire desired body part is scanned, the collimator is closed, preventing x-ray photons still generated inside the tube from reaching the patient.

3.1.2 X-Ray Generation

The x-rays that are emitted from the tube are generated by two main mechanisms: Bremsstrahlung and characteristic radiation. The x-ray beam consists of photons at multiple energy levels (polychromatic spectrum), in the neighborhood of several dozen keV to a couple of hundred keV. The maximal energy is set by the tube voltage, referred to as the kV (kilovolts) or kVp (kilovolts peak) setting. This is selected by the user according to the clinical need.

3.1.2.1 Bremsstrahlung Radiation

Bremsstrahlung radiation means "braking radiation." It is the radiation emitted from the deceleration of the electrons due to interactions with other charged particles in the metal target. During an inelastic collision between the energetic electron and the nucleus of the metal target, the electron slows down and changes direction (deflections). The electron's lost kinetic energy is converted to electromagnetic radiation, namely, bremsstrahlung radiation. The spectrum of emitted bremsstrahlung x-ray photons is a continuous one, ranging from nearly zero, when only small amount of kinetic energy has been lost, to the maximal kinetic energy level of the electron (Fig. 3.4). The maximum energy arises when the electron has practically stopped and all of its kinetic energy has generated an x-ray photon. High Z materials (high atomic

Fig. 3.4 A schematic illustration of x-ray tube spectrum, with Bremsstrahlung and characteristic radiation

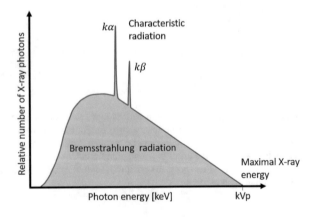

number materials like metals) yield more deflections than lower Z materials and consequently produce comparatively more x-ray bremsstrahlung radiation.

3.1.2.2 Characteristic Radiation

In the Bohr model of the atom, the electrons occupy "orbits" around the nucleus in different energy shells, where the innermost one is called the K-shell, the next one is the L-shell, and so on. The K-shell electrons are bound more tightly to the atom than the outer shells' electrons and are said to be in a lower energy state.

An ionization interaction happens when there is an inelastic collision between a charged particle (e.g., a high kinetic energy electron) and one of the electrons bound within a shell. The charged particle may lose energy in such collisions, transferring it to the bound electron such that it is used to overcome the binding energy and shift the target electron to a higher energy shell. This is called excitation. In such case, where electron is removed from one of the inner orbitals, another one, from an outer shell, will replace it by "jumping" to the inner shell, now missing an electron, to refill it. The energy released in this last transition from higher to lower shells will manifest as a photon with exactly the same energy as the difference in binding energy levels between the two shells. This in turn can cause a cascade of released photons as electrons from even higher energy shells fill lower energy shells with vacancies. Since the shells have quantized binding energy levels, the released photons will possess only those values of the "allowed" energy levels. Different elements have different shell structures with different binding energies that are characteristic of the element, hence the name characteristic radiation or characteristic x-rays.

> The x-ray beam is polychromatic. The maximal energy of the x-ray spectrum is set according to the selected tube voltage.

3.1.3 X-Ray Interaction with Body Tissues

There are several types of interactions between x-ray and matter for the clinical diagnostic imaging x-ray range. In most cases, interactions can result in the local deposition of energy, but in some cases an x-ray will continue after the initial interaction, as a scattered x-ray. The relevant types of interactions are the photo-electric effect, Rayleigh scattering, and Compton scattering. All of them are inter-actions between a photon and one of the electrons that orbits the atom involved in the interaction.

3.1.3.1 The Photoelectric Effect

With the photoelectric effect interaction, the x-ray photon is completely absorbed and all of its energy is transferred to the electron. In such a case where the electron is bound to its parent atom with binding energy E_1 and the energy of the incident x-ray is given by E_0, the kinetic energy E_k of the photoelectron equals $E_k = E_0 - E_1$. If the energy of the incident x-ray photon is less than the binding energy of the electron, i.e., $E_0 < E_1$, photoelectric interaction with the specific electron is energetically not possible and will not take place. When $E_0 = E_1$, then photoelectric interaction is most probable, whereas the probability is decreases with increasing E_0 thereafter.

As mentioned earlier, in the Bohr model, the electrons are assigned to different orbitals (e.g., K, L, M, and N). If a photoelectric interaction with the K-shell electrons is not feasible because of insufficient incident x-ray energy, such interaction might still happen with electrons from one of the outer shells. The binding energy E_1 that is associated with the K-shell is called the K-edge and so on for each electronic shell. The term "edge" refers to the sharp rise in the probability of photoelectric interaction when the process becomes energetically possible.

The photoelectric effect results in the ionization of the atom, and a single ion pair (a positive charged atom and a negative charged electron) is initially formed. When the parent atom and the bound electron become separated, it creates a vacancy in one of the orbital shells. This in turn starts a cascade of electrons from higher to lower orbitals and produces radiation characteristic to the element of the atom involved in the interaction (characteristic radiation). It is important to note that though this effect in many ways resembles the mechanism for generating characteristic x-rays, there is a major difference. Since different materials possess different binding energies of the electrons' shells, the characteristic radiations they emit are different as well. Usually x-rays are produced by electrons hitting a high atomic number metal target like tungsten ($Z = 74$) and the characteristic x-rays energies are therefore quite high.

However, the imaging x-rays interact with human tissue which, unlike a metallic target, consists mostly of low atomic number, such as carbon ($Z = 6$), hydrogen ($Z = 1$), and oxygen ($Z = 8$). Such compounds have low K-shell binding energy. For example, the K-shell binding energy of oxygen is 0.5 keV, which is very small compared to energies of incident x-rays (up to 150 keV). Such low-energy x-rays cannot travel very far before being totally absorbed locally.

3.1.3.2 Rayleigh Scattering

This type of scattering consists of an elastic (coherent) scattering of x-rays by atomic electrons. The energy of the scattered x-ray equals the incident x-ray energy. There is no exchange of energy between the x-ray and the medium, and ionization does not occur. However, the scattered x-ray has a new path compared to its original trajectory. This has an undesired effect on the imaging outcome, in that scattered x-ray may cause image blur if detected. As expected, higher-energy x-rays are subjected to much a smaller angle scattering compared to lower-energy x-rays. The interaction with the involved atom cannot result in ionization due to the laws of energy and momentum conservation. As a result of this mechanism, Rayleigh scattering is most common for low-energy x-rays hitting high Z materials. As the energy range for diagnostic x-ray is in the order of 100 keV, and the tissue is made of mostly low Z materials, the coherent scatter cross section (i.e., the probability of Rayleigh scattering occurring) is relatively small and is usually less significant for x-ray imaging.

3.1.3.3   Compton Scattering

Compton scattering involves the incoherent scattering of an x-ray photon by an atomic electron. It typically takes place in cases where the atomic electron binding energy is significantly smaller than the energy of the x-ray photon. This is the main reason that the Compton effect mostly occurs at the outer electron shells where the electrons are essentially free in the medium. In such interactions the incident x-ray photon with energy E_0 is scattered by the medium. The products of the interaction include a scattered x-ray photon, with its new energy value E_1, an electron with kinetic energy T, and an ionized atom. In Compton scattering, a relationship between the fractional energy loss and the scattering angle θ is observed according to

$$\frac{E_1}{E_0} = \frac{1}{1 + \alpha \cdot (1 - \cos \theta)} \tag{3.1}$$

where $\alpha = E_0/(m_0 \cdot c^2)$ in which m_0 is the rest mass of the electron and c is the speed of light in a vacuum, such that $m_0 \cdot c^2 = 511$ keV. Qualitatively this equation implies that with an increased scattering angle, there is a reduced energy of the scattered x-ray photon. This effect is amplified at higher incident photon energies. Higher-energy x-ray photons have increased chance of forward scattering, whereas photons with low energy are more likely to backscatter. This kind of scattering is the most probable interaction of x-rays within tissues at energies used for diagnostic imaging.

3.1.3.4   X-Ray Attenuation

Planar x-ray and computed x-ray tomography measure the attenuation of the x-ray photon beam by either absorption or scattering. In case that a beam of N x-ray photons is incident at a thin plate of material with thickness dx and probability μ of interaction (Fig. 3.5), the reduction of photons from the beam is given by

$$dN = -\mu \cdot N dx \tag{3.2}$$

Fig. 3.5 Schematic
depiction of x-ray photons
through a plate with an
attenuation coefficient μ and
thickness dx

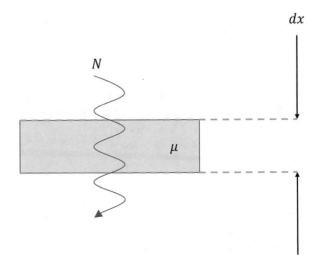

Rearranging and integrating Eq. (3.2), with N' as an index of integration, results in

$$\int_{N_0}^{N} \frac{1}{N'} dN' = -\mu \cdot \int_{0}^{l} dx \tag{3.3}$$

Solving this equation yields

$$N = N_0 \cdot e^{-\mu l} \tag{3.4}$$

Equation (3.4) is known as the Lambert-Beers equation.
For a uniform path and a known original energy I_{x_0}, it has the form

$$I_x = I_{x_0} e^{-\mu \Delta x} \tag{3.5}$$

The units of the thickness of x are usually (cm), so the units of μ would be (cm^{-1}). The probability of absorption, μ, is called the linear attenuation coefficient and sometimes is marked by μ_l. It is the probability of interaction from all interaction mechanisms and is therefore the sum of the probabilities of all the interaction types $\mu_l = \tau + \sigma_R + \sigma_C + \pi$, where τ is the attenuation coefficient for the photoelectric effect; σ_R and σ_C are the Rayleigh and Compton scatter attenuation coefficients, respectively; and π is the pair production attenuation coefficient. At x-ray energies greater than 1.022 MeV, interaction with matter can produce an electron/positron pair with probability π. In diagnostic x-ray the used energy is not high enough for pair production, so in reality $\mu_l = \tau + \sigma_R + \sigma_C$.

Equation (3.5) is also true in the case of a non-uniform material in which each layer, Δx_i, has an attenuation coefficient μ_i, up to n layers of material:

$$I = I_0 e^{-\mu_1 \Delta x_1} e^{-\mu_2 \Delta x_2} \ldots e^{-\mu_n \Delta x_n} = I_0 e^{-\sum_{i=1}^{i=n} \mu_i \Delta x_i} \qquad (3.6)$$

The resulting exponent is the sum of all the different attenuations along the path of the photon.

The linear attenuation coefficient μ_l describes the attenuation properties of a specific material, for a specific x-ray energy. Since it depends on the electron density of the material, water vapor, liquid water, and ice will have different values of μ_l for the same energy. Since μ_l is changing proportionally with the density of materials, a way to compensate for density is to normalize μ_l by the density ρ, giving what is called the mass attenuation coefficient μ_ρ:

$$\mu_\rho = \frac{\mu_l}{\rho} \qquad (3.7)$$

Consequently, the mass attenuation coefficient for water vapor, liquid water, and ice is the same for a specific energy.

Since the units of μ_l are (cm^{-1}) and the units of ρ are $(g\ cm^{-3})$, according to Eq. 3.7, the units of μ_ρ, which is the mass attenuation coefficient are $(cm^2\ g^{-1})$.

Now Eq. (3.5) becomes

$$I_x = I_{x_0} e^{-\mu \cdot \rho x} \qquad (3.8)$$

where instead of the length x a new quantity ρx is being used, which is the density times the thickness of the material. This so-called "mass thickness" ρx is defined as the mass per unit area. Just like the linear attenuation coefficient, the mass attenuation coefficient is the sum of its components: $\mu_\rho = \frac{\tau + \sigma_R + \sigma_C}{\rho}$. The mass attenuation coefficient μ_ρ is the coefficient used for all calculations in conventional x-ray imaging today, where the temperature remains constant during all diagnostic scans, clinically taking into consideration as 37 °C.

The density ρ defines the ratio between the linear and the mass attenuation coefficients, and the density of each element differs depending on its state and purity. In general, for most elements, there is a trend of increasing densities with the increasing of the atomic number Z. In addition to the atomic number, the photon energy is also a factor that influence μ. This is summarized in Table 3.1.

Since different materials and tissues are made of a different combination of elements and molecules, each has a different attenuation. A higher value of μ means higher attenuation.

For example, adipose (fatty) tissues have a lower x-ray attenuation than water, which has a lower attenuation than soft tissues. The most attenuating body parts are the bones, although if there are metal implants within the body, they are usually even more attenuating than the bones.

Table 3.1 Summary of factors affecting the attenuation

	Photoelectric effect	Rayleigh scatter	Compton scatter
Approx. μ_ρ dependence on Z	$\propto Z^3$	$\propto Z$	σ_C is independent of Z
Approx. energy dependence	$\propto E^{-3}$	$\propto E^{-2}$	Decreases with energy
Contribution to the total attenuation	At low photon energies, this is the main interaction for photons with the absorber	Have relatively small contribution and amounts to only a few percent of the total attenuation	This is the predominant type of photon interaction in the diagnostic energy range

According to the attenuation of the x-ray beam, each pixel in the image is assigned a different gray-level (shade of gray). The difference in pixel values between the different body parts is the source of the image contrast. It provides the physician with the ability to distinguish between different elements that are being scanned with x-ray.

The difference in attenuation between the different tissues is the source of contrast in x-ray-based imaging.

Current common practice is to display more attenuating tissues with lighter shades, while less attenuating tissues appear with darker shades.

3.2   Projectional Radiography

3.2.1   Detection and Data Acquisition

For projectional radiography and fluoroscopy, the x-rays that pass through the patient body are usually collected by a flat panel detector or film, which is placed across from the tube with the patient, or the examined body part in-between.

3.2.1.1   Film Screen Radiology

In previous years, screen film radiography was the only method to detect the x-ray photons and form an image. It used films made of a very a thin transparent base, coated with a special silver halide suspension. When this suspension is exposed to x-rays, it produces a silver ion and an electron pair. The electrons get attached to "sensitive spots" which are irregular shaped crystals and attract the silver ion. As a result of enough exposure, the silver ions aggregate, forming blobs of black metallic silver to form the latent image. A second step involves the processing of the film in a dark room and developing it with alkaline solution. During this process, these silver blobs become opaque and make the image visible and insensitive to further exposure. Highly attenuating structures like bones will appear white, while the areas of the film that were more exposed to radiation will appear with darker shades.

For this reason, another parameter that should be taken into consideration when using film is the total exposure intensity and exposure time. If the exposure time is too short, not enough photons can reach the film, and the resultant image might be too "bright," which would impair the image contrast. This is less of an issue when working with non-film detectors, as the image brightness can be manipulated by means of post-processing; however, extremely photon-poor x-ray radiographs will be noisy. Conversely, if the exposure time is too long, too many photons reach the film or detector, saturating it and making the image too dark with reduced contrast.

The film method had many limitations, therefore, computed radiography (CR) and digital radiography (DR) are currently the common means of detecting and generating the output image for x-ray imaging. Some of the advantages that CR and DR offer are: larger dynamic range, no need for chemically development of the films, and increased sensitivity. However, their main advantage over the old-fashioned film images is the ability of digital clinical images to be retrieved from a computerized database such as a PACS (picture archiving and communication system). This facilitates storage, distribution, comparison, and fast digital processing of the image, to improve the image quality, and enhance diagnostic capabilities, according to the clinical needs.

3.2.1.2 Computed Radiography and Digital Radiography

In the CR system, a plate is used to detect the x-rays which pass through the scanned body. It is placed inside a mechanical opaque housing called a cassette. This is to prevent its exposure to visible light in the examination room, which might also be detected by the plate, and interfere with the x-ray photon detection.

The plate is coated with a photostimulable crystal made of halogenides called the phosphor. Once scanned with x-rays, the energy of the incident x-ray photon causes the stimulation of the electrons to a higher energy level within the crystal. This energy is then temporarily stored within the phosphor. After which, one has to process the plate in order to convert the data into a digital image. For this purpose, red light laser is shone separately at each pixel, initiating the release of the stored energy by photostimulated luminescence: bluish visible light photons are emitted. This is performed pixel by pixel so each time the emitted light is in correlation with the stored energy of that specific pixel. The emitted light photons are detected and transformed into electronic signals, either by silicon photodiodes or, more commonly, by photomultipliers. From such a signal-per-pixel system, it is possible to generate a digital image and to acquire the output image. Since the stored energy within the phosphor might be released slowly over time, the readout process usually happens straight after the scan, to prevent any loss of data.

Sometimes, the laser does not stimulate the release of all the previous high state electrons and some of the stored energy remains as residual energy. This is usually referred to as "trapped" energy. In order to reuse the plate, one must "clean" it first, by wiping all of the residues, to avoid "contamination" in the next use of the plate. This is usually achieved by subjecting the plate to intense white light. Because reading CR requires two steps, this method is used for projectional radiography, but cannot be used during fluoroscopy, which requires real-time imaging.

DR technology, which is becoming more common compared to the CR technology, can be divided into two types: direct and indirect conversion systems. Indirect systems include two steps: a scintillator that turns x-rays into visible light and a photodetector that converts the luminosity measurements into electric current (Fig. 3.6). The main interaction between the x-ray photons and a scintillator is the photoelectric effect. While traveling a short distance within the scintillator, the resultant photoelectron may excite other atoms by imparting energy to their electrons and moving them to a higher energy shell. When these electrons return to their lower energy states, they emit characteristic radiation in the range of visible light and UV radiation, for scintillators. On the side of the scintillator opposite to the incoming x-ray beam, there is a silicon detector array consisting of many submillimeter pixels with a very thin-film transistor (TFT) array that forms a grid on a silicon wafer. Each detector also contains a photodiode. The photodiode converts the visible light that was generated in the section of the scintillator layer in front of the pixel to electrical current. This conversion is of a correlative manner, where more visible light produces more signal. In order to generate an accurate digital representation of the x-ray

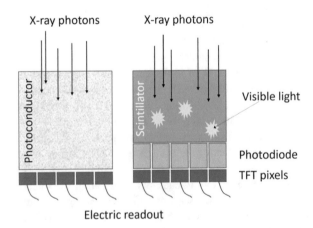

Fig. 3.6 Detector panels can indirectly (right illustration) or directly (left illustration) convert the energy of the incident x-ray photons to electric signal. While the direct method is using a "single step" photoconductor to convert the incident energy to charge, the indirect method has two steps: first it has to transform the x-ray into visible light, using a scintillator, only then it uses a photodiode to convert it into electric charge. Both methods are using an array (pixels) of thin-film transistor for the final readout

image, the electric current from the photodiodes is amplified and processed. This additional electronics is placed behind the sensor array so it does not attenuate the incoming signal.

The direct conversion systems are built with photoconductors that can convert incident x-ray photons directly into electric current, without the need to any intermediate stage. When x-ray photons hit the layer of the photoconductor, it generates electron-hole pairs due to the internal photoelectric effect. A voltage difference applied to the photoconductor layer draws the electrons and holes to corresponding anodes and cathodes, respectively. The resulted electric current is correlative with the energy of the incident x-ray photon. An array of thin-film transistors reads the signal and creates the grid of pixels.

> X-ray detectors are based on converting the energy of the x-ray photons either to a measurable amount of light (indirect conversion method) or directly to a measurable electric current (direct conversion method).

An important difference between the two methods is the spatial resolution. Even though both methods can have very-small-size pixels, with the indirect method, the visible light spreads laterally across the wafer of the photodetector. This results in a "blob" of light that can be detected by several adjacent pixels. Registration of some signal in pixels adjacent to the pixel recording the main event is sometimes referred to as "cross-talk." Such a blob will result in a measurement of a smaller amount of energy than the original energy of the incident photons. This can degrade spatial resolution and increase the noise within the image. Such impediments can be improved by using pixelization of the scintillator with guard rings around each detector pixel that reduces the spread of scintillating light. Regardless, one of the main benefits of the direct conversion method is its superior spatial resolution due to each photon being detected by a single pixel to begin with.

Another important benefit of the direct conversion over the indirect conversion is the specific measurement of energy of each single photon. This is in contrast to the indirect conversion approach, where the detectors do not detect each photon incident individually, but rather yield a signal that is proportional to the total amount of scintillated light from all of the incident x-ray photons at a specific pixel. This means that with the indirect method, the system cannot differentiate between a case with one very energetic photon and several less energetic photons if the same luminosity is generated in both cases. With the direct method, each incident is measured separately; hence, it is easier to differentiate incoming x-ray photon energies. This information can be used to differentiate between tissues with similar total attenuation but different atomic composition.

To optimally match the clinical needs (Table 3.2), each different method has different advantages that should be taken into consideration, such as the cost, ease of use, and time to operate.

Table 3.2 Parameters influencing the resultant image quality

Parameter	Description	Desired
Quantum efficiency	The probability of an interaction between the detector medium and the incident x-ray photon	As high as possible
Electronic noise	Variations in signal that are not a result of x-ray attenuation	As low as possible
Dynamic range	The range of photon intensity that can be detected (above the electronic noise and below the saturation level)	As high as possible
Spatial resolution	The ability to separate two adjacent objects, which depends on the pixel (detector element) size of the grid	As high as possible
Data acqui- sition time	The time it takes to read all the pixels and acquire the image	As short as possible (especially for real-time fluoroscopy)

One of the causes of blur in the image is scattered radiation). This happens when an incident photon reaches a detector after being scattered: instead of traveling along a straight path inside the body and hitting the detector pixel along that path, a photon might be scattered and hit another adjacent detector pixel. Such incidents reduce the measurement accuracy for both pixels and harm the contrast and resolution of the final image.

In order to reduce this effect, an anti-scatter grid might be used. This grid is placed after the patient, just before the detector, and is made of a very attenuating material, such as lead. Photons along paths deviating too much from the focal spot origin within the x-ray tube will be attenuated by the grid before actually reaching the detector, lowering their negative impact on image quality. This process is referred to as collimation. On the other hand, since fewer photons hit the detector with colli- mation, more radiation is needed to maintain image intensity compared to the situation without an anti-scatter grid.

A different current from each detector pixel is measured providing the contrast needed for image formation. Typically, the pixelated detector needs to be calibrated in order to correct for different sensitivity and performance between detectors.

3.2.2   Image Formation

In planar imaging, the image formation and reconstruction process can be thought of as straightforward. However, some factors have a significant impact on the resultant image that should be taken into consideration.

One of these factors is the magnification. Three main terms are used to describe the geometry of the source, scanned object, and detectors (i.e., image): source-object distance (SOD), object-image distance (OID), and source-image distance (SID). The magnification factor describes how much bigger than the scanned object the image appears on the detector:

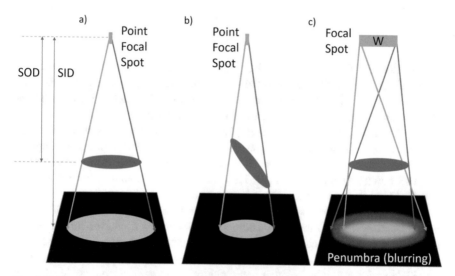

Fig. 3.7 Illustration of magnification and penumbra. The left illustration (**a**) demonstrates the magnification of the object projection on to the detectors, relative to the object's original size. The middle illustration (**b**) shows that the angle of the object, compared to the path between the source and detector, also impacts the size of the projection. The right illustration (**c**) demonstrates an addition to the "shadow" of the projection called the penumbra giving "blurred" edges to the object projection. A smaller focal spot offers sharper edges and therefore better spatial resolution

$$M = \frac{\text{SID}}{\text{SOD}} \qquad (3.9)$$

The magnification factor M is what causes objects that are far from the detectors to look larger than their actual size on the resultant image (Fig. 3.7a). The effect is similar to making "hand shadow puppets" with a candle flame in a darkened room. A shadow cast on a wall (image) will appear larger as one's hand (object) approaches the candle (source).

This could lead to ambiguity. For example, when objects with similar size are located at different distances from the source, they will appear as different sizes in the x-ray radiograph. Also, changes in the angle of the object with respect to the source and detectors will result in different projections yielding different images due to parallax (Fig. 3.7b).

The design of the focal spot as well as the distances within the imaging system are also very important, as they have a significant effect on the image resolution and blurring. Since the focal spot is not infinitely small, the width of the effective focal spot results in an area on the detectors, at the edge of an object image, which has a varying density of radiation (Fig. 3.7c). This in turn causes an edge gradient blurring which is termed the penumbra. A smaller magnification factor helps in minimizing the width w of the penumbra and the blurriness of the image according to

$$w = W\left(\frac{SID}{SOD} - 1\right) \tag{3.10}$$

Equation (3.10) is used to calculate the size of the penumbra, where W refers to the focal spot width, SID refers to the focal spot-detector distance, and SOD is the focal spot-object distance. To reduce the penumbra and its deleterious effects, the effective focal spot size or the magnification should be reduced.

Another important step that takes place during the image formation process, especially in the non-film techniques, is the detector calibration. For example, a calibration might level out variations in the sensitivity and performance between different components of the system. This is performed automatically to make sure that the image is even and to reduce the appearance of artifacts.

The selection of the tube voltage (kV) also has a major impact on the final image. For most body tissues, the attenuation μ monotonically decreases for increasing energy in a non-linear manner. The difference in attenuation between the different tissues (i.e., the image contrast) is less prominent as the energy level of the photons gets higher (Fig. 3.8). On the other hand, photons with very low energy level might not be able to pass through the body, as they would be completely attenuated and absorbed before reaching the detectors. This reduction in photon intensity might result in a very low signal that would yield a higher noise level in the image. Consequently, the kV must be chosen at a level high enough to ensure that the

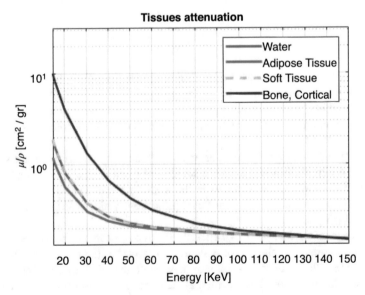

Fig. 3.8 Attenuation curves of tissues. Each tissue type has a different attenuation curve, resulting the various brightness levels in the final image. This is the source of the image contrast, which allows the distinction and differentiation between different tissues within the image. Note the differences in attenuation values for bone and soft tissues at 60 kV vs. 120 kV. The bone possesses high attenuation values, so it appears much brighter than soft tissue

Fig. 3.9 Clinical comparison between higher and lower voltage. While the x-ray CT image on the left (**a**) was taken with 120 kVp and exhibits lower noise level, the image on the right (**b**) was taken with 80 kVp, demonstrating higher noise level, but better contrast. Note that the white objects (bones and contrast agents) appear whiter and more noticeable when the contrast is increased. The display grayscale settings between the two images are the same

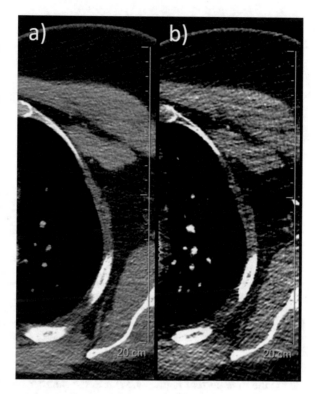

x-ray beam has enough energy to penetrate the body with sufficient intensity, but with a low enough energy to maintain image contrast.

The kV selection must be tailored for each specific patient and the scanned organ (Fig. 3.9). For example, when scanning the breast during mammography, a typical voltage setting is about 30 kV, since there is a need for high contrast, and the organ itself is quite small and less attenuating. On the other hand, when scanning the abdomen, the selected voltage might be around 120 kV, in order to make sure that the photons are energetic enough to travel through the pelvic bone without being completely attenuated and to ensure the image is not too noisy. For obese patients the selected voltage might be 140 keV, while for pediatric, it might be only 80 keV, as the photons have less "body" to travel through, before reaching the detectors. The clinical needs ultimately dictate the selected voltage that would yield optimal image quality while keeping acceptable radiation dosage.

The noise level in the image is also controlled by the tube current selection (sometimes called mA or mA·s selection). In theory, an infinitesimally isotropic x-ray beam energy would be inversely correlated with the square of the distance from the source. $I = a \cdot h^{-2}$, where is I the intensity, a is a constant factor, and h is the focal spot-object distance. This is still a very good approximation for most clinical systems design, if starting to measure the energy levels from about 50 cm away from the focal spot (the actual x-ray source), as the air attenuation is usually negligible.

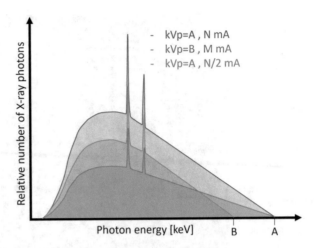

kVp=A , N mA
kVp=B , M mA
kVp=A , N/2 mA

Fig. 3.10 A schematic illustration of different spectra for different combinations of tube current (mA) and tube peak voltage (kVp) selection. The kVp selection sets the maximal energy of the x-ray photons (the blue and green lines). The only difference between them is that the current differs by a factor of 2. Note that the photon energy distribution itself remains exactly the same. For the case with the lower kVp (orange line), the photon distribution is different and has a different photon energy maximum. The characteristic radiations (the peaks of the three graphs) are always at the same energy, as they are defined by the tube's target material and are not kVp or current dependent

Since the tube current is proportional to the number of photons, and the signal-to-noise ratio increases with the square root of the number of photons, the physician can reduce the noise level, by increasing the tube current.

Note that even though both tube current and kV selection impact the noise level, the tube current does not impact the contrast level of the image. Increasing the tube current only increases proportionally the total number of photons detected (Fig. 3.10). The spectrum of the polychromatic beam (i.e., the distribution of the energies for a specific kV selection) remains constant regardless of the current selection.

The relation between the patient radiation exposure dose and tube current is approximately linear. The patient dose is also approximately proportional to the square of tube voltage. For the detectors, the relations are different. When film screen radiography was used, it was shown that an increase in 15% in kV reduced the contrast but increased the image detector dose by a factor of 2, just like doubling the tube current. This rule of thumb still applicable today with digital radiography and is called the "15% rule." It works both ways, so a reduction of 15% in the kV is equivalent to a reduction of half of the current, in terms of detector dose.

 Lower kVp increases the contrast between different tissues, but also the noise level, for a constant current. Higher current decreases the noise level but does not change the contrast of the image.

According to the Rose criteria, an estimation of the required number of detected photons n per unit area to show clearly an object of diameter d and at a contrast C is

$$n = [k/(C \cdot d)]^2 \qquad (3.11)$$

The value of k is usually taken to be equal 5. Consequently, a decrease in the contrast between an object to be imaged and its background will require a greater number of photons to be detected compared to a case with more contrast. For example, an increase in kV will decrease the contrast and could result a lower detectability.

The technician controls the setting that would result a good image quality, taking all factors into consideration. For example, when using a current (mA) setting I, and a kV setting of V, it can be estimated that patient dose D varies as $D \propto I \cdot V^2$. According to the 15% rule, the user can retain similar image quality by changing the settings I_0 and V_0 to $I_1 = 0.5I_0$ and $V_1 = 1.15\,V_0$. This will approximately maintain the detector dose but will decrease the patient dose: $D = I_1 \cdot V_1^2 \cong 0.66I_0 \cdot V_0^2$. Because of the increase in kV, the contrast-to-noise ratio would be less with the new settings, but if it is still acceptable, then a significant dose reduction can be achieved.

3.2.3 Projectional X-Ray Types and Modalities

3.2.3.1 Planar Imaging

Projection radiography or planar imaging is a type of medical imaging that generates two-dimensional images using x-rays. The x-rays pass in a specific direction from the x-ray tube through the patient or the scanned body part and hit the detectors positioned on the other side of the patient (Fig. 3.11). This process generates a single static image: the x-ray source and detectors are fixed during the scan. The image itself shows the difference in attenuation and absorption as different shades of gray. The value of each pixel in the resultant image is based on integration of the effect of all the tissues along the path of the photons that reach the detectors.

The fact that the image presents a superposition, i.e., a two-dimensional representation of a three-dimensional object, is the source of an inherent defect. This can cause a shadow effect where data from different tissues, one in front of another, overlap and possibly conceal a clinical problem. This superposition also exacerbates poor soft tissue contrast. With the absence of depth information, it is difficult to distinguish several poorly attenuating objects from one that is very attenuating. The development of CT scanners resolved these issues.

Despite these limitations, 2D x-ray radiography is still one of the most widespread imaging methods for clinical diagnostics, because it is fast, cost-effective and involves little ionizing radiation exposure to the patients. These scans include chest x-rays, extremity x-rays, and bone x-rays, among others.

Fig. 3.11 An illustration of chest x-ray procedure. The patient is positioned next to the detector panel, while the x-ray tube is on the opposite side

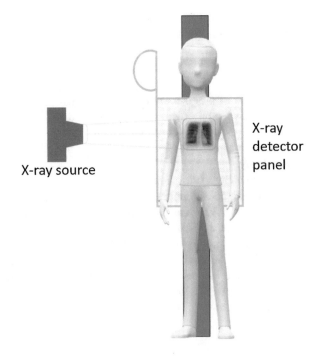

X-ray source

X-ray detector panel

Chest x-rays are one of the most routine types of medical images as they can detect a variety of clinical conditions such as pneumonia or lung cancer. The ionizing radiation dose is sufficiently low for chest x-rays to be used as a lung cancer screening test.

These planar x-ray radiographs are also routinely used for clinical cases involving the skeletal system. This method is a very easy and quick tool to diagnose conditions such as dislocated joints or fractured bones. It is also very useful in presenting foreign objects inside the body.

Dual-energy x-ray absorptiometry (DEXA) is a technique frequently used to estimate bone mineral density and to assess osteoporosis. With this scan technique, the scan is made with two x-ray beams of different energies. By utilizing the unique x-ray attenuation of bone and soft tissues, at given energy levels, it is possible to separate the two tissue types in the scanned area. Based on that separation, an assessment of the mass of the bones is made, and the resultant images can be used to generate a score for comparison to known reference values.

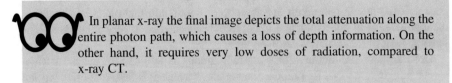

In planar x-ray the final image depicts the total attenuation along the entire photon path, which causes a loss of depth information. On the other hand, it requires very low doses of radiation, compared to x-ray CT.

In addition to planar imaging, there are a few special types of projectional radiography techniques that are unique to specific tissue types and clinical needs.

3.2.3.2 Mammography

A very specific type of projection radiography is the mammography. With this type of imaging, x-rays are used to generate 2D images of the internal anatomy of the breast and the mammary gland, hence the name of the resultant images: mammograms.

The breast is made mostly out of fatty tissues, with relatively low x-ray attenuation coefficients. In addition, the diameter of the breast itself is fairly small. Therefore, the x-ray photons have only a short distance to cover in a medium that is not very attenuating. Since the beam is not hardening significantly inside the medium, soft x-ray photons (i.e., lower energy) can be used. The maximum of the energy spectrum, set by the tube voltage, is usually low, approximately 30 kV, which results in a comparatively high image contrast (Fig. 3.12).

For this reason, a mammography exam requires a very low ionizing radiation dose. It is low enough to be used as a routine screening test for detecting breast cancer. With the help of mammography, it is possible to detect breast cancer in very early, more treatable stages, even before the tumor or lump is large enough to be detected by palpation, or before any other symptoms are manifest.

During the exam the breast is compressed between two sheets. The x-ray tube is placed quite close to the breast on one side, and the detection panel is adjacent to the breast at the opposite side. Usually images from two directions are taken, from the superior and lateral positions.

Fig. 3.12 An example of an x-ray mammography image of the breast. Most of the breast is composed of fatty tissue. It is a less attenuating tissue and so it is depicted with very dark shades. There are some scattered areas of fibroglandular density, resulting some "structures" within the breast. The muscle tissue, which is part of the chest wall, is even more attenuating and is depicted with brighter shades (at the upper right part of the image)

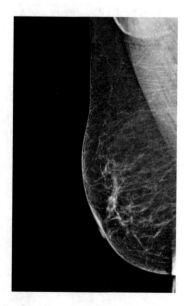

In recent years, digital breast tomosynthesis, which is a new form of mammography, has also been used, instead of traditional mammography. In this procedure, the breast is also compressed between two sheets, but instead of two single projections, the x-ray tube moves along a section of an arc. The system acquires many different projections, from which a three-dimensional model of the breast is generated, for better visualization and clinical diagnostics.

3.2.3.3 Dental

Dental x-rays can be divided into two main types, extraoral X-ray, when the film (or plate) is located outside the mouth, or intraoral x-rays, when the film is located inside the mouth.

Intraoral radiography is used to find cavities, look at the tooth roots, and see developing teeth (Fig. 3.13). The most familiar are the bite-wing images, where the patient bites a device with a "wing-like" shape that holds the small film unit, while the tube radiates. This gives very detailed images of individual teeth.

Extraoral radiography is used to exam the entire mouth, as they usually present most of the teeth, but they might also show the jaw and part of the skull. Extraoral radiography includes techniques such as panoramic x-rays, showing the entire set of teeth. With this technique the x-ray tube travels along an arc around the patient's head, to get a shot of the entire set of teeth. Alternatively, dental cone-beam CT may be used. This provides volumetric data of the teeth. With this method the tube and detectors rotate a full circle around the patient's head. It uses a much lower ionizing radiation dose, compared to conventional medical CT, but a considerably greater dose than the standard dental x-ray.

In recent years the use of CR and DR is becoming more widespread in dental x-rays, and the CR method is currently the common practice for most dental application. This is due to a couple of reasons, mainly for being less costly. Another advantage of CR is that it obviates the need for chemical development of film and

Fig. 3.13 An intraoral dental x-ray image. The metal fillings are bright white, as they attenuate much more than the enamel and dentine that makeup the dental bone

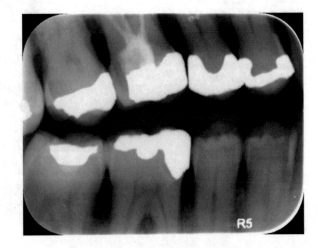

offers a better dynamic range than film. In addition, the CR plate is very thin, almost as thin as film; therefore, it is easy to use by the technician and is well accommodated by the patient. It is more difficult to position the DR sensor accurately, which is much thicker in comparison to the CR plate. In both methods, small electronic pads replace the older film units, allowing the images to be saved in a digital format, which is a very important advantage of both approaches.

3.2.3.4 Fluoroscopy

Fluoroscopic imaging is applied when the physician would like to see a "movie" of the x-ray images in real time. Typically, the patient is laying on top of the system table, while the tube radiates the desired organ from above, and the detectors are positioned below. The tube and detectors are connected by a "C-arm" system that can revolve around the patient to acquire images from different angles as needed. The images are transmitted to a monitor located just near the patient table for an easy review by the physician during the procedure.

In contrast to the static x-ray images of planar imaging, fluoroscopy offers the ability to monitor different dynamic processes such as showing a beating heart, food traveling inside the pharynx during swallowing, or the spread of a contrast agent in the gastrointestinal tract or blood vessels. The availability of real-time imaging of dynamic internal processes facilitates diagnosis as well as treatment.

The live motion ability combined with high spatial resolution is especially important in procedures involving instrument insertion, where the physician needs to see at every moment where the instrument is located. Examples of such procedures include orthopedic surgery, the placement of stents or of other devices within the body.

Since the "movie" is presented in real time, the patient is being irradiated during the fluoroscopy. Since this procedure involves relatively long exposure, the dose levels are higher in comparison to those used in planar imaging. Typically, fluoroscopes have hands-free methods of being turned on and off so that the irradiation occurs only as needed, thereby limiting the radiation dose.

3.2.4 X-Ray Contrast Agents

In medical imaging, the difference of the measured physical property of each tissue, is the source of the image contrast, and provides the ability to separate the tissues in the final image. In order to enhance the visibility of inner organs, structures, and dynamic processes, a designated substance, which usually generates a much stronger signal compared to the one of the tissues, is introduced to the body often orally or by intravenous injection. The differences between the native tissue and the substance, in the resultant image, are significant enough to generate visible contrast; hence, they are referred to as "contrast agents."

Each imaging modality has its own unique agents appropriate for the measured physical attribute. In addition, contrast agents have to be biocompatible, to minimize risk to the patient.

Given that in x-ray imaging x-ray attenuation is the physical phenomenon providing image contrast, common contrast agents are highly attenuating at the working photon energy range. This corresponds to x-ray tube voltages of about 30–150 kV.

> X-ray contrast agents are solutions containing materials with relatively high x-ray photon attenuation that causes them to appear brighter in a standard radiograph, compared to the darker shades of the less attenuating tissues. Blood vessels, tumors, or tissues that are colored by a contrast agent are much more noticeable in the resulting image.

3.2.4.1   Iodine-Based Angiography

One of the most common contrast agent procedures is angiography, which refers to the imaging of the circulatory system (Fig. 3.14). This is performed usually with an iodine-based contrast agent that enters the bloodstream via an intravenous (IV) catheter. In this case, the contrast agent is usually injected during the scan.

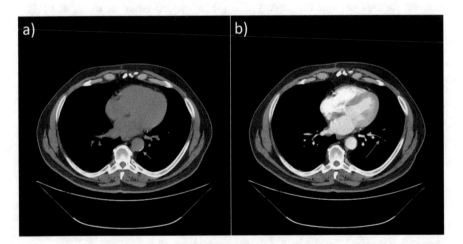

Fig. 3.14 Iodine contrast agent comparison, using two axial CT images of the heart. The left image (**a**) was taken without any contrast agent. A vertebra and the bones of the ribs appear white, while the heart itself has an almost uniform darker shade of gray. It is possible to see three white points, demonstrating calcifications, in the aorta (note arrow). The right image (**b**) was taken with a contrast agent. The blood within the heart and aorta has a much brighter shade due to the iodine solution, while the muscle tissue of the heart is still darker. A narrow blood vessel can easily be identified (note arrow)

Fig. 3.15 Attenuation
curves of contrast agents.
The high attenuation values
of the contrast agents,
compared to the soft tissue,
result in their brighter
appearance in a standard
x-ray radiograph. Note the
k-edge (sharp increase in
attenuation) of iodine,
barium, and gold, at 33 keV,
37 keV, and 80 keV,
respectively. Such increases
in attenuation add to the
visibility of those materials
in the final image

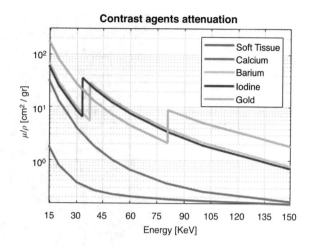

Since the iodine in the contrast material has a much larger atomic number
($Z = 53$) in comparison to atoms such as hydrogen, carbon, oxygen, nitrogen,
phosphorus, and calcium, which compose most of the body, it attenuates much
more than soft tissues and looks very bright in the x-ray images, just like bones
(Fig. 3.15). This is called a radio-opaque contrast agent.

Angiography is usually performed in combination with the fluoroscopic technique. With dynamic presentation of the contrast agent and blood dynamics, the
physician can detect an aneurysm or stenosis in arteries or veins during the imaging
procedure. If needed, the procedure can become a therapeutic one, where the
physician can place medical inserts, such as stents, very accurately, given that the
real-time imaging enables the physician to guide the catheter within the body.

With fluoroscopy, it is possible to have different work setups: pulse mode or
continuous mode. With pulse mode, a higher x-ray current is used, but the frame rate
is quite low usually, a few frames or "snapshots" per second. In most cases, this
results in a lower accumulated radiation dose to the patient. The continuous mode
works with much lower tube current, to prevent the tube from heating due to the
continuous and sometime long activity. This increases the total exposure to the
patient but also enables the fast frame rate to a couple dozen per second, which is
needed to produce a real-time movie.

Digital subtraction angiography (DSA) is a common technique to "extract" the
blood vessels in an image from the background. At the beginning of the procedure,
pre-contrast images of the region of interest are taken. With some image manipulation, these can serve later as "masks" for image subtraction. Once the contrast agent
is administrated, the imaging system automatically aligns the contrast image with the
previous image. By subtracting the image with contrast from the original mask, the
vasculature with contrast becomes evident. Due to the subtraction, even though
vessels with contrast agent traditionally appear white on x-ray images, on DSA
images, they have darker shade of gray than the background. This way, most of the
background tissues are removed from the resultant image, leaving essentially only

Fig. 3.16 An example of iodine DAS image during brain fluoroscopy. Due to the subtraction, the blood vessels with iodine appear black, while the rest of the background is gray. The arrow indicates a narrower part in the main blood vessel

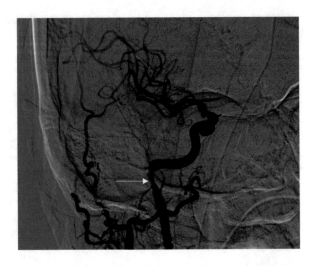

the blood vessels clearly visible (Fig. 3.16). DSA images usually have very good SNR and high resolution that show even very thin vessels.

3.2.4.2   Barium Contrast

Another common radio-opaque contrast agent is a barium-based suspension. Like iodine, barium has a high atomic number ($Z = 56$) compared to most elements found in the body; hence, it is also considered to be a radio-opaque contrast agent. It is usually used for examining the digestive system. One of its advantages is that it can be used in different forms and can be administrated to the body by drinking or as enema. For this reason, it can be used to diagnose pathologies in many different regions of the gastrointestinal tract and is used in esophagus testing and colon exams. Combined with fluoroscopy, it is possible to detect abnormalities during swallowing, as the physician tracks the barium dynamics from the mouth to the stomach. It is also possible to see the colon clearly when it is filled with contrast agent in order to detect abstractions, such as polyps, blockages of the large intestine, and ulcerative colitis (inflammatory bowel disease).

3.2.4.3   Nanoparticles and Microbubbles as Contrast Agents

In recent years the research of nanoparticles as contrast agents has made significant progress [1]. With this approach, elements such as gold and bismuth, with very high atomic numbers, are used. Due to their very high attenuation, they provide effective contrast, even in small amounts. Unlike the iodine or barium contrast agents, which were dispersed freely within the bloodstream or along the digestive system, nanoparticles can be made with special markers that makes them preferentially accumulate at targeted tissues, such as cancerous tumors. Though not yet in a commercial use, this promising technique might assist in cancer diagnosis and screening in the future.

Another relatively new approach that is being researched is the use of microbubbles as contrast agents. In contrast to the previous presented agents, which were all radio-opaque, the microbubbles are radiolucent: they allow x-ray photons to pass more readily than much of the surrounding tissue. This means that they generate contrast by appearing darker, in the final image, in the same manner as air. Microbubbles have already been used for many years as contrast agents for ultrasound imaging and might be effective for x-ray imaging as well.

3.3   X-Ray CT

3.3.1   Basic Concept

X-ray computed tomography (CT) is a digital process where multiple projections from different angles, acquired around a single axis of rotation, are used to generate a three-dimensional volumetric representation showing the internal structure of the scanned body (Fig. 3.17). Such volumetric data can later be manipulated in order to

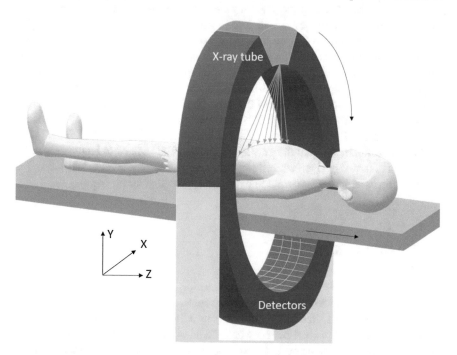

Fig. 3.17 A diagram of a CT scanner. The patient is on top of a table, while the x-ray tube and detectors are rotating simultaneously around the inferior/superior axis. The table can move in and out of the gantry, during the scan

define and illustrate structures and organs of patients, with image contrast based on differences in x-ray attenuation.

> In CT imaging, the data is volumetric and not a summation of values; hence, each pixel represents the x-ray photon attenuation of a specific volume element in the body. Volumetric data allows the generation of images from different directions and volume rendering of the scanned body.

The first clinical CT scanner system was developed by Godfrey N. Hounsfield during the early 1970s. The pixel values assigned to a CT image, which represent the radiographic brightness, are named Hounsfield units (HU), in his honor. He shared the Nobel Prize in Medicine in 1979 with Allan M. Cormack for the development of computer-assisted tomography.

Ever since its introduction, CT has become a useful and important tool in medical imaging to supplement planar x-ray imaging. In recent years, it has also been used for screening for disease as well as preventive medicine.

The key difference, and one of the main advantages, of CT over projectional radiography is the fact that it does not calculate a pixel's value based on the superposition of all the different mediums along the beam path. For example, if there are two objects along the photon path, with medium attenuation, they might look the same as a single object with higher attenuation in projectional radiography, due to the summation. In CT, on the other hand, each voxel displays the value of a volume element in space. Consequently, it is possible to generate 3D representations, or volume rendering, of the scanned body from CT data, which is not possible in projectional radiography (Fig. 3.18). In the former example, a physician will be able to see the different locations of the two objects, in the volume rendering of the CT, no matter what was their "order" with respect to the photon path. It is also one of the reasons that CT offers a much higher image contrast.

Another difference is that the gray levels of the CT are scaled to Hounsfield units. This means that for a given set of scan parameters, the gray level (pixel values) of specific organs in the image would always be approximately the same.

3.3.2   CT Evolution

The first CT scanners had only a single detector, with a narrow "pencil"-shaped x-ray beam. Once the detector took a reading, both detector and beam changed their lateral location, until a "line" was acquired (Fig. 3.19). With this acquisition method, the source generated parallel beams when scanning each "line." After such a "line" was completed, both detector and beam rotated around the patient a small amount (e.g., 1°), and a new "line" was acquired. This process was repeated until the rotation covered 180°. Such a manner of movement of the tube-detectors duo was called

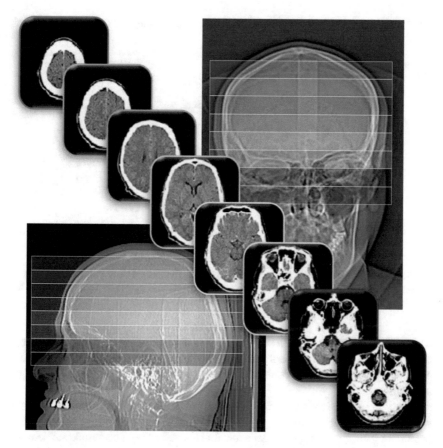

Fig. 3.18 Many adjacent axial images cover a large volume. Each image can be thought of as a two-dimensional plane, which together generate a three-dimensional volume. This is in essence a matrix of the voxel's values. With a matrix at hand, one can look at the entire volume, from each desired direction

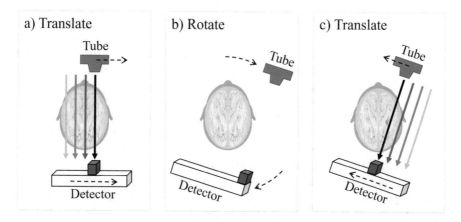

Fig. 3.19 An illustration of a single detector CT. The tube and detector would first translate simultaneously after each reading was acquired (**a**). Once a "line" was formed, both the tube and detector would rotate a little (**b**), and then a new translation stage would start to acquire another "line" (**c**). This would continue until a half circle rotation was accomplished

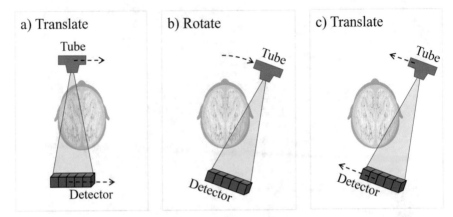

Fig. 3.20 Second-generation CT, with a few detectors, to form a fan beam. The movement was still termed rotate-translate, but since several detectors were involved at each "reading," the scan time was significantly reduced, compared to the first-generation CT design

a "translate-rotate" acquisition. Each slice was scanned independently, generating a cross-sectional image, i.e., axial image, of the inner body's structures. It was a time-consuming process that required several minutes to complete a single-slice acquisition.

The second generation of CT tried to handle the time-consuming aspect of scanning by having several detectors, instead of a single one (Fig. 3.20). At the beginning there were only three detectors that formed a line. The number of detectors grew over time until the detector line was composed of several dozen detectors. Now the beam shape was no longer a "pencil beam" but was changed to a narrow fan beam shape. Each "frame" was comprised of multiple readings, as the detectors yielded simultaneous readings from each x-ray snapshot. The movement of the main components was still considered "translate-rotate," but the acquisition time was reduced significantly. Along with this major improvement, new challenges emerged: Now, due to the fan beam, the reconstruction needed to handle the challenge of photons scattered to adjacent detectors, which impaired the image resolution and image quality. In addition, with the fan beam shape, the readings of the detectors are not generated by parallel straight lines, but by several diagonal lines having a conjoined apex at the tube.

With second-generation CT scanners, full-body scans yielded very good results, as scan time was short enough so that most patients were able to hold their breath during the entire scan. Still, for cardiac scans or chest scans, this was not fast enough. The most time-consuming step was still the mechanical "translate" movement of the system of the rotate-translate configuration. In addition, the system configuration of each scan was very sensitive to changes. Even small changes in the system's tube/detector position relative to its original position caused a misalignment of x-ray beams and detectors. This yielded image artifacts because now x-ray beams were not reconstructed with their "matching" detector and pixel, giving inconsistent projections in the sinogram data.

In order to eliminate the need to perform the "translate" motion, the short line of detectors was replaced with a much longer array assembled from hundreds and even thousands of detectors. With it, a matching x-ray beam that was wide enough to encompass the entire width of the patient was used. This way, each "frame" covered the scanned body, and the system only needed to rotate between each frame, in order to take views from multiple directions. The detector and x-ray tube were tightly linked opposing one another. They rotated together around the superior/inferior axis location, called the isocenter, while remaining stationary with respect to each other. This is referred to as "rotate-rotate" motion. In CT, unlike planar radiography, the acronym SID stands for the constant Source-Isocenter Distance, whereas the distance between the source and the detector array, is called SDD (Source-Detector Distance).

With a "rotate-rotate" configuration, the main time-consuming part became the delay between consecutive slices. Electric cables supplied the power and connected the detector array and the x-ray tube to the gantry. The cables enabled the control and data transfer between the detectors and the gantry. They were rolled out when the scan took place, but after each full rotation, the motion had to stop, and reverse direction, so the cables could be reeled back. Now each axial scan was composed of three steps: scan-break-reverse. Such a scan could take 10 seconds, of which only a couple of seconds were needed for the actual data acquisition. The time gap between the data acquisition steps was responsible for the long acquisition times and entailed low temporal resolution. This was particularly problematic when performing contrast-enhanced scans, with bolus tracking.

In third-generation scanners, the continuous rotation of the x-ray tube/detector assembly provides a solution (Fig. 3.21a). Generally, this is enabled via the use of slip rings. It uses a rotating metal ring, while stationary "brushes" contact on the outside of the ring, enabling the supply of power to the rotating assembly and transmission of electrical signals from the assembly to the stationary gantry. The slip ring method provides the connection between the gantry to the x-ray tube and

a)
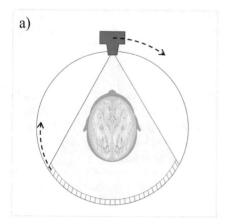

b)
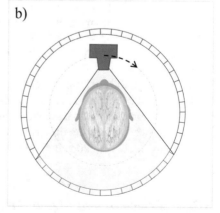

Fig. 3.21 Illustration of third (**a**) and fourth (**b**) CT generations. In the third-generation design, the tube and detectors rotate together around the patient. In the fourth-generation design, only the tube rotates around the patient, while the detectors cover a full circle of 360°

detectors, without the need for cables, allowing for nonstop rotation. Combined with the ability of smoothed and controlled table movement within the gantry, this enabled the scanning of an entire body as a series of rapid consecutives scans.

In parallel to the third generation, the fourth-generation CT presented another concept which tried to minimize the image artifacts that happened due to movements and detectors' instability (Fig. 3.21b). In this configuration the detectors cover the entire circumference, forming a circle around the patient and tube. During the data acquisition, the tube rotates around the patient, while the ring of detectors remains static. Since the detectors didn't move and for part of the time were subjected to direct x-ray exposure, without the patient interference, it helped with the detectors' calibration process and stability.

It is important to note that even though in both cases, of third and fourth generations, there is a fan beam (or cone beam in more recent designs), the tube and detectors have the "opposite" roles, in the reconstruction process. While in the third generation, the apex of the fan beam, for reconstruction, is the tube, in the fourth generation, the apex, for the reconstruction, is actually a single detector, collecting data from multiple positions of the tube, which forms the fan beam. This means that unlike the third generation, where the limits of spacing between samples are determined by the detector size, in the fourth generation, the sampling rate sets the limit of the spacing. If the rate is fast enough, the spacing can be much smaller than the detector size. A higher sampling density is a significant advantage, as it can reduce artifacts and increase image resolution.

One of the major drawbacks of the latter configurations was the extremely high number of required detectors, since one had to establish a full circle, which was quite large in diameter (>60 cm), as it included the patient, the table, and the x-ray tube inside. More detectors also meant more electronics and other hardware elements, which made the fourth-generation design rather costly. Ultimately, partly due to practical and financial reasons, this type remains uncommon in the clinic.

In further development to the third-generation type CT, more detectors were added in later designs, which were not arrange in a "line," but a wide and long arc, consequently changing the x-ray beam shape, from a "fan beam" to "cone beam." By widening and subdividing the detectors in the superior/inferior axis direction, several transverse slices could be acquired simultaneously resulting "multi-slice" CT.

As expected, the combination of large detector arrays, combined with slip rings, reduced the total procedure time drastically, from a few seconds per rotation, in early third-generation models, to less than a third of a second per rotation, in later models. With its subsecond rotation time and submillimeter resolution, the third-generation design is the most common configuration used in clinics today.

3.3.3   Detection and Signal Sources

Unlike the flat panel detectors used in planar x-ray radiography, in CT scanners the detectors are arrayed in an arc shape. The array is made of many solid-state detectors and each one is a small block of scintillator material. They operate as indirect conversion detectors, in which x-ray photons are absorbed by the scintillator to produce visible light. The resultant light is then measured by a photodiode, which converts it to electrical current. Each detector cell results in an independent measurement.

Whereas the x-ray tube is typically radiating continuously during acquisition, the detectors are not "detecting" continuously. Instead, they have a "reading window" duration. This is the integration period, or IP, used to generate the measurement, and it is the factor that determines the number of angular readings in a single rotation. For example, if a single rotation takes 1 second, and each IP takes 1 ms, there can be approximately 1000 frames per rotation, and the angular readings are approximately every 0.36°. The signal measured during a single IP by all of the detector elements is called a projection.

There is a tradeoff between the number of projections and SNR. More projections require shorter IPs, which result lower SNR. The SNR is reduced in measured projection since during short IP fewer events would be detected. Since the system also has some inherent electric noise, if the measured signal from the scintillators is too low, then the final output might be too noisy to be reliable.

The minimum duration of an IP is also limited by the afterglow phenomenon. The detection process depends on the interaction between the x-ray photon and the scintillator producing visible light. The light is the result of fluorescent emissions from excited electrons returning to lower energy states. This decay has a specific time constant that can be modeled as an inverse exponential process, so after a certain amount of time, it is assumed that there are no more excited electrons. This can be thought of as a "reset" point for the detectors before a new IP start. However, due to impurities in the scintillating material, some electrons get "stuck" in an excited mode and produce delayed emission. As a result, it takes a longer period of time for the light emission process to effectively finish. This is the afterglow. If it continues too long, these delayed emissions might be counted in the signal of the next projection, which can lead to image artifacts.

Closely packed, small detector elements are subject to additional sources of interference. With cone-beam CT, photon scatter is a more prominent factor than in earlier CT generations. Scattered radiation may hit a detector element different from the one expected from a direct ray path from the x-ray tube, introducing errors to the projections. An anti-scatter grid at the detector face is often used to lessen this effect. Essentially, the incoming beam paths are partially collimated at the detector face to define direct paths to the focal spot in the x-ray tube. The anti-scatter grid must be coupled closely with the detector array even during rapid revolutions, as even small misalignments will introduce errors in the projection data.

Cross-talk can also introduce projection error. This happens when there is a "leakage" of the signal of an x-ray photon signal from one detector cell to its neighbor, since the visible light doesn't have a preferred direction and spreads within the scintillator. Cross-talk to neighboring detector elements can decrease the spatial resolution. To reduce this effect, each detector cell is isolated with a reflective coating that aids in directing the light toward its corresponding photodiode.

The measured current from each detector follows Eq. (3.8), and the reference current is measured by a reference detector. The different values of every row in the detectors array result in a "profile" of attenuation. Mapping the different profiles along all the viewpoints, results a sinogram. During reconstruction, the sinograme is then processed to generate a single 2D image or a 3D volume. The image reconstruction process is explained in more details in Chap. 2.

3.3.4   Image Formation

There are several main steps when generating a CT image for clinical diagnostics:

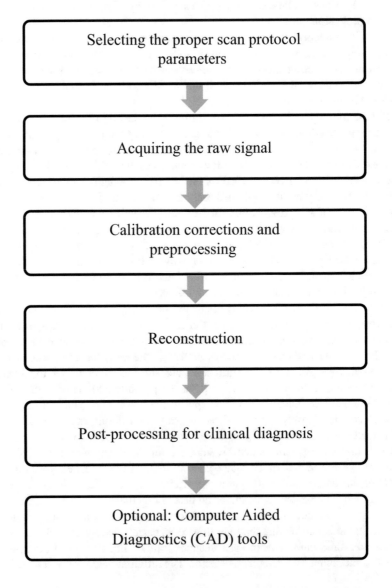

3.3.4.1   Selecting the Proper Scan Protocol Parameters

Selecting the proper scan and reconstruction parameters appropriate for the clinical needs of the diagnostic CT image is the first step of the procedure. Typical CT studies start with a very-low-dose scan, taken without the revolving of the gantry, in order to produce a "projection-like" image of the body. This image aids in the planning of the start and end points of the actual tomographic scan (as seen in Fig. 3.22). Selection of scanning parameters such as the kV_p and the tube current effect the contrast, noise, and patient ionizing radiation dose. The diameter or width of the slices in the transverse (x–y) plane is called the field-of-view (FOV). The FOV and bowtie filter selection are also important parameters, as for smaller scanned objects, such as the heart or extremities, a smaller FOV could be preferred and a suitable bowtie filter selection might reduce the patient dose.

Care must be taken when choosing suitable scan parameters, since errors might entail a supplementary scan with its additional radiation dose to the patient. However, the reconstruction parameters can be altered after the scan is finished, so long as the raw data is still available for reconstruction. Reconstruction parameters are implemented only after the data is already acquired. Similarly, post-processing parameters (e.g., related to filtering, segmentation, or display) can be altered after the acquisition. If required, a variety of reconstruction and post-processing parameters may be tested to obtain optimal results, without any need for additional scans.

Fig. 3.22 The patient is scanned with very low-dose planning scan, during which the tube is static. Based on that scan, the physician can define very accurately the exact start and end locations of the axial/helical scan, for example, only the lung volume (marked in yellow)

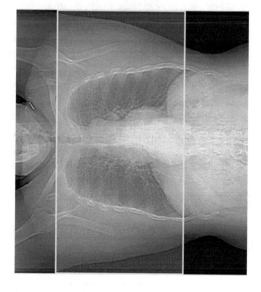

Regardless, default reconstruction and post-processing parameters are usually assigned to specific acquisition protocols.

3.3.4.2   Acquiring the Raw Signal

Once the parameters are selected, the radiating phase (actual scanning) takes place, which is typically only a couple of seconds per scan. During this step, the system acquires the raw signal I from each detector, for each IP during the scan.

A special detector, sometimes called reference or normalization detector, measures the tube's output level giving the reference intensity, I_0, of the scan. This is the base-level signal which the system uses for the normalization.

3.3.4.3   Calibration Corrections and Pre-processing

During the pre-processing step, the signal is tested to see if there are any problematic or missing readings. For example, if there is a missing reading from a specific detector, it can be corrected by means of interpolation.

In addition, some corrections for known physical phenomenon are applied before the reconstruction step, based on calibration data. Calibration data is the name for data colledted from scans that are taken with known conditions and setup, such as scans of known obejcts, called phantoms, with specific scan parameters. They yield system performance and measurement data for a given setup that are taken prior to clinical scanning. For example, "air-scans" are taken with no object within the beam path, where the current, voltage, and bowtie filter are predefined and known. Analysis of such "air-scan" signals, can be used to calculate needed adjustments for differences in detector element sensitivity. In addition, it measures the unique sensitivity of each detector element. This calibrated data is used to "normalize" the signal between the different detector elements and serve as baseline. This is in addition to the reference detector that helps normalize the entire signal between different frames and compensate for tube output discrepancies within the scan.

Another example might be scanning specific phantoms using the different bowtie filters. This helps in gathering the needed information regarding the changes in the spatial distribution of the x-ray spectrum each of the bowtie filter inflicts.

Some of the calibration data is used during the calibration correction step and some during the reconstruction step. The calibration data is used to improve the accuracy of the measured signal and help adjusting the result of every scan, making sure the final CT values are according to scale.

3.3.4.4   Reconstruction

Given a baseline intensity, I_0, that is measured by the air-scans and by the reference detector, one can alter Eq. (3.6) to isolate the summation of product of the attenuation coefficient μ_i and the incremental thickness Δx_i for each ith discrete step of n steps along an x-ray beam path:

$$\ln\left(\frac{I_0}{I}\right) = \sum_{i=1}^{n} \mu_i \Delta x_i \qquad (3.12)$$

In CT scanners, the images are composed of square matrices with a specific number of elements (pixels) the user can select. For example, a matrix of 512×512 pixels for a transverse slice is common. The user also defines the desired transverse FOV, such as 50 cm diameter. This sets the pixel size. In this example it would be 500 mm/512 pixels, so the pixel width is just under 1 mm. By selecting a larger matrix size or smaller FOV, the user can reduce the size each pixel represents. Note, not to confuse the image pixels with the pixels of the detector array. The former is an attribute the user can control, when choosing different reconstruction parameters, while the latter is a part of the scanner's hardware.

Upon acquisition, the data can be normalized, and the summation of the difference attenuation is obtained. This is the attenuation of the different tissues μ_{tissue}^i along the beam path i.

In CT images, reconstructed image pixels are assigned values called Hounsfield units (HU) and are scaled according to Eq. (3.13):

$$\text{CT number}\,[\text{HU}] = \frac{\mu_{\text{tissue}} - \mu_{\text{water}}}{\mu_{\text{water}} - \mu_{\text{air}}} \cdot 1000 \qquad (3.13)$$

where μ_{tissue} is the measured attenuation coefficient of a particular pixel and μ_{water} and μ_{air} are the reference attenuation coefficients of water and air, respectively. Since it is considered that air is not attenuating, i.e., $\mu_{\text{air}} = 0$, Eq. (3.13) can be now written as

$$\text{CT number}\,[\text{HU}] = \frac{\mu_{\text{tissue}} - \mu_{\text{water}}}{\mu_{\text{water}}} \cdot 1000 \qquad (3.14)$$

It is easy to see that according to this scaling method, air has a CT value of -1000 HU. As part of the calibration process, μ_{water} for different scan parameters is measured. Remember that μ_{water} is a constant that is energy dependent, so the system needs to calibrate this value, regardless of the scanner parameters: water is always assigned a CT value of 0 HU. It is important to remember though that the actual measured value is not precisely 0 HU for water and -1000 HU for air, but a value very close to that. This is due to noise, scatter, and other imperfections that may shift the value of specific pixels from their "true" value. From Eq. (3.14) it can be seen that as the attenuation of the medium increases, so does the CT value.

Note that this scaling method sets the water value and air value to be constant, regardless of the energy of the x-ray beam. Assigning pixel values to a known and calibrated scale, regardless of the different mediums along the x-ray path, is a major factor in the consistency of the CT values assigned to each organ of the display. It has an enormous impact on the accuracy of the final diagnosis.

During reconstruction, the output images are generated, by finding the specific value for each pixel that is consistent with the summation of values along each beam path represented by the acquired projection data. Image reconstruction techniques are explained in more detail in Chap. 2. The resultant image is in fact a map of the attenuation of each pixel, presented in HU values. These original images are the basis for post-processing.

> CT values are always calibrated to a known scale (HU); hence, the values of different tissues remain approximately the same, for scans taken with similar conditions, even between different patients.

3.3.4.5   Post-Processing for Clinical Diagnostic

During this step, the basic output images are processed and altered. Some of these parameters are included within the reconstruction parameter options and should be chosen as needed. For example, the user can choose the reconstruction kernel that controls the sharpness of the images. A "sharp" kernel will enhance the edges, while a "softer" kernel will smooth the image, in order to reduce the image noise.

The user can also generate additional sets of images, based on the original set, when needed. For example, choosing to display thin slices might be appropriate for viewing lung pathologies, whereas larger features may be more appropriately viewed with thicker slices, reducing image noise.

By manipulating the reconstructed volumetric data (like a three-dimensional matrix of voxels with known values), it is possible to present any plane, from any direction. Transverse (also called "transaxial" or "axial"), sagittal, coronal, oblique, and even curved views can be sampled from the volumetric data. The general name for such technique is MPR (multi-planar reformation). This is very helpful, for example, in the case of the backbone vertebrae, where it's often helpful to produce images that are exactly perpendicular to the vertebrae main axes. All of the different reconstructions, using the same scanned data, but different reconstruction parameters, optimize the physician's ability to detect any clinical findings and give an accurate diagnosis.

3.3.5   Axial Scans and Helical Scans

Modern CT scans can be divided into two modes: axial mode and helical (spiral) mode.

Axial scans were the original mode of scans but are still used today for some specific clinical needs. When working with this mode, the tube and detectors are rotating around the patient during data acquisition, but the table and the patient location are static.

A single axial scan "width" along the z-axis (longitudinal or superior/inferior axis) is called coverage and is usually expressed in millimeters. This is based on the tube aperture and controlled by a collimator at the beam exit of the x-ray tube. It is described as the z-axis "length" of the beam at the isocenter. It means that the z-axis coverage is the product of the slice thickness and the number of slices. The maximal coverage is limited by the actual size of the detector array in the z-axis direction.

As the acquisition is performed with a circular motion, evenly around the isocenter, the image space (the support) is a circle (though often it is cropped to have a rectangular shape). The FOV is usually measured in centimeters, and the maximal FOV is limited by the design of the detector array and the bowtie filter. Still, the user can reconstruct a smaller FOV upon request.

With axial scans, the slices are acquired at specific discrete locations based on the detector array, where each single row of the detector array generates the data for the reconstruction of its matching slice. Assuming that there is no gap or overlap between the slices, the interval between two slice's centers in the z-direction dictates and equals the original slice thickness. The slice thickness itself (width in the z-direction of each slice) is usually measured in millimeters (Fig. 3.23). When there is no overlap between slices, it is simply the coverage divided by the number of slices. For example, a scanner with coverage of 20 mm and 32 slices has a slice thickness of 0.625 mm at the isocenter.

Originally it is assumed that the reconstruction plane (i.e., the "axial" slice) is identical to the acquisition plane, hence the data from different detector rows did not "mix" with data from other rows during the reconstruction process. This assumption was applicable for narrow coverage and is sometimes called "2D axial" reconstruction. When the beam is not a fan beam but a wide cone beam, and the coverage

Fig. 3.23 Volumetric data. Each volume element, called a voxel, has a volume of $\Delta x \Delta y \Delta z$. The slice thickness is determined by the coverage and the number of slices. The user can also set the number of pixels in each slice by selecting the matrix size. The actual size of each pixel is determined by the total FOV and the number of pixels that are used for display

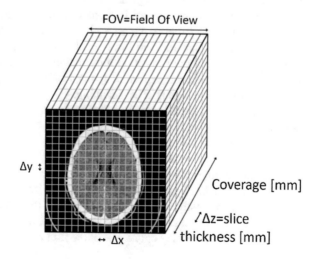

becomes large, the assumption of parallel planes defined by each detector row is no longer valid. In such case, image artifacts might appear, without proper handling. Reconstruction that accounts for this is sometimes called "three-dimensional axial" reconstruction.

Sometimes there is a need to cover a volume that is longer than the maximal axial coverage. In such a case, only after the data acquisition of the full circle is finished, the table incrementally moves forward by a step equaling the coverage, in order to take another scan. The process is repeated until the entire volume is covered by enough axial rotations. The axial scans are closely adjacent, but not overlapping one another to ensure that no body part is scanned twice. During the table movement, the tube is turned off and radiates only once the table is stationary in its new location.

As described in the previous section, the "dead time" between the consecutive scans was one of main motives for the development of "nonstop rotation" (i.e., helical mode).

With the helical scan mode, the tube radiates and rotates continuously while the table translates the patient along the z-axis at a constant speed. This way, one can imagine that the focal spot is moving around the patient in a spiral fashion, while projections are continually being acquired.

The constant speed at which the table is translating during the tube rotation is usually not measured by the usual units of speed of centimeters per second (cm s^{-1}). It is commonly expressed in terms of pitch. The pitch describes the ratio between the distance the table is covering during a single rotation and a single rotation coverage.

If the coverage is c and the distance the table travels with one revolution is d, then the pitch is calculated according to Eq (3.15):

$$p = \frac{d}{c} = \frac{v \cdot t}{c} \tag{3.15}$$

Here v is the table speed (cm s^{-1}), and t is the time it takes for the gantry to complete one rotation (seconds). If the coverage is 80 mm, a single rotation takes 0.5 seconds, and the table speed is 6 cm s^{-1}, then the pitch is 1.33. Higher pitch enables a faster scan of larger volumes, but reduces the number of samples for the same volume, in comparison to lower pitch. Lower pitches are typically used for protocols requiring more projection information to get finer details and are associated with higher radiation exposure. A pitch of 1 samples the volume at approximately the same coverage as axial scanning.

With helical CT, the table and patient are constantly moving during data acquisition, so there isn't a specific "detector location" for a specific slice, as with axial scanning. It can be said that the data acquisition is isotropic for all the sampling within the selected scanned volume. Therefore, any position along the z-axis, within the scanned volume, can be selected as a slice to be reconstructed.

This is especially important when the feature of interest is very small, particularly in the Z direction, with a width that is equal the slice thickness or smaller. When scanning in axial mode, it might be seen by two detector rows, so each one of the two matching slices will display only a part of the feature's true intensity and the contrast would be lessened. This is called the partial volume effect: the signal is shared between two volumes. In such case, the signal is shared between two volumetric slices. With helical scanning, it is possible to select the reconstruction such that the feature plane is in the center of a plane of a specific image. This would mean that the full feature is included in a single slice, which would result a display of its full intensity and contrast.

Since the reconstruction z-position can be selected arbitrarily from helical scanning data, it is possible to reconstruct overlapping slices. This means that unlike the case of axial scans, the interval between two consecutive slices can be smaller than the slice thickness and that data is "shared" by adjacent slices. For example, 50% overlapping means that half of the current slice is covered by the preceding slice and the other half by the next one. A 0% overlap means no overlap at all, as is the case with the axial scans. One of the useful advantages that helical overlapping images offer is the improved visualization of 3D, as the images are redisplayed from other viewing angles. A small "gap" between two consecutive axial images can generate artifacts when rendering 3D images, which is done by means of interpolation, due to the fact that the data is less continuous. This problem is minimized when interpolating the overlapping images.

In axial scans, the table and patient are static during the scan. In contrast, during helical scans there is a nonstop smooth motion of the table during the scan that enables a faster acquisition with longer scan volume.

3.3.6   Image Manipulation and Clinical Examples

3.3.6.1   Gated Scans

For third-generation CT scanners, motion of the heart and lungs presented one of the challenges in generating diagnostic images. This is analogous to trying to photograph a moving object: if the shutter is not quick enough, moving items appear blurred in the resultant image. Even though third-generation CT scanners had a coverage of a few centimeters and rotation times less than 1 second per rotation, this was not enough to overcome this obstacle.

For scanning lungs, the solution was quite simple. Since the breathing rate is not very fast to begin with, approximately 12 breaths per minute, the patients were asked to hold their breath and remain still. This gave enough time to perform the scanning of the entire lungs' volume, without significant motion artifacts. This is known as the "breath-hold" technique.

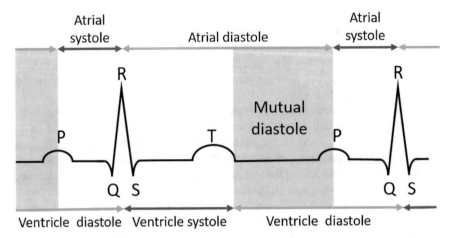

Fig. 3.24 Illustration of an electrocardiogram, with systole and diastole times. The mutual diastole occurs when the heart is not contracting. Using data only from that period minimizes possible motion artifacts in the image

For cardiac scans a more sophisticated solution was needed. It was accomplished by taking advantage of the periodic motion of the beating heart and using gated scanning with gated reconstruction algorithms.

The heart does not move in the same manner during the entire cardiac cycle. Each heartbeat has a more "static" period, and a more "dynamic" period, which are called diastole and systole, respectively. During the diastole, the heart muscle relaxes as the chamber fills with blood. During systole, a strong contraction of the cardiac muscles occur, and the blood is pumped out of the chamber.

The ventricles and atria each have different systole and diastole times, but since these times partially overlap, it is possible to find a short window when they are both at their "static" periods (Fig. 3.24). This window can be found for each patient from an electrocardiogram.

Another challenge was that the size of the heart is roughly 12 cm, which was too large to be volumetrically imaged by a single axial scan due to insufficient z-axis coverage.

Prospective gated axial scans (also called "step-and-shoot") provided one solution (Fig. 3.25). With this approach the scans are meticulously timed, and taken automatically, in accordance with an ECG (electrocardiogram). The patient is irradiated only during the diastole with a single axial scan. After that, the table and patient are moved slightly along the z-direction, so during the next heartbeat, a new section of the heart can be scanned. Since the diastole is shorter than the duration of a full rotation, the patient is irradiated only about a half of the rotation during each axial scan. Several scans are taken, until the full length of the heart is scanned. This approach reduces unnecessary irradiating during systole and significantly minimizes the overall radiation exposure of the patient, compared to older methods such as retrospective gating.

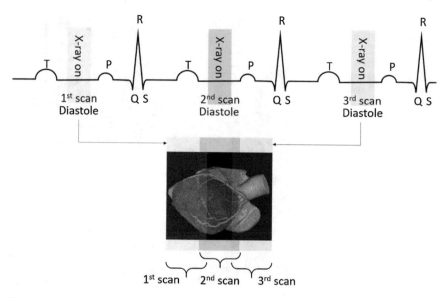

Fig. 3.25 An illustration of prospective gated axial procedure. Three axial scans are performed, during approximately the same phase of the diastole, of three consecutive heartbeats. There is a small overlap between the scans, to make sure that the entire heart volume is scanned without any gaps

In retrospective gating, the heart is irradiated and data acquired during several diastole/systole cycles. The gating is performed retrospectively, and most of the data acquired is typically not used, resulting in much higher radiation dose than prospective gating methods. With retrospective gating the patient is scanned continuously with a helical scan using a very low pitch synchronized with an ECG recording. Several heartbeats are included within this continuous scan. After acquisition, the user can select a specific heart phase, as per the ECG signal, and reconstruct images based only on data acquired close to that specific phase (e.g., 70% through a cycle). The low-pitch scan enables the collection of a sufficient number of projections from different viewpoints to reconstruct a good-quality CT image, even if only a short specific phase is of interest. One benefit of this approach is the option to reconstruct every phase during the cardiac cycle, even systole, since the data throughout the cycle is acquired. This is in contrast to the step-and-shoot approach, where only the scanned phase can be reconstructed.

Fortunately, current scanner designs offer much wider coverage, of up to 30 cm coupled with a faster rotation time as short as a quarter of a second. This allows the scanning of the whole heart during a single heartbeat. They result a much clearer final image, showing subtle details such as small blood vessels.

In combination with iodine as a contrast agent, coronary computed tomography angiography (CCTA), also called virtual angiography, has become one of the main tools for physicians to assess the condition of patient with chest pain [2]. By using a bolus "test" scan, with only a small amount of iodine, the technician can time the "real" scan, with the full amount of iodine, in a way that the iodine would be at its

Fig. 3.26 An example of CCTA volume rendering of the heart. The coronary arteries are filled with iodine-based contrast agent and appear with much brighter shades in the final image

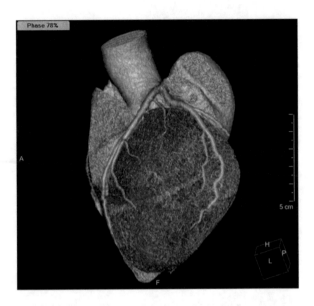

peak concentration in the coronary arteries during the scan. Such an exam enables the physicians to view contrast-enhanced coronary arteries from many different directions and even look "inside" the lumen of the arteries in order to detect occlusions or severe stenosis that might impair blood supply to the myocardium (Fig. 3.26). By timing several successive scans, it is also possible to observe the blood vessels (with the iodine) during different arterial phases. From such an exam the physician can obtain a functional assessment in addition to imaging the physiology.

This exam can provide a more accurate diagnosis of coronary artery disease, and since it is noninvasive, this procedure is being used more frequently.

3.3.6.2   Image Presentation and Post-Processing

One of the main advantages of CT images is the fact that they are digital images. CT images are almost invariably stored in a DICOM (Digital Imaging and Communications in Medicine) format. This means that no matter which scanner model or brand generated the images, they are always kept in a very specific format. This format of reconstructed image data is public and known. Hence, there are numerous software applications that offer the physicians either manual or automatic tools that can change the appearance of the images, perform different measurements, or manipulate the images, in order to reach a better clinical diagnosis.

Some of the basic image manipulation is setting the window level and window width (also called "windowing"). Currently, standard CT scanners generate 12-bit range image pixels. This is a 4096 HU range spanning from the minimal value of −1024 HU to a maximal value of 3071 HU. It is very difficult for the human eye to analyze correctly the subtle differences between adjacent HU values; therefore, setting the windowing is a crucial step for better clinical diagnosis.

The window width (*WW*) controls the range of CT values of the image that are displayed, while the window level (*WL*) controls the mid-value of this range. The maximal (white) and minimal (black) limits are calculated according to Eq. (3.16):

$$\text{Max value} = \text{WL} + \frac{\text{WW}}{2} \,; \text{Min value} = \text{WL} - \frac{\text{WW}}{2} \qquad (3.16)$$

For example, brain windowing with WW = 80 and WL = 40 means that the span of visible HU values is between 0 and 80. Any pixel with a HU value greater than or equal to the maximum gray level value (80 in this example) appears white, while any pixel less than or equal to the minimum gray level value (0 in this example) appears black.

Wide windowing, usually WW > 400, is best for diagnostic cases where the variations in attenuation are very distinct, for example, bone fractures. Narrow windowing is better when the attenuation coefficients among pixels in the region of interest are quite similar and harder to distinguish, for example, when looking at the difference between white and gray matter in the brain. By expanding the dynamic range of displayed pixels that otherwise would appear similar, it is easier for the human eye to make such distinctions (Fig. 3.27).

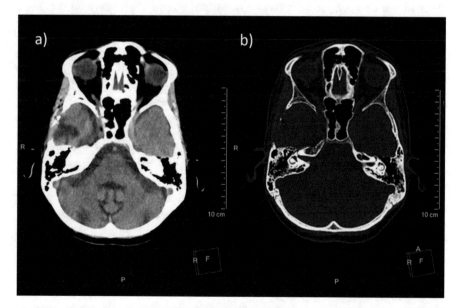

Fig. 3.27 Two combinations of window level and width. On the left image (**a**), the window width is quite narrow, enabling a very good visualization of soft tissues and hemorrhage, but the bone volume appears completely white, as most of the values are above the upper threshold of the display window. On the other hand, the image on the right (**b**) demonstrates a very wide windowing, where the soft tissues might seem homogenous, but it is much easier to see the bone structure such as the bones of the inner ear

It is important to remember that even though the appearance of the image as it is displayed is different, the HU of each pixel remains the same. This means that the physician can still make quantitative measurements on the image, regardless of the window level.

Another helpful display option, for diagnostic purposes, is changing the black & white color-scheme. For example, it is possible to use a brownish-red color-scheme instead of gray. Note that the image is still only "gray-scaled", i.e., only one value per pixel, but the colormap is different. With this display method, each gray level is presented with its matching rad-brown shade of color. This result a more realistic appearance of the rendered volume. The heart in Fig 3.26 is an example of using such color palette. Since this method only changes the display, and not the pixel values, any kind of measurements on the images, will remain unchanged.

Another image processing tool is the image filter. Some filters can enhance the edges or make the image much smoother, according to the clinical need. If applied to the reconstructed volumetric data itself, those filters might sometimes alter the HU values, as they inherently change the pixel values of the image, not only its appearance. In some instances, filtering and interpolation are applied only for display purposes, and the underlying pixel values remain unchanged. It is important to have some understanding of the image reconstruction, post-processing, and used display techniques, especially when making quantitative measurements.

Since CT data are actually volumetric and not only two-dimensional, one can think of the data arrayed as a three-dimensional matrix, with many voxels. Each voxel represents a small volume element. With that in mind, it is easy to perform surface or volume rendering. Rendering is the process of creating a synthetic image that looks three-dimensional from the two-dimensional data (Fig. 3.28). This can help the physician to see the entire scanned organ as a whole. It also allows the physician to "play" with the three-dimensional model, rotate and translate it, in a very intuitive way, for a better view of the region of interest.

In addition, it is possible to generate a maximum intensity projection (MIP). With such an image processing technique, the volumetric data is projected on the visualization plane, where only the most intense voxel from each of the ray paths orthogonal to the selected plane is displayed (Fig. 3.29). It can help highlight small or fine elements, such as nodules in the lungs, or small blood vessels with contrast agents (Fig. 3.30). It is important to note that when looking at the MIP image, just like in a conventional projection, the information of the order of the objects along the projection is lost.

With the evolving abilities of computers, one of the most helpful tools is the computer-aided detection or computer-aided diagnosis (CAD). CAD tools can highlight suspected findings, such as lung nodules. They can also automatically perform segmentation of designated areas such as liver segmentation or heart chamber segmentation and add relevant measurements on each segment (Fig. 3.31). Such automatic techniques are helpful as they save the physician time, replacing manual labor, but also give more consistent measurements, which are beneficial when trying to evaluate the progress of a prolonged clinical condition.

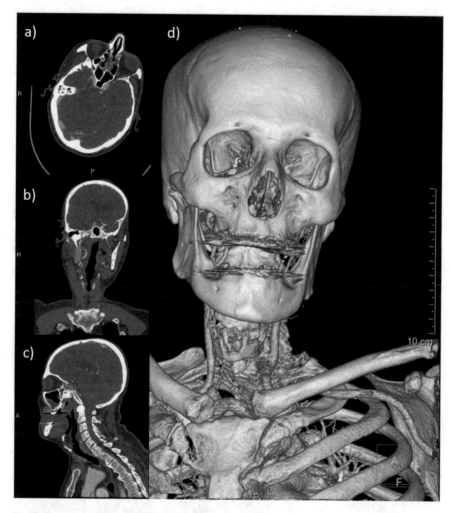

Fig. 3.28 Volume rendering and different image planes. The physician can review the patient from many different directions, since the CT data is volumetric data. Picture (**a**) presents the original axial (transverse) images, picture (**b**) presents the coronal images, and picture (**c**) presents the sagittal images. Picture (**d**) demonstrates a volume rendering of the data such that the skeleton and iodine-filled blood vessels are visualized as surfaces

3.4 Dosimetry and Safety

3.4.1 Ionizing Radiation and Health Risks

When atoms inside living cells are ionized, there are three possible outcomes: sometimes the cell dies; in other cases, the cell repairs itself completely and continues to function as before; lastly, the radiation can cause cellular degradation where the cell fails to repair itself. In the first case, if a sufficient number of cells die,

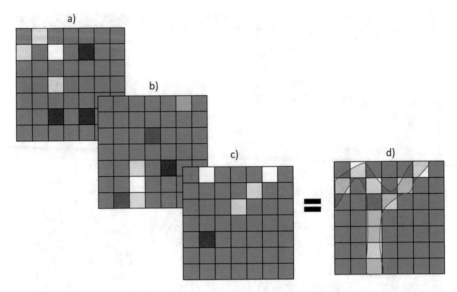

Fig. 3.29 MIP example. Assume that there are three selected planes (**a**, **b**, and **c**) with pixels depicted by several gray levels. Darker shades represent lower intensities, and brighter shades represent higher intensities. Each pixel in location (m, n), of the MIP image (**d**), has the value of the maximal intensity from the available three values of that same location (m, n) in the different planes. The result could demonstrate a blood vessel with contrast matter (marked in color). Otherwise the blood vessel branching is less evident because the details are shared among several planes

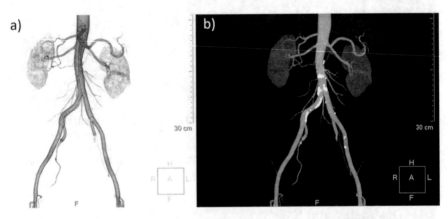

Fig. 3.30 A clinical example of 3D volume rendering image (**a**) vs. MIP image (**b**). As the iodine solution has a relative higher attenuation, the MIP image helps to depict even very small blood vessels that cannot be seen in the 3D rendering. In addition, it is much easier to identify small calcifications within the blood vessels (white). On the other hand, since the MIP is only a projection of the volume, there is loss of information. For instance, some of the blood vessels that are located in front of the aorta are visible only in the 3D rendering

Fig. 3.31 An example of CAD, automatically depicting and coloring the different heart chambers, coronary arteries, and the aorta

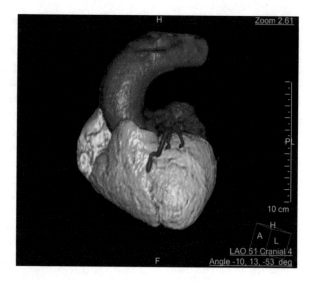

tissues or organs can be damaged, perhaps sufficiently to imperil the organism. Effects of this magnitude are called deterministic effects because the damage can be correlated to the amount of ionizing radiation received. The amount of radiation received during diagnostic imaging is a couple orders of magnitude below the threshold known to cause deterministic effects in humans. If only a small number of cells die, organisms are generally able to recover without biological damage. The third case listed above is of concern when estimating health risks due to diagnostic scanning. A cell damaged by radiation that does repair may mutate and pass on mutations upon cell division, possibly generating cancer in the future. Because this undesired and dangerous outcome is not certain, and very likely stochastic by nature, these effects are called indeterminate or stochastic risks associated with ionizing radiation. The risks of diagnostic scanning with x-rays due to ionizing radiation are known to be small, but their actual value is unknown. Therefore, at the levels of radiation used for diagnostic scanning, health risks are estimated and assigned according to consensus within the health physics community.

Different cell types differ in their susceptibility to x-rays. Cells that are least specialized are more likely to be affected by ionizing radiation, as well as cells that frequently reproduce. For this reason, fetuses, babies, and children are more susceptible to ionizing radiation damage than adults.

The damage that tissue or organs might suffer from ionizing radiation depends mostly on the amount of the absorbed radiation (total dose D) which is measured in units of gray (Gy). The radiation dose is the amount of radiation energy deposited in an absorber, per absorber mass:

$$1 \text{ Gy} = 1 \text{ joule energy deposited per kg of absorber (tissue)}$$

The assumption is that the biological radiation damage is proportional to the amount of energy deposited by ionization within a given mass. This is reasonable since the number of ions produced will be proportional to the energy deposited. However, the potential of harming a tissue or organ is a combination of the amount of radiation (Gy), the type of radiation, and the sensitivity of the tissue toward radiation damage. For example, alpha-particle radiation (see Sect. 4.1.1) can be 20 times more biologically damaging per Gy than x-rays (or beta-particle radiation). Consequently, a quality factor of 20 is used for alpha-particles, whereas it is just 1 for x-rays. The product of the quality factor and the absorbed dose in Gy is called the equivalent dose with units called sieverts (Sv). Additionally, different regions of the body are more susceptible to radiation damage than others. For example, extremities such as hands are much less sensitive to radiation damage than a number of the internal organs of the abdomen, such as the stomach. When estimating the risk to whole-body radiation exposure in humans, an effective dose (H_T) is calculated (also in Sv) that accounts for this by assigning a weight to different body parts in proportion to their respective sensitivities. Such an estimate may be calculated as

$$H_T = \sum W_T W_R D_{T,R} \tag{3.17}$$

The factor $D_{T, R}$ is the dose of type R (e.g., alpha-particles or x-ray) that tissue T (e.g., hand or stomach) absorbed. The factor W_T is the weighting factor of the radiated tissue: it is a factor that represent how sensitive is the tissue to radiation. The factor W_R is the weighting factor of the radiation type itself (e.g., 20 for alpha-particles, 1 for x-rays).

This measure is used to estimate the potential for causing biological damage caused by radiation and is tissue specific, as each tissue has a different weighting factor. For calculation purposes, 1 Sv is often assigned a 5.5% risk of developing cancer. Since an exposure of 1 Sv can cause deterministic effects in certain scenarios, it is more common to express indeterminate or stochastic risk with smaller exposure values, such as assigning a risk of 5.5×10^{-4} to an exposure of 10 mSv. Diagnostic CT scans may have radiation exposures to the patient in this order of magnitude. This risk says that 1 out of $1/5.5 \times 10^{-4}$ or 1 out of 1818 patients may eventually develop a cancer from an exposure to a 10 mSv diagnostic scan. However, there is no consensus that there is any proven risk for effective doses below 50 mSv. The risk is assigned to assist in the proper care of patient populations. Since the risk is assigned in proportion to the exposure, and since there is no level below which the assigned risk is zero, this is referred to as the linear no-threshold model (LNT).

3.4.1.1 Projection Radiography Dose Measurements

In projection radiography such as fluoroscopy, the dose area product (DAP) is used for the assessment of the radiation risk from the x-rays. As the name suggests, it is the absorbed dose, multiplied by the radiated area. It is measured in units of

(Gy cm^2). Different parameters, such as the kVp, the tube current, the exposure duration, and the exposure type, influence the DAP and the potential risk. To aid in optimizing diagnostic imaging protocols, today all commercial systems are equipped with built-in DAP meters. These are an ionization chambers that are placed between the x-ray collimators and the patient and cover the entire x-ray field, to provide accurate measurements of DAP. Since the x-ray beam starts with a point source, but diverge, creating an x-ray field, the irradiated area increases with the square of distance from the source. This means that $A \propto d^2$, where A is the area and d is the distance from the point source. On the other hand, the radiation intensity decreases according to the inverse square of distance. This means that $I \propto d^{-2}$ where I is the radiation intensity. Since the absorbed dose is proportional to the intensity, it means that the DAP is independent of distance from the source.

3.4.1.2 Computed Tomography Dose Measurements

Compared to planar x-ray imaging, for CT it is somewhat more difficult to measure the effective dose to the patient for a diagnostic scan. It is a function of many different parameters, such as scanner geometry, x-ray beam filtering, beam divergence, and beam uniformity. Additionally, the patient size will affect the dose. For example, children would generally receive a greater dose if scanned with the same protocols as adults. Modern acquisition protocols typically vary the radiation exposure, which also affects the estimated dose.

To reduce these variables in order to compare scanners and protocols, radiation exposures related to standard setups provide indices of measurement. One standard is called the computed tomography dose index (CTDI). Such indices are based on the dose that is measured on a specific well-defined homogenous phantom. The CTDI$_{100}$ is calculated according to Eq (3.18), where the number of slices is n, T is the slice thickness, and $D(z)$ is the radiation dose that is measured at point z along the scan, at the isocenter. The dose itself is measured using a 100 mm standard pencil dose chamber that is slid into holes within a standard acrylic phantom.

$$\text{CTDI}_{100} = \frac{1}{nT} \int_{-50\text{mm}}^{50\text{mm}} D(z)dz \qquad (3.18)$$

The amount of dose that reaches the skin and the peripheral surrounding is higher than the amount that actually reaches the center of the patient. This is due to the attenuation of the photons, once entering the body. To account for this, a similar value, CTDI$_w$, can be defined that includes the placement of the dosimeter at the peripheral parts of the phantom at anterior/posterior and left/right positions in addition to the central measurement. The results are then weighted according to

$$\text{CTDI}_w = \frac{2}{3} \left\langle \text{CTDI}_{100}^{\text{peripheral}} \right\rangle + \frac{1}{3} \text{CTDI}_{100}^{\text{central}} \tag{3.19}$$

where $\left\langle \text{CTDI}_{100}^{\text{peripheral}} \right\rangle$ is the average of the four peripheral CTDI_{100} measurements. This is the dose index for a single axial scan. For helical scans, the index is further refined as CTDI_{vol}, which includes the pitch as a factor that impacts the dose:

$$\text{CTDI}_{\text{vol}} = \frac{\text{CTDI}_w}{\text{pitch}} \tag{3.20}$$

Intuitively, all other parameters being equal, a pitch of 0.7 will collect twice the number of projections as a pitch of 1.4 and will take twice as long; therefore, one would expect the radiation exposure to double, too. Lastly, the dose length product, *DLP*, is a quantity defined for use in CT which is in a way analogous to the DAP used in planar radiography.

$$\text{DLP} = nT \cdot \text{CTDI}_{\text{vol}} \tag{3.21}$$

This implies that if all other parameters remain equal, a scan length that is twice as long is associated with doubling the dose and therefore double the *DLP*. Higher dose rates or higher total dose correlates with a higher assigned health risk, according to LNT. Lowering the rate of exposure, along with lowering the total exposure, is generally assumed to reduce the risk, due to the fact that it gives the cell more time to repair itself correctly and lower the likelihood of long-term effects. However, total exposure serves as the basis for the assigned health risk.

Above a threshold that varies with biology, radiation can impair the healthy functioning of tissues and lead to physiological effects. These are the deterministic effects described above. Minor effects can be redness of the skin, as in burns, or loss of hair. Significant effects can be part of acute radiation syndrome, where the DNA (deoxyribonucleic acid) is harmed or the tissue becomes necrotic. Deterministic effects can include symptoms such as nausea, vomiting, diarrhea, and headaches. Severe cases include life-threatening symptoms such as neural problems, decrease in white blood cells, and death. To avoid being alarmist, it is important to remember that the threshold of such deterministic effects is a couple orders of magnitude higher than the radiation exposures associated with common radiological exams. Radiation exposures from diagnostic procedures, even from CT, are small (Table 3.3). Regardless, gross negligence in the use of x-ray equipment has been known to cause some of the minor deterministic effects.

Table 3.3 Effective dose estimation for clinical procedures [3]

Diagnostic procedure	Typical effective dose (mSv)
Chest x-ray (PA film)	0.02
Lumbar spine	1.5
Upper GI exam	6
Barium enema	8
CT head	2
CT chest	7
CT abdomen	8
Coronary artery calcification CT	3
Coronary CT angiogram	16

3.4.2   The ALARA Concept

In order to ensure that patients are not overexposed to ionizing radiation, the ALARA concept has been adopted by the health physics community. Within the bounds of regulatory limits or good practice guidelines, this principle states that radiation exposure should be *As Low As Reasonably Achievable*. For radiological exams, this means that equipment should be chosen and procedures should be adjusted to always try to reduce the patient radiation dose to the minimum needed in order to maintain a good image quality required for an accurate clinical diagnosis. The same principle is applied to workers, such as x-ray technicians, who might be exposed to radiation fields on a daily basis.

In addition, it is important to avoid irradiating body parts that are not clinically relevant for an accurate diagnosis and protect parts that may be more sensitive to radiation damage. For example, in some procedures, it is standard practice to shield organs (e.g., with a 2-mm-thick lead sheet) like the thyroid gland or the ovaries if they fall within the exposed FOV but their images are not relevant to the diagnosis.

It is imperative to remember that the risk of not properly diagnosing an indicated condition far outweighs the assigned risk of any diagnostic imaging procedure. Special care should be taken when assessing the risk of screening a largely healthy population for a specific disease. The prevalence and prognosis of the disease, for which the screening is performed, should be used to estimate a reasonable radiation exposure for the exam. For example, mammograms performed on a general population to screen for breast cancer have very low levels of exposure. Given good clinical practice, the risks associated with radiation exposure are negligible compared to the benefits of medical x-ray imaging.

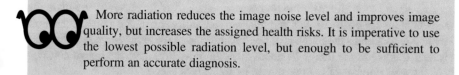 More radiation reduces the image noise level and improves image quality, but increases the assigned health risks. It is imperative to use the lowest possible radiation level, but enough to be sufficient to perform an accurate diagnosis.

3.4.3   Quality Versus Safety

Quantum noise is often used to describe the cause of noise in x-ray imaging, since most of the noise is a result of the random variations in detected x-ray intensity across the scanned plane. A less significant factor is the electronic noise, and it is usually negligible in comparison.

Fluctuations in the detection of the x-ray photons follow Poisson statistics. For example, if during each IP of a CT scan, a detector should measure $N_0 = 100$ photons, then it will actually measure a number around 100, within each of the IPs of the scan, but probably not exactly 100, due to statistical fluctuations.

Remember that some of the properties for Poisson distribution are

$$E\{N\} = \text{Var}\{N\} = N_0 \tag{3.22}$$

Quantifying these random fluctuations of x-ray detection, they can be represented as the standard deviation, σ. Here, this is given by the square root of N_0:

$$\sigma = \sqrt{\text{Var}} = \sqrt{N_0} \tag{3.23}$$

Therefore, in the above example, we can expect to measure in most of the cases 100 ± 10 photons at each IP. Obviously, due to the nature of the Poisson distribution, a smaller part of the IPs might result significantly higher or lower measured values.

From here we can calculate that the detector signal-to-noise ratio as

$$\text{SNR} = \frac{N_0}{\sqrt{N_0}} = \sqrt{N_0} \tag{3.24}$$

Since the number of x-ray photons is proportional to the tube current, doubling the current will result in about a 40% increase in the SNR of the detected photons.

Note that the standard deviation, σ, is by itself a random variable. One way to estimate its randomness is by using the coefficient of variation also known as relative standard deviation (RSD). It is defined as the ratio of the standard deviation to the mean.

$$\text{RSD} = \frac{\sqrt{\text{Var}\{N\}}}{E\{N\}} = \frac{\sqrt{N_0}}{N_0} = \frac{1}{\sqrt{N_0}} \tag{3.25}$$

This means that although the absolute variance of N increases when there are more photons (increase of N_0), the relative variability, or image noise, decreases. Practically, doubling the number of photons used to make an image will reduce the visualized noise in the image by about 40%. In order to reduce the image noise by half, the total number of photon should be four times bigger.

Equation (3.23) might be a little confusing, as at first glance, it might look like more photons increases the noise level since increasing N_0 increases the standard deviation $\sqrt{N_0}$. However, recall that the measured signal from the N_0 photons is always normalized and scaled, so it is the relative deviation from that value that is the more meaningful one.

On the one hand, more photons are better for image quality, as they result less image noise; on the other hand, more radiation increases the assigned health risk to the scanned patient.

According to the ALARA concept, the optimal solution is to choose the scan parameters specifically to the clinical need in order to maintain good image quality, with the lowest dose possible. As previously discussed, altering the voltage and current changes the photon's energy and beam intensity. Therefore, by choosing lower voltage, or by lowering the current when possible, one can reduce the dose.

The continuous x-ray exposure used during helical CT scans subjects the patient to a higher absorbed dose, compared to exposures due to the momentary x-ray pulse used in planar radiographic imaging. Consequently, CT has been the subject of much design effort to decrease the absorbed dose.

There are two main approaches to achieve reduce CT radiation dose.

The first approach to reduce patient dose is the use of automatic exposure control (AEC) tools. Traditionally, the entire CT scan was taken with a constant current. With more modern AEC methods, the tube output is a profile with respect to the z-axis that can change during the scan itself [4, 5]. With this technique the tube current modulates during the scan. It is a function that changes along the tube z-axis position and is larger when scanning an attenuating part and lower when scanning less attenuating objects. An example could be a scan that starts at the hips with higher current and gradually decreases along the thighs and shins, until it reaches the much less attenuating feet (Fig. 3.32).

An even more complex profile would be one where the modulating function is a combination of tube position in z-axis and the tube angle. With this kind of approach, the current changes are fast enough that when scanning something non-symmetric,

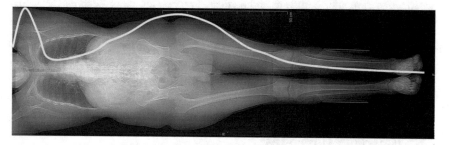

Fig. 3.32 Current profile along the scanned body (yellow line with vertical axis in relative units). Left to right in the image, at first, the current increases due to the attenuating shoulders, followed by a decrease over the length of the much less attenuating lungs. As the profile continues to the abdomen area, which is quite attenuating, there is a second increase, followed by a monotonic decrease along the thighs, shins, and feet, as the legs are getting narrower and less attenuating

in terms of attenuation, even for the same z-axis location, the current is changing during the rotation itself. For example, when the tube is facing the shoulder area laterally (e.g., right to left), the x-ray path is very attenuating; hence, the current will be quite high. On the other hand, when the tube is facing the same shoulder area from above (e.g., the tube is anterior to the patient), the x-ray path is much less attenuating; therefore, the current will be much lower. With this approach the image quality stays approximately the same throughout the entire acquisition, while the total dose is much lower than using a conventional constant current.

Traditionally, the tube voltage of CT scanners was 120 kVp. As the tube design improved, new options for kVp selection were added, and today common options are for 80, 100, 120, and 140 kVp, yet 120 kVp is the most common selection. Another aspect of the AEC tools is the automatic recommendation and selection of specific kVp, according to the patient size and clinical needs. For example, pediatric scans can usually use a much lower kVp, such as 100 kVp or 80 kVp, which also helps in reducing the total dose. Generally, one or two very-low-dose planning scans may be performed so that such AEC tools can estimate the attenuation before the full-dose CT is performed.

The second approach to reduce patient dose is by applying sophisticated reconstruction algorithms, instead of the traditional filter back projection [6–8]. By applying, for example, iterative reconstruction methods, it is possible to maintain or improve the image quality for a decreased x-ray beam intensity.

With iterative reconstruction methods, a model is built, according to a statistical metric, and the image estimate keeps improving from one iteration to the next. These approaches are much more robust to noise in projection data and subsequently can maintain image quality for acquisitions with reduced radiation exposure.

Filtered back projection reconstruction can take several seconds to complete the formation of an entire set of images. In comparison, the iterative methods of reconstruction are much slower to produce the images and could take even 30 minutes per scan on older scanner consoles. However, such methods can be used to reduce the dose by about 50% while maintaining good image quality. Such dramatic dose saving is preferred whenever clinically possible. With the new advances of computer hardware and dedicated processing units, such as CPUs and GPUs, such computation is performed much faster. Therefore, this technique is becoming the standard for reconstruction in the clinic.

Those two approaches, dynamic radiation profile and iterative reconstruction, can be used individually or combined together for optimal results, based on the clinical needs, in order to minimize the required dose to the patient.

3.5 Emerging Directions

3.5.1 Dual-Energy CT (DECT)

In CT, the combination of the attenuation of elements within tissues or materials set the final CT value. Hence, materials with different composition might have similar

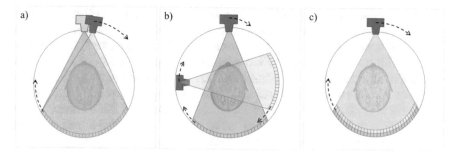

Fig. 3.33 Dual-energy approaches: (**a**) interleaving beam spectra, (**b**) dual tube system, (**c**) dual-layer detectors

CT values, which makes the differentiation between them very difficult. With the help of dual-energy CT, it is sometimes possible to classify the pixels that belong to different groups of materials or elements.

There are several methods that offer such ability in today's scanners [9, 10].

The first approach relies on interleaving beam spectra (Fig. 3.33a). With this approach, a fast switching between the kVp of the tube is applied, so one frame has high kVp (e.g., 140) and the next one has low kVp (e.g., 80), repeatedly. This way the sampling density for each energy is reduced, but with high rotation frequency, there are still enough samples to generate two sets of images (high and low energy) when the scanned area is static. This technique is less effective when scanning the heart, for example.

Another approach presents a dual tube system (Fig. 3.33b). With this concept, two different tubes, positioned at 90° relative to each other along the gantry's arc, are used. One radiates with only high kVp, and the other, with only low kVp. In addition to the dual tubes, there are two matching sets of detection arrays, also positioned at 90° relative to each other, each facing a different tube. With this technique, the sampling rate is not affected at all, as both tubes can radiate simultaneously. On the other hand, scattered photons from one tube might be detected by the array of the other tube, which could lead to erroneous signals and impair the final image.

The last approach uses a different type of detector (Fig. 3.33c). This new type of detector has a "sandwich-like" structure. The upper part can detect the low energy photons of the x-ray spectrum. The middle part serves as a barrier, removing a significant portion of the lower energy photons in the spectrum, that were not attenuated by the upper part. The lower part is sensitive to the remaining higher energy photons. This way the detector yields two signals simultaneously that discriminates the energy of the low energy photons and the high energy photons of the given spectrum. This approach may suffer from inaccurate photon discrimination, and because the process in part relies on absorption for discrimination, it reduces the signal strength.

Each approach has pros and cons, but ultimately each of them allows the user to benefit from the tissue and material separation that can be achieved from the low and high energy data. This discrimination is reliant on the differing energy dependence of the absorption coefficient for different materials. As mentioned earlier, this is very applicable, especially for distinguishing iodine-based contrast agents and calcium, as those two elements have K-edge of approximately 33 keV and 4 keV, respectively.

Since the attenuation of the iodine solution is a function of its concentration, it means that in some cases, both of the materials, iodine solution and calcium, might have a comparable white appearance in the final CT image. It can be difficult to tell if a bright area is plaque, which might clog blood vessels, or in fact a contrast agent flowing through the vessels. For a given spectrum, their CT value might be the similar, depending on the density of the iodine and plaque.

With the additional data from two energies, it becomes easier to classify the pixels in the image, as each of them behaves differently. The pixel values of the iodine-based agent would change dramatically between the lower (e.g., 80 keV) and higher (e.g., 140 keV) images. In contrast, the pixels of the calcium-based tissues would shift much less significantly.

In addition, usage of the "raw data," prior to the image formation, can also be useful. From the "higher energy" data and the "lower energy" data, it is possible to generate a virtual mono-energetic (monochromatic) image, to any selected energy [11]. When generating such monochromatic image according to a specific K-edge energy of a selected material (e.g., 33 keV for iodine), the differences between iodine and calcium are even more evident, than in the original, multi-energetic image.

> Since the divergence in attenuation for tissues and contrast agents is larger for lower energies than for higher energies, it is possible to differentiate between materials when scanning with dual-energy techniques.

Based on this difference, many clinical applications benefit from the material separation.

One example is automatic bone removal in CT angiography (Fig. 3.34). With this application, it is possible to automatically identify and remove the bones, allowing a much better display of the iodine-filled vessels. Another clinical application is atherosclerotic plaque removal. This can be thought of as an advanced option of the automatic bone removal application. Here, in addition to the bones, which are quite massive part of anatomy, the application also delineates plaque blobs. Once removed, it enables the physician to detect more easily narrow parts of the blood vessels and even measure and quantify the obstruction.

3.5.2   Multi-energy CT and Photon Counting

A type of detector that would be able to completely discriminate and identify the energy of each detected x-ray photon would naturally enable the most robust

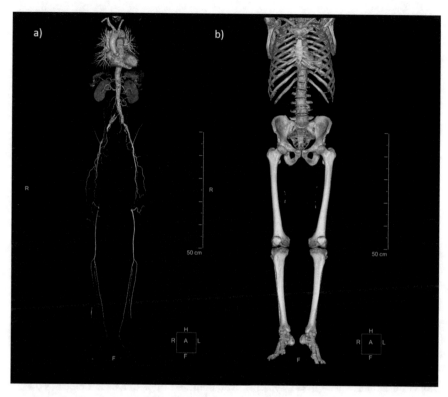

Fig. 3.34 An example of iodine/bone differentiation by means of dual energy: (**a**) organs with contrast agent (e.g., blood vessels, kidneys), (**b**) only bones are displayed in the image

solution for material decomposition. This can be achieved with a direct conversion detector. When the x-ray photons interact with the detector, it results an electric current, which allows the classification of the photon's energy [12].

By setting a single-energy threshold of discrimination, such a detector can serve as a dual-energy detector, as it sorts the photons into two groups: one of high energy (above the threshold) and another of low energy (below the threshold). But by setting multiple thresholds, it can serve as a multi-energy detector, as it classifies the photons to multiple groups. One can think of this as analogy to a "normal" color image where each color represents an energy level, instead of only having an "intensity" image, of the current black and white CT technique. Such "colors" can make material decomposition possible.

In CT imaging, the x-ray photon energy is usually under 150 keV, and the attenuation of x-rays by matter is mainly due to the Compton scattering and the photoelectric effect. It is possible to model these two interactions for many materials with a combination of two variables, the mass density ρ and the atomic number Z, and generate a curve for specific material information. Since most of the body tissues are composed of elements with a very low K-edge (around 0.5 keV or even less), their attenuation can be thought of as a monotonic descending function.

Once calibration data for predefined specific materials is obtained, and the physical interactions are modeled, the attenuation of any material can be expressed as a linear combination of any pair of two basic predefined materials, from the pool of calibrated data. Then, material-specific information can be obtained after scanning, such as mass density, or effective atomic number, which allows for better discrimination between different objects. There are several approaches for material decomposition. Some work on the pre-reconstructed data and some on the reconstructed data (i.e., the volumetric image itself).

With this additional data, two types of clinical application were developed. The first quantifies the concentration of a specific component, within a mixture, for example, to calculate the concentration of contrast agent uptake, in order to optimize its administration.

Another type of clinical application is the classification of materials into predefined groups, such as in a clinical application that performs kidney stone characterization. With this technology it is possible to separate uric acid from non-uric acid stones. This is a very useful classification, since if the physician knows, based on the CT images, that the patient has a uric acid stone, a relatively simple treatment of urinary alkalization could be started. This can minimize the need for additional medical tests and prevent the much more invasive procedure of stone removal.

While photon counting detectors are used in nuclear medicine, currently they are not commercially used in CT systems. The main reason is that such detectors do not yet offer the capability of handling the high photon flux that is usually used in clinical CT. When the input flux is too high for the detector to discriminate different events, it might treat multiple photons as a single event. This is called the pile-up effect, where multiple photons hitting the detector might be counted as a "single event" of a high-energetic photon, instead of several photons, each with different, lower energies.

Even though currently there aren't any commercial photon counting CT systems, it holds much promise for future scanners, newer applications, and better diagnoses.

Lastly, it is important to mention deep-learning-based applications [13, 14]. This emerging direction is trying to offer an alternative to the current CAD mechanisms. Current CAD methods are usually based on an ad hoc process involving several sequential steps of image processing and computer vision techniques along with predefined conditions and limitations based on assumptions and clinical experience. As each step relies heavily on its predecessor, any change or tuning in parameters or algorithm structure within the process is very laborious and difficult to perform. The deep-learning approach implements the automatic exploitation of different features of the CT images and, according to the performance, adjusts parameters or variables within the algorithm. This is done in a ubiquitous manner, correlating CT image features to known clinical findings in a process of self-improvement. Such new techniques are used in radiography [15] as well as CT analysis and are gaining popularity.

Acknowledgments The authors wishes to thank Dr. Jacob Sosna, Dr. Alexander Benshtein, Dr. Dany Halevi, and Nathalie Greenbaum from Hadassah Medical Center, Jerusalem, for their assistance in gathering the clinical images for this chapter.

Bibliography

1. De La Vega JC, Häfeli UO. Utilization of nanoparticles as X-ray contrast agents for diagnostic imaging applications. Contrast Media Mol Imaging. 2015;10(2):81–95.
2. Budoff MJ, et al. Diagnostic performance of 64-multidetector row coronary computed tomographic angiography for evaluation of coronary artery stenosis in individuals without known coronary artery disease. J Am Coll Cardiol. 2008;52(21):1724–32.
3. McCollough CH, Bushberg JT, Fletcher JG, Eckel LJ. Answers to common questions about the use and safety of CT scans. Mayo Clin Proc. 2015;90(10):1380–92.
4. Kalra MK, et al. Techniques and applications of automatic tube current modulation for CT. Radiology. 2004;233(3):649–57.
5. Kalra MK, Maher MM, Toth TL, Kamath RS, Halpern EF, Saini S. Comparison of Z-axis automatic tube current modulation technique with fixed tube current CT scanning of abdomen and pelvis. Radiology. 2004;232(2):347–53.
6. Hara AK, Paden RG, Silva AC, Kujak JL, Lawder HJ, Pavlicek W. Iterative reconstruction technique for reducing body radiation dose at CT: feasibility study. Am J Roentgenol. 2009;193 (3):764–71.
7. Beister M, Kolditz D, Kalender WA. Iterative reconstruction methods in X-ray CT. Phys Med. 2012;28(2):94–108.
8. Silva AC, Lawder HJ, Hara A, Kujak J, Pavlicek W. Innovations in CT dose reduction strategy: application of the adaptive statistical iterative reconstruction algorithm. Am J Roentgenol. 2010;194(1):191–9.
9. Forghani R, De Man B, Gupta R. Dual-energy computed tomography: physical principles, approaches to scanning, usage, and implementation: part 1. Neuroimaging Clin N Am. 2017;27 (3):371–84.
10. Forghani R, De Man B, Gupta R. Dual-energy computed tomography: physical principles, approaches to scanning, usage, and implementation: part 2. Neuroimaging Clin N Am. 2017;27 (3):385–400.
11. Yu L, Leng S, McCollough CH. Dual-energy CT–based monochromatic imaging. Am J Roentgenol. 2012;199(5_supplement):S9–S15.
12. McCollough CH, Leng S, Yu L, Fletcher JG. Dual- and multi-energy CT: principles, technical approaches, and clinical applications. Radiology. 2015;276(3):637–53.
13. Litjens G, et al. A survey on deep learning in medical image analysis. Med Image Anal. 2017;42:60–88.
14. McBee MP, et al. Deep learning in radiology. Acad Radiol. 2018;25(11):1472–80.
15. Lakhani P, Sundaram B. Deep learning at chest radiography: automated classification of pulmonary tuberculosis by using convolutional neural networks. Radiology. 2017;284 (2):574–82.

Chapter 4
Nuclear Medicine: Planar and SPECT Imaging

Synopsis: In this chapter the reader is introduced to the physics of nuclear medicine imaging and methods of forming functional planar and single photon emission computed tomography (SPECT) images.

The learning outcomes are: The reader will know the fundamental components of a conventional gamma camera, including the concept of a radiopharmaceutical, and be able to compare this to solid-state gamma camera designs. Students will be able to implement reconstruction algorithms for SPECT systems and characterize factors affecting image quality in SPECT cameras

Nuclear medicine (NM) techniques provide a unique method for imaging human physiology. Typically, a biologically functional molecule (ligand) is labeled with a radioisotope and referred to as a radiopharmaceutical or, more specifically, a radiotracer. Usually introduced by intravenous injection, the radiopharmaceutical becomes incorporated throughout the body over a short time period (e.g., 1 hour). The resultant radiotracer distribution is dependent on the body's functional use of the molecule (Fig. 4.1). Radioisotopes are chosen such that they undergo nuclear decay processes that produce high-energy photons similar to x-rays. The photons are emitted by the radiotracer from the body and are detected by the NM camera, providing data on the location and paths of the functional molecule. NM is a powerful diagnostic tool providing true molecular imaging: nano- and picomolar concentrations of radiotracers can be detected. These small quantities of radiopharmaceutical do not perturb the patient's biochemistry, nor is the patient exposed to unreasonable amounts of ionizing radiation. Since physiological changes typically precede any measurable anatomical change, pathologies often may be detected with NM before any anatomical change can be detected by other modalities.

The most common clinical diagnostic imaging procedures on nuclear medicine cameras are bone scans and myocardial perfusion imaging (MPI or heart scans). For

© Springer Nature Switzerland AG 2020
H. Azhari et al., *From Signals to Image*,
https://doi.org/10.1007/978-3-030-35326-1_4

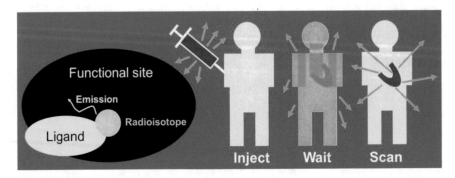

Fig. 4.1 Emission imaging: A radiopharmaceutical is injected and the biologically active ligand is taken up by a functional site. Radioisotopes attached to the ligands emit high-energy photons suitable for imaging

example, the former can be used to detect stress fractures before they are evident in other imaging modalities, and the latter may assess stenosis in the coronary arteries. A related technology is addressed in the next chapter: positron emission tomographic scanners (PET) are used primarily for oncological studies. Also, within many nuclear medicine departments, special radiopharmaceuticals are used for radiation treatment of disease. These include radioiodine NaI pills for hyperthyroidism and thyroid cancer; radioactive yttrium microspheres for liver cancer; radioactive radium $RaCl_2$ solution for bone cancer; and radioactive lutetium Lu-PSMA (prostate-specific membrane antigen) solution for metastatic prostate cancer. However, only diagnostic imaging in nuclear medicine departments is described here.

4.1   Physical Phenomena

4.1.1   Physics of Radioisotopes and Radioactive Decay

Whereas for x-ray imaging the energy for signal generation is supplied by the x-ray tube, the source of the signal in nuclear medicine imaging techniques is the nucleus of unstable isotopes. The nuclei of radioisotopes release energy to achieve lower energy, stable (non-radioactive) states by emitting particles that can then be used for signal generation for imaging. Table 4.1 lists many of the radioisotopes commonly used for diagnostic imaging. The initial radioisotope is called the "parent" and the resulting isotope is called the "daughter." Since most decay modes listed involve the change in the number of protons in the nucleus (the atomic number), the daughter is not only a different isotope; it is a different element from the parent. The exception in Table 4.1 is technetium, in which a metastable state of technetium, ^{99m}Tc, decays to the radioisotope ^{99}Tc.

The notation ^A_ZX is a shorthand method of listing isotopes, such that X is the chemical symbol of the element, Z is the atomic number, and A is the atomic weight

Table 4.1 Common radionuclides used in nuclear medicine imaging

Radionuclide	Relevant decay mode[a] (and branching ratio)	Half-life		Principle imaging photons	Modality
^{11}C – carbon	β^+ (100%)	20.39	minutes	511 keV	PET
^{13}N – nitrogen	β^+ (100%)	9.97	minutes	511 keV	PET
^{15}O – oxygen	β^+ (99.9%)	2.04	minutes	511 keV	PET
^{18}F – fluorine	β^+ (96.7%)	109.8	minutes	511 keV	PET
^{64}Cu – copper	β^+ (17.4%)	12.70	hours	511 keV	PET
^{67}Ga – gallium	EC (100%)	3.26	days	93, 184, 394 keV	SPECT[b]
^{68}Ga – gallium	β^+ (89.9%)	67.6	minutes	511 keV	PET
^{90}Y – yttrium	β^- (100%) β^+ (0.0032%)	64.0	hours	Bremsstrahlung e.g., 60–170 keV, 511 keV	SPECT/ PET/ therapy
^{82}Rb – rubidium	β^+ (100%)	1.273	minutes	511 keV	PET
^{99m}Tc – technetium	IT (100%)	6.01	hours	141 keV	SPECT
^{111}In – indium	EC (100%)	2.80	days	171, 245 keV	SPECT
^{123}I – iodine	EC (100%)	13.2	hours	159 keV	SPECT
^{124}I – iodine	β^+ (22.8%)	4.18	days	511 keV	PET/ therapy
^{131}I – iodine	β^- (100%)	8.02	days	364 keV	SPECT/ therapy
^{177}Lu – lutetium	β^- (100%)	6.7	days	113, 208 keV	SPECT/ therapy
^{201}Tl – thallium	EC (100%)	3.04	days	60–80 keV x-rays	SPECT
^{223}Ra – radium	α^{++} (100%) and decay chain	11.44	days	82, 154, 269, 351, 402 keV	SPECT/ therapy

[a]See Table 4.2 for decay mode explanation
[b]SPECT in this table includes non-tomographic planar imaging as an option

(the number of protons and neutrons in the isotope). Since the chemical element X is determined by its atomic number and the resultant electron shell structure, Z is often omitted in this notation as redundant information.

There are five decay modes most relevant to nuclear medicine (Table 4.2). In alpha (α)-emission, a helium nucleus consisting of two protons and two neutrons is ejected from the parent nucleus $^A_Z X$. The nuclear equation can be written as $^A_Z X(,\alpha)^{A-4}_{Z-2} J$, where J represents the daughter element and the bracket notation is of the form "(incident reactant, emitted product)." Note that the atomic number and atomic mass number of the parent and the products balance since the α-particle could

Table 4.2 Radioactive decay schemes for radioisotopes used in nuclear medicine

Unstable parent $_Z^A X$	Decay mode	Emissions	Daughter
	Beta (negative)-emission	β^-, antineutrino	$_{Z+1}^A E$
	Isomeric transition (IT)	γ-Ray	$_Z^A X$
		Conversion e$^-$	
	Positron emission	β^+, neutrino	$_{Z-1}^A G$
	Electron capture (EC)	Neutrino and characteristic x-ray or Auger e$^-$	
	Alpha-emission	α^{++}	$_{Z-2}^{A-4} J$

be written as $_2^4$He. Since all the nuclear reactions in Table 4.1 are spontaneous decays, there is no incident reactant particle initiating the transition, so that space is left blank in the equation. An example is the decay of radium-223 into radon-219: $_{88}^{223}$Ra$(,\alpha)_{86}^{219}$Rn. Radon-219 is also an alpha-emitter, leading to a decay chain of radioactive daughters. Primarily, radium-223 is used for radiation therapy in the treatment of prostate cancer metastasized to the bone, because, as a calcium analog, it can deliver radiative energy lethal to cancer cells active in the bone. Since only 1.1% of the energy is released as high-energy photons, ^{223}Ra is limited as an imaging tracer, although such imaging has been moderately successful [1].

Beta-particles are high-energy electrons emitted from the nucleus, including anti-electrons called positrons. An example of a radioisotope in which virtually all the decays are beta-emission is yttrium-90: $_{39}^{90}$Y$(,\beta^-)_{40}^{90}$Zr. Here, the beta-negative symbol (β^-) indicates that this is a decay by the emission of a standard negatively charged electron and not by the emission of a positively charged positron (β^+). Yttrium-90 is also used primarily for radiotherapy, for example, the treatment of liver tumors by means of radioactive microspheres (resin or glass beads <70 μm in diameter containing ^{90}Y). The beta-emissions are only indirectly imaged in nuclear medicine. These high-energy electrons slow down in patient tissue and release energy via the bremsstrahlung emission of x-rays (Sect. 3.1.2), which can then be imaged with a nuclear medicine camera. A very small percentage (0.0032%) [2] of the yttrium-90 decays involve the emission of a positron via the decay of a metastable state of the daughter zirconium-90. Because of the relatively high activities used in therapy, these positrons can generate enough of a signal for PET imaging [3].

The most common positron (β^+) emitter used in medical imagine is fluorine-18: $_9^{18}$F$(,\beta^+)_8^{18}$O. As an ionic aqueous solution (e.g., Na[^{18}F] dissolved in water), the radioisotope can be used for PET bone scans since fluorine is quickly absorbed by the skeleton [4], but it is usually employed attached to a ligand. A positron emitted from within a body can travel up to about 2 mm, depending on its energy, before it slows down sufficiently to interact with the surfeit of electrons in most matter. The subsequent matter/anti-matter annihilation usually produces two 511 keV that travel in virtually opposite directions. The coincidence detection of this pair is used in PET to reconstruct 3D radiotracer distributions. The annihilation photon energy (E_γ) of

511 keV is the energy of the rest mass of an electron (m_e) converted to electromagnetic radiation as per Einstein's well-known equation $E_\gamma = m_e c^2$, where c is the speed of light. A competing reaction to positron emission is electron capture (EC) in which an inner orbital electron is captured by the nucleus, reducing its atomic number, giving an identical daughter element. For example, only 96.7% of all ^{18}F decays are positron emission, whereas the remaining 3.3% are via electron capture. In nuclear physics or nuclear engineering jargon, one says that the branching ratio of positron emission is 96.7% for ^{18}F decay. Also in the jargon, annihilation photons are occasionally included in the category "gamma rays," although historically the term refers to photons emitted from a nucleus.

Decay by isomeric transition (IT) provides no change in atomic mass number or atomic number since the nucleus is in a metastable state and releases energy to a lower energy state by emitting a photon which is not a nucleon (proton or neutron) and carries no electric charge. All radioactive nuclei are in a metastable state in that they will exist temporarily at a higher energy state, before transitioning to a lower energy state, like a skier who has stopped briefly while gliding down a hill. Technetium-99m (^{99m}Tc) is a metastable state of technetium-99 (^{99}Tc) that decays by IT and emits γ-rays at energies similar to a very energetic ("hard") x-ray, 140.5 keV, with a branching ratio of 89%. As a salt solution (^{99m}Tc-pertechnetate), ^{99m}Tc can mimic iodine and is commonly used for thyroid imaging, but it is usually employed attached to a ligand. Because the energy of its emissions is released solely as photons useful for imaging, ^{99m}Tc can provide for emission imaging with minimal radiation exposure to the patient. Other radioisotopes may emit charged particles followed by the IT decay of their daughters, with the IT γ-rays providing imaging photons. For example, the beta-emissions of iodine-131 are biologically damaging, useful for ablating remnant metastatic thyroid cancer tissue, and the IT decay of its daughter (xenon-131) provides 364 keV useful for nuclear medicine imaging. A competing nuclear reaction with gamma-ray emission during isomeric transition is internal conversion: the energy associated with gamma-ray emission is instead transferred to an inner orbital electron called a conversion electron. Like a beta-particle, this electron has high energy, the energy released by IT minus the binding energy of the electron. But unlike beta-particle emission, the nucleus does not change its total charge and remains the same element. Also, the energy of the conversion electron is discrete, not a continuous spectrum, dependent on the nuclear energy level transitions and the energy levels of the electronic shell transitions. Characteristic x-rays may be emitted after internal conversion, as higher energy orbital electrons release energy in discrete units as they fill a lower energy orbital left vacant by the conversion electron.

Historically, the decay rate of isotopes was measured relative to the decay rate of 1 g of ^{226}Ra, and this was called 1 curie (Ci). This unit is named after the only person to have received a Nobel Prize in both Chemistry and Physics, Marie Curie, and for her husband Pierre. The curie has been redefined as 3.7×10^{10} decays per second. A decay per second is called a becquerel (Bq) such that a commonly used injected dose of 10 mCi of a radiotracer is equal to 370 MBq. The becquerel is named after Henri Becquerel who shared the Nobel Prize in Physics with Marie and Pierre for the

Fig. 4.2 Radioactive decay. One-half of the parent nuclide decays into a daughter nuclide (or daughter nuclides) in one half-life. After two half-lives, the remainder of the parent is 1/4 and after three, it's 1/8. If there is only one daughter and it is stable, after six half-lives, >98% of the parent nuclide has decayed to become the daughter nuclide

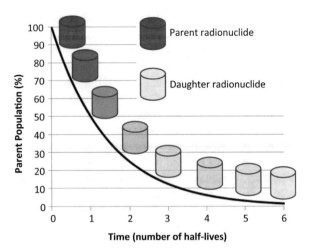

discovery of radioactivity. Because many branching ratios are not 100%, and because radioactivity can follow extended decay chains, the activity in becquerels should not be interpreted as emitted particles per second.

Radioactive decay is a stochastic process. The rate of decay is proportional to the number of parent nuclei (N). This is intuitively understood since doubling the number of parent nuclei would double the rate of decay, $dN/dt = -\lambda N$, where t is time and λ is the decay constant or decay rate constant and the negative sign accounts for depletion. Given initial conditions $N(t = 0) = N_0$, this can be integrated to give

$$N = N_0 e^{-\lambda t} \tag{4.1}$$

Commonly, the decay of a radioisotope is characterized by a half-life (τ), which is the time needed for the parent population to deplete to half its original number, given no additional production of the parent. By this definition, $\tau = \ln(2)/\lambda$, and Eq. 4.1 can also be written as

$$N = N_0 (\tfrac{1}{2})^{t/\tau} \tag{4.2}$$

which aids in mental calculations since one can often estimate the percent remaining population by approximating the number of half-lives that have past. By definition the activity A is given by $A = |dN/dt|$, so the activity is proportional to the parent population by a factor of the decay constant: $A = \lambda N$. Figure 4.2 shows a graphical representation of the relationship between activity (or parent population) and half-life.

After one half-life, one-half of the population of a radionuclide, and one-half of its activity, remains due to random radioactive decay.

4.1.2 *Propagation and Attenuation of Ionizing Radiation*

Ionizing radiation consisting of charged particles, like alpha and beta radiation, interacts with the orbital electrons and nuclei of matter through which it travels, via Coulomb forces. These collisions can be elastic, but the particles can pass energy to the stopping material through inelastic collisions. With inelastic collisions, orbital electrons may be excited to a higher energy state or be removed from their bound state (ionization), thus reducing the incident particle's energy. Collisions with nuclei are often elastic; however, high-energy electrons that are slowed by a nucleus' Coulomb forces will lose energy by means of radiative losses called bremsstrahlung: a high-energy photon is generated. The probability of incident alpha-particles producing bremsstrahlung radiation is typically several orders of magnitude less than that for beta-particles. The distance traveled by high energy charged particles can be characterized by an average range, dependent on the stopping medium and the particle energy, charge, and mass. Examples of ranges of charged particle ionizing radiation in water (a tissue surrogate) for some isotopes used in nuclear medicine are given in Table 4.3.

Similar to x-rays, at the photon energies used in nuclear medicine, gamma rays (γ-rays) and annihilation photons interact with patient tissues and radiation shields primarily through the photoelectric effect and Compton scattering (Sect. 3.1.3). As a high-energy photon beam passes through homogenous material, the probability of absorption by any layer of equal thickness remains the same. Consequently, the change in intensity (photon flux) of the beam is proportional to the material thickness, $dI/dx = -\mu I$, where I is the beam intensity, x is the material thickness, and μ is a proportionality constant called the absorption coefficient. Given an initial beam intensity $I(t = 0) = I_0$, this can be integrated to give

$$I = I_0 e^{-\mu x} \tag{4.3}$$

as indicated by Eq. 3.5. By analogy to half-life, the absorption can be characterized by a half-value layer (HVL), which is the material thickness needed to reduce the beam intensity by half. By this definition, $\text{HVL} = \ln(2)/\mu$, and Eq. 4.3 can also be written as

Table 4.3 Example charged particle ionizing radiation and their average range in water

Isotope	Particle emitted	Maximum energy (MeV)	Mass (atomic mass unit)	Electronic charge	Average range (mm)
^{18}F	β^+	0.6	0.00054	+1	0.3
^{68}Ga	β^+	1.8	0.00054	+1	1.2
^{131}I	β^-	0.9	0.00054	−1	3.5
^{223}Ra	α^{++}	8	4	+2	<0.09

$$I = I_0(\tfrac{1}{2})^{x/\text{HVL}} \tag{4.4}$$

The half-value layer does not scale linearly with energy. For example, HVL for ^{99m}Tc γ-rays (141 keV) is 0.3 mm of lead (Pb), but for 511 keV annihilation photons from PET materials such as ^{18}F, HVL is 6 mm Pb (see Fig. 4.3).

The half-value layer depends on the material, the type of radiation, and the energy of that radiation. Because Eq. 4.3 is modeled on the probability of a discrete interaction between the particle and matter, the concept of HVL is appropriate for radiation that does not carry charge, such as photons and neutrons. As mentioned above, radiative particles carrying charge, such as α- or β-particles, undergo continual slowing via Coulomb forces, and as such their interaction with matter is best characterized by range.

With respect to absorption by matter, γ-rays are the most penetrating, β-rays are less penetrating, and α-rays are the least penetrating. A sheet of paper is an effective shield against α-radiation, Perspex or acrylic can provide a good shield against β-radiation, and γ-radiation is usually shielded with lead or other dense metals such as tungsten or steel (Fig. 4.4). The low average atomic number of Perspex or acrylic reduces the amount of bremsstrahlung radiation produced while stopping β-particles.

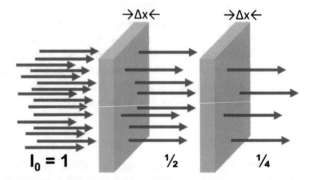

Fig. 4.3 Half-value layer (HVL). Each addition of one HVL (here Δx) of a material will reduce the radiation intensity (number of photons) by half

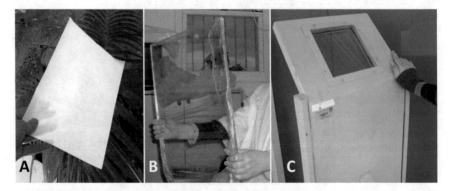

Fig. 4.4 Example shields. (**a**) Paper can stop α-rays. (**b**) Plastic stops β-particles producing minimal bremsstrahlung x-rays. (**c**) Lead, tungsten and/or steel for γ-rays

Since material HVLs have a non-linear relationship with photon energy, a typical shield for 141 keV γ-rays from ^{99m}Tc is 3 mm of lead, whereas a typical shield for 511 keV annihilation photons from PET materials is 50 mm of lead, even though the energy of the latter is less than four times larger. Shielding in a nuclear medicine facility is usually determined by the need to keep the radiation exposure of personal small and well within regulatory limits.

4.1.3 Radiation Safety with Open Sources

Nuclear medicine departments present a challenge for radiation safety design because the primary sources of radiation are mobile and have free will: the injected patients. Unlike x-ray tubes used with other radiological exams, these sources cannot be turned off. Fortunately, most patients are compliant and radiation exposure to staff and the general public can be kept low.

Within a nuclear medicine department, there are typically several sealed sources used to check machine calibration or image quality. Sealed in plastic or epoxy, these sources containing relatively long-lived isotopes do not present a hazard of contamination unless the seal is broken. For example, most departments have a small sealed ^{137}Cs source (half-life: 30 years) that is NIST-traceable (referenced to a source at the National Institute of Standards and Technology) for a daily check that the dose calibrator is giving accurate measurements of radiopharmaceutical activity. Large, flat ^{57}Co sources (half-life: 272 days) are used for daily quality control checks of the nuclear cameras. These sources are checked periodically and are sequestered safely if any cracks in the seal or leaks are found.

However, most radioisotopes in the clinic are not permanently sealed: they are open sources. Radiopharmaceuticals are usually in solution such that they can be intravenously injected into the patient. There are exceptions, such as ^{131}I that is usually given as a solid capsule of NaI salt but even this is an open source for practical purposes since the capsule is designed to be easily broken and easily dissolved after ingestion. Additionally, radiopharmaceuticals are usually eliminated from the body in the urine, but may also egress via saliva, feces, or perspiration. Consequently, the biological waste from patients must be considered to be open sources of radiation.

In a similar manner to x-ray tubes, sealed sources present the possibility of external exposure of personnel or the general public to radiation fields. External exposure means that the source of the ionizing radiation is external to the individual. External fields provided by charged particles are typically characterized by a range in air, but common sealed sources found within nuclear medicine departments do not give off significant α- or β-fields beyond their sealed container. The external field due to γ-rays emitted can be characterized by a γ-constant or exposure rate constant. For a given isotope, a given activity, and a given distance, the exposure rate (mSv/hour) will be constant since the photon flux and energies will be constant (Fig. 4.5). Since there is no preferential direction for gamma-ray emission, a small

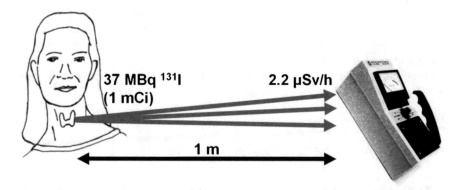

Fig. 4.5 Exposure rate constants (Γ constants) give the radiation field for a fixed distance from a point source of a fixed activity of an isotope. Here, a field measurement can be used to deduce the remaining uptake of ^{131}I in a thyroid patient. The point source approximation is suitable since almost all ^{131}I concentrates in this small organ. There is little attenuation of the γ-ray beam. The approximation is better at 2 m, where the field is expected to be ¼ the field at 1 m

source (ideally a point source) will radiate photon flux homogenously in all directions, leading to an inverse-square dependence of the photon flux and consequently the exposure rate. So the γ-constant (Γ) can be given in units of μSv h^{-1} MBq^{-1} m^2, and the exposure rate $\dot{D}$ from a small source can be written:

$$\dot{D} = \Gamma A / r^2 \tag{4.5}$$

where A is the source activity and r is the distance from the source. When uptake within a patient is highly localized, as in many ^{131}I treatments, Eq. 4.5 can be rearranged in order to estimate the activity A remaining from a radiation exposure rate from the radiation field at a known distance. In practice, a distance of at least 2 m is often used to ensure that the small source approximation is valid.

Internal exposure is radiation exposure from sources within the body. Open sources provide for the real possibility of internal exposure (Fig. 4.6). Liquid sources might be inadvertently ingested if, for example, personnel eat with contaminated hands. Although most radiopharmaceuticals in nuclear medicine are not volatile, Na [^{131}I] solutions, ^{133}Xe gas, and vaporized graphene containing ^{99m}Tc (Technegas) can contaminate the air if not properly vented and subsequently contaminate the lungs.

The risks associated with internal exposure can be characterized by the annual limit on intake (ALI). An ALI is the amount taken internally that would expose the worker to the annual maximum average radiation dose (20 mSv according to ICRP, 50 mSv for US Nuclear Regulatory Commission). To be clear, an internal exposure of 1 ALI does not indicate that any harm has been done, but it does indicate a consensus opinion on the maximum annual risk due to internal exposure acceptable for radiation workers. A higher ALI indicates that there is less risk associated with the radioisotope. Diagnostic materials have comparatively high ALI values since

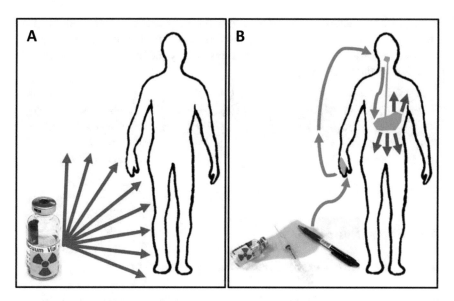

Fig. 4.6 External and internal radiation exposure. (**a**) External radiation fields cause exposure as a flux of particles from a source external to the body. Removing the source stops further exposure. (**b**) Internal radiation exposure occurs by ingestion, inhalation, or penetration through the skin by a radioactive source. The exposure continues until biological and physical processes eliminate the source. Spills of radioactive material can contaminate equipment such as pens, which in turn can contaminate the hand and then the mouth leading to ingestion

risks associated with diagnostic scans of patients also need to be minimized. Conversely, materials used for treatment have comparatively low ALI values since the goal of treatment is to cause deterministic effects by fatally damaging pathological tissue. For example, the ALI (ICRP) for ingestion of ^{99m}Tc (used for imaging) is 888 MBq, whereas the ALI for ^{131}I (used for treatment and imaging) is one one-thousandth of that activity (0.888 MBq). Iodine-131 inventories and disposal must be very carefully controlled in nuclear medicine, as well as for other treatment materials such as ^{89}Sr, ^{90}Y, ^{177}Lu, and ^{223}Ra. Because of this difference, waste with moderate contamination is typically sequestered safely for 10 half-lives for diagnostic material and 20 half-lives for treatment materials, reducing remaining activities by one one-thousandth and one one-millionth, respectively. Regardless, waste is checked before disposal according to national regulation. Alpha-emitters often have small gamma constants, presenting relative low risks from external fields, but often have very small ALI values, indicating a relatively high risk if contamination is present. For example, the gamma constant is 0.02 μSv h^{-1} MBq^{-1} m^2 for ^{223}Ra compared to 0.16 μSv h^{-1} MBq^{-1} m^2 for ^{18}F, a PET tracer, but ^{223}Ra's ALI is just 0.185 MBq compared to 410 MBq for ^{18}F. Although a nuclear medicine department's activities and radiation fields are comparatively small, care must be taken to keep exposures to a minimum for personnel since such exposures are part of the daily routine. Generally, personnel maintain a distance of 1 m from a patient, when close interaction is not required, to reduce external exposure, and protective clothing such as smocks, gloves, and shoe covers reduce the risk of contamination. External exposure to the hands can be mitigated by the use of tongs, pincers, and radiation shielding of syringes.

 Radiation safety aims to reduce potential harm by reducing exposure from high energy particles generated outside the body and from radioactive materials that might contaminate the body.

4.2   Signal Sources

4.2.1   Emission Versus Transmission Scanning

In a transmission image, internal structure is revealed by the transmission and absorption of photons that originate from outside the object being imaged (Fig. 4.7a). X-ray radiographs and CT scans are transmission images: the x-ray tube transmits photons across the object. Transmission images primarily give information about the anatomy. In an emission image (Fig. 4.7b), the internal structure is revealed by the emission of photons originating from within the object, such as occurs in nuclear medicine imaging. Consequently, the term "emission" is used in the acronyms SPECT (single photon emission computed tomography) and PET (positron emission tomography) (Fig. 2.3).

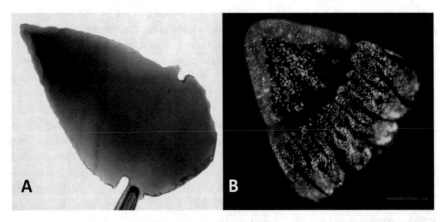

Fig. 4.7 Transmission vs. emission imaging. (**a**) The internal structure of a backlit arrowhead is revealed by photons from outside passing through the object as in a transmission image. (**b**) In an emission image, the internal structure is shown by photons originating from within the object such as with this bioluminescent jellyfish (Periphylla) (used with permission by Steve Haddock)

Nuclear medicine techniques image only high-energy photons that are generated within a body and emitted due to radioactive decay processes.

4.2.2 Concept of a Radiopharmaceutical

A radiopharmaceutical is a biologically active ligand labeled with a radioisotope that will eventually decay radioactively and emit a signal. In nuclear medicine the signal is a high-energy photon (γ-ray). The signal modulation is the intensity, or flux, of these photons, not their energy: a higher flux from a region of the body indicates a higher concentration of the biologically active ligand. The radiopharmaceutical is introduced into the patient usually by intravenous (IV) injection, swallowing, or breathing. After some biological uptake time (1 second to 1 week; often ~1 hour), the signal from radioactive emission is used to make an image (Fig. 4.8). For example, the ligand methylene-diphosphonate (MDP) can be preferentially taken up by bone

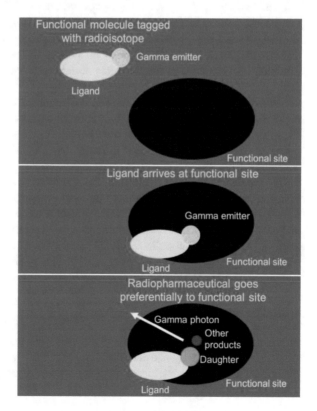

Fig. 4.8 Emission imaging and the radiopharmaceutical concept

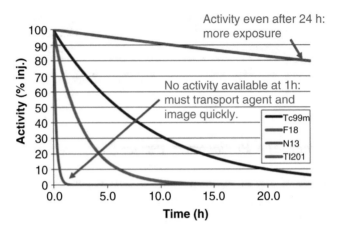

Fig. 4.9 Activity vs. time curve showing the radioactive decay of several isotopes

after IV injection when MDP is tagged with ^{99m}Tc that emits 141 keV gamma photons. In another example, a patient's leukocytes (white blood cells) are removed and labeled with ^{99m}Tc to track infection after IV injection (lymphoscintigraphy). A standard meal containing scrambled eggs with some ^{99m}Tc-pertechnetate solution can be used to track gastric emptying for several hours after swallowing. Lung ventilation scans are typically performed after the patient breathes small ^{99m}Tc metal crystals that have each been trapped in a shroud of six member graphitic rings (Technegas). Special tests may require special means of introducing the radiophar-maceutical. For example, in lymphoscintigraphy a ^{99m}Tc-sulfur colloid is injected subdermally about 1 cm from a known breast tumor to help identify the sentinel node: the first lymph node in the lymph bed to receive drainage from the lesion. Some radioisotopes serve as their own ligand. For example, ^{201}Tl is chemically a potassium analog and thus is used to image cardiac viability after IV injection as a salt solution.

The choice of ligand depends on its biological activity. The choice of radioisotope is not arbitrary. Chemically, the isotope must be capable of being attached to the ligand without negatively affecting the ligand's desirable properties. Metals like technetium do not attach readily to biochemical materials and often must be che-lated: a pair or more coordinate bonds form a "pincer-like" structure to hold the metal ion. The isotope or its daughter must be able to emit or generate photons with energies sufficient to be emitted from the body in useful intensities, but with low enough energies to interact with practical detectors. For a gamma camera, this range might be from 25 to 600 keV. The half-life of the isotope needs to be sufficiently long to enable imaging after an uptake period, but long half-lives may proportionally increase patient radiation exposure if the radiopharmaceutical is not biologically cleared from the body quickly. For example, the PET tracer ^{13}N-ammonia needs to be immediately delivered to the imaging room, usually by pneumatic delivery systems, for cardiac scanning because its half-life is 10 minutes (Fig. 4.9).

Table 4.4 Physical parameters of isotopes related to external radiation exposure

Isotope	Half-life (hour)	Photon energy for imaging (keV)	Imaging photons per decay	Radiation field at 1 m (Γ) (mSv/MBq)
^{201}Tl	73.1	69–80, 167	1.04	0.185×10^{-4}
^{99m}Tc	6.01	141	0.89	0.224×10^{-4}
^{18}F	1.83	511	1.94	1.58×10^{-4}
^{13}N	0.166	511	2	1.63×10^{-4}

Conversely, cardiac studies using ^{201}Tl have the highest exposure for typical diagnostic nuclear medicine scans, in part, because the half-life is comparatively long at 73 hours. With a half-life of 6 hours, ^{99m}Tc can provide a strong signal after 1 or 2 hours of uptake, but after 2.5 days, the activity of any remaining in the body is negligible.

Regardless, the radioactive emissions from a patient after a scan is complete do not usually warrant special instructions to reduce exposure to others because the fields are small [5]. The external fields produced are dependent on the activity remaining in the patient and the radiation field constant (Γ-constant) as described in Eq. 4.5. The Γ-constant is primarily a function of the total photon energy being emitted. Table 4.4 details this relationship for some radiotracers used in cardiac imaging, showing higher Γ-constants for higher total photon energies. Breastfeeding mothers provide an exception to this rule because of the possibility of passing some of the radiopharmaceutical to the baby as a breast milk contaminant. If cessation is warranted, it is usually just a few hours. However, Na[^{131}I] and Na[^{123}I] studies usually require permanent cessation [6].

4.2.3 Gamma Sources

As mentioned in the previous section, the choice of isotope is constrained both by the biological properties of the isotope and/or its compounds and by its physical characteristics such as half-life, particles emitted, and Γ-constant. For imaging, high-energy photons must be emitted from the region of radiotracer concentration. In the case of positron emitters (see Table 4.1), these photons are generated via positron/electron annihilation, and this mechanism is discussed in more detail in Chap. 5. Gamma cameras only use the beam intensity of high-energy photons as a signal; they do not directly image charged particle emissions. The gamma emitters listed in Table 4.1 are the isotopes used in standard nuclear medicine diagnostic procedures.

By far the most common radioisotope used in nuclear medicine is ^{99m}Tc, a gamma emitter that decays from a metastable state of technetium-99 into a ground state ^{99}Tc (lowest nuclear energy level). Technitium-99 itself is radioactive, $^{99}_{43}$Tc $(,\beta^-)^{99}_{44}$Ru, but it has a comparatively long half-life of 21,110 y. As discussed in Sect. 4.1.1, activity, A, is proportional to the decay constant, λ, for a given population of a nuclide, N, according to $A = \lambda N$, and therefore inversely proportional to the half-life. Given a half-life of ^{99m}Tc of 6.01 hours, a dose with an activity of 370 MBq of ^{99m}Tc decays rapidly to just $370 \times 6.01/21110/365.25/24 = 1.2 \times 10^{-5}$ MBq of ^{99}Tc, after accounting for the conversion of years to hours. This provides an external field a couple of orders of magnitude less than background and is 11 orders of magnitude below an ALI for ^{99}Tc [7]. Consequently, with the level of activities used in a nuclear medicine clinic, the activity of ^{99}Tc does not have to be considered.

Isotopes of iodine, ^{131}I and ^{123}I, are commonly used in diagnostic imaging. Iodine-131 emits gammas via an excited nuclear state of its daughter, xenon-131, after $^{131}_{53}$I$(,\beta^-)^{131}_{54}$Xe decay. After gamma emission, this daughter, ^{131}Xe, is stable. Iodine-123 decays primarily by electron capture, $^{123}_{53}$I$(e^-,\gamma)^{123}_{52}$Te, and in a similar manner emits a gamma via a metastable state of its daughter, tellurium-123, which is then effectively stable (half-life of $>1 \times 10^{13}$ years). Iodine is used for thyroid and thyroid cancer imaging. Iodine-131 is used for treatment as Na[^{131}I] and as ^{131}I-MIBG (metaiodobenzylguanidine). Iodine-123 MIBG is used for imaging neuroendocrine tumors and enervation of the heart.

Thalium-201, used for cardiac imaging, is not technically a gamma emitter, but emits high-energy photons suitable for nuclear medicine imaging. After ^{201}Tl undergoes electron capture, the daughter product, mercury-201, emits x-ray photons primarily in the energy range of 68.9–80.3 keV. The ground state of ^{201}Hg is stable. Gallium-67 citrate has been used to image infection, inflammation, melanoma, and lymphoma [8] among other pathologies. This EC decay to a stable daughter, $^{67}_{31}$Ga $(e^-,\gamma)^{67}_{30}$Zn, produces a number of γ-rays of which 93, 184, and 296 keV are typically used in imaging. Leukocytes labeled with indium-111 have been used to image infection and inflammation, and ^{111}In-octreotide is a radiotracer for neuroendocrine tumors. Indium-111 can be imaged using the 173 and 247 keV γ-rays emitted via EC decay to a stable daughter product, $^{111}_{49}$In$(e^-,\gamma)^{111}_{48}$Cd. Additionally, cobalt-57 with a half-life of 272 days provides a source of γ-rays for quality control imaging or for flooding the background field during imaging to outline the contours of a patient's body during lymphoscintigraphy. The gamma peak used is 122 keV produced by EC decay to a stable daughter, $^{57}_{27}$Co$(e^-,\gamma)^{57}_{26}$Fe. To be clear, ^{57}Co is not used as a radiotracer but as a source of γ-rays external to the body, when needed.

The gamma imaging of alpha-emitter treatment materials such as ^{223}Ra is not typically performed clinically, but has been used for initial dosimetry studies [1]. Decay of alpha-emitters with clinically useful half-lives often involves a chain of daughter products emitting α-, β-, and γ-rays. After four α- and three β-emissions, ^{223}Ra becomes ^{207}Pb which is stable. Since only 1.1% of the energy released is in the form of high-energy photons and since clinical activities use are small (<7 MBq), the intensity of photons available for imaging is small, making quality images difficult to obtain. Radium-223 distribution after treatment has been imaged using 82, 154, 269, and 402 keV photons which are emitted primarily during the decay of ^{223}Ra and its first daughter ^{219}Rn.

Some isotopes, such as chromium-51, may be used as radiotracers in a nuclear medicine department, but not for γ-ray imaging. Also, preclinical scanners for

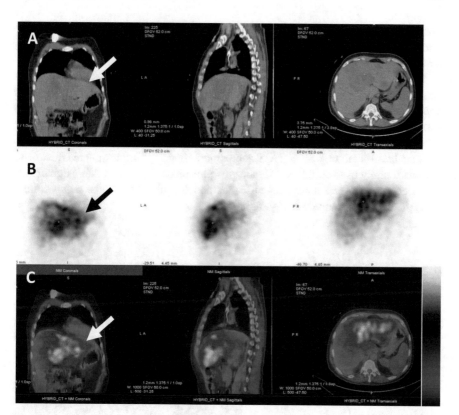

Fig. 4.10 Bremsstrahlung imaging. (**a**) CT acquisition. (**b**) SPECT acquisition of ^{90}Y microspheres treatment. (**c**) Fused SPECT/CT images. Images show a greater average concentration of ^{90}Y in the left lobe (arrows) compared to the right lobe of the liver

laboratory animals use a broader array of isotopes for γ-ray imaging since there are fewer practical constraints, such as those related to radiation exposure.

4.2.4   Bremsstrahlung Sources

Imaging isotopes used in therapy can be useful for patient dosimetry or to track response to treatment. Although some beta-emitters used in radiotherapy may also produce gamma-ray photo peaks useful for imaging (e.g., ^{131}I, ^{177}Lu), others, like ^{90}Y, do not. Given sufficient beta-emission energy (e.g., $E_\beta = 2.28$ MeV maximum for ^{90}Y), the bremsstrahlung radiation produced by the beta-particle interaction with tissue can be used for nuclear medicine imaging. The relatively high activities used in treatment facilitate bremsstrahlung imaging because the efficiency of bremsstrahlung production is low for practical acquisitions at activities more typical of diagnostic imaging. For example, activities for treatment of liver lesions with ^{90}Y microspheres can range up to 3 GBq, whereas a cardiac scan can be performed with under 0.3 GBq ^{99m}Tc sestamibi.

Bremsstrahlung imaging for the clinic can be optimized heuristically or by simulation for a specific tracer. Since this radiation is broad-spectrum, the range of photons accepted for imaging is comparatively broad. For ^{90}Y imaging, ranges from about 55–285 keV have been used although simulation suggests that 60–170 keV provides better image contrast without giving too much image noise [9]. Figure 4.10 shows a bremsstrahlung SPECT/CT acquisition after a ^{90}Y microspheres treatment. Although rarely used currently, ^{32}P chromic phosphate ($E_\beta = 1.71$ MeV maximum) has been used to treat malignant effusions and can also be imaged using similar energy ranges for bremsstrahlung radiation.

4.2.5 Radiotracer Production

The fundamental signals in nuclear medicine arise from nuclear decay. The energy released in decay must have at one time been stored via nuclear reactions. Although there are many means of inducing nuclear reactions, the two workhorses of medical radioisotope production are nuclear reactors and cyclotrons.

Fig. 4.11 A light-water uranium-235 nuclear reactor used for research and for the production of iodine-125 at McMaster University, Canada

Fig. 4.12 ^{99m}Tc generator
(**a**) and schematic (**b**). Saline
solution (blue) preferentially
dissolves the ^{99m}Tc
produced by ^{99}Mo in a
ceramic column (yellow/
red) and removes it as a
pertechnetate solution (red)
to a vacuum vial in a process
called elution. Hash marks
indicate radiation shielding

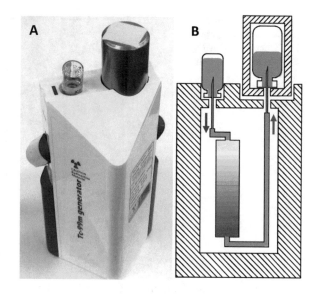

Uranium-235 targets can be placed in nuclear reactors and exposed to the neutron flux there (Fig. 4.11). This induces fission of the uranium nuclei: they split into smaller nuclei. One product is molybdenum-99. The target is removed for chemical processing to preferentially remove the ^{99}Mo which is then fixed in a ceramic matrix and mounted in a portable device called a generator. With a half-life of 65.94 hours, ^{99}Mo is constantly producing ^{99m}Tc, with a branching ratio of 0.876. Flooding the matrix with a saline solution will preferentially remove ^{99m}Tc as ^{99m}Tc-pertechnetate solution while leaving the remaining ^{99}Mo behind. Typically, a vacuum vial attached to one end of the system draws saline attached to the other end of the system through a series of tubes and through the ceramic matrix (Fig. 4.12). This process is called elution. The generators are shipped to radiopharmacies, often within a hospital or nuclear medicine clinic. The ^{99m}Tc-pertechnetate solution may be used as a radiotracer itself (e.g., for thyroid imaging) or may be chemically reacted to form other tracers such as ^{99m}Tc-methy-lene-diphosphonate (MDP) for bone scans. Although the half-life of ^{99m}Tc is relatively short, rendering the radiotracer ineffective within a few hours, the relative long half-life of ^{99}Mo means that more ^{99m}Tc is constantly being generated over a number of working days. This ensures that ^{99m}Tc is available within the radiopharmacy to make radiotracers with sufficient activity for diagnostic imaging.

The production of ^{99m}Tc from ^{99}Mo can be tracked by rate equations:

$$\frac{dN_2}{dt} = k\lambda_1 N_1 - \lambda_2 N_2 \tag{4.6}$$

where N_1 and N_2 are the ^{99}Mo and ^{99m}Tc populations, respectively, with corresponding decay rate constants of λ_1 and λ_2. The branching ratio of ^{99m}Tc from ^{99}Mo is accounted for by $k = 0.876$. It is straightforward to show that the

Fig. 4.13 Theoretical activity curve for a Mo^{99}/Tc^{99m} generator showing Tc^{99m} buildup after four elutions to zero Tc^{99m} activity spaced every 24 hours

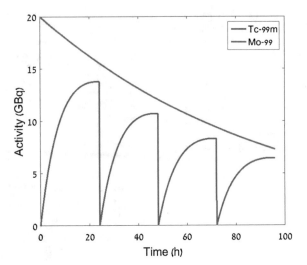

amount of ^{99m}Tc available in terms of activity A (recalling $A = \lambda N$) at some time t after it has all been removed is

$$A_2(t) = kA_1(0) \frac{\lambda_2}{\lambda_2 - \lambda_1} \left(e^{-\lambda_1 t} - e^{-\lambda_2 t} \right) \tag{4.7}$$

where $A_1(0)$ is the initial ^{99}Mo activity immediately after an elution removing all ^{99m}Tc and $A_2(t)$ is the activity of the ^{99m}Tc at time t after elution. Figure 4.13 shows a theoretical plot of these activities for these elutions every 24 hours from a 20 GBq ^{99}Mo generator. In practice, the activity of ^{99m}Tc remaining in the generator is not zero. It is also straightforward to show that the activity of ^{99m}Tc in the generator will maximize at about 1 day after elution.

Devices capable of accelerating protons and nuclei to high energies can impart sufficient energy that these particles can overcome the Coulomb barrier between them and the similarly positively charged target nuclei during collisions. Energy stored in these nuclear reactions is the source of the signal when this method is used to produce diagnostic medical radioisotopes. The unstable radioisotopes constantly decay, releasing gamma rays or other particles ultimately providing high-energy photons for diagnostic imaging.

In cyclotrons, protons, or other positively charged ions, are generated and accelerated across a high voltage potential. The potential is maintained between two "dees" which resemble a large metal hollow disk-shaped cylinder that has been separated along a diameter. Parallel to the axis of this cylinder, which is perpendicular to the flat planes of the dees, a static magnetic field is maintained. Because of the Lorentz force, the ions move in a circular path. The frequency of circulation about this path can be calculated by equating the Lorentz force with the centripetal force exerted on the ion. This is the cyclotron frequency. By reversing the polarity of the high voltage at the cyclotron frequency, the ions are repeatedly accelerated across

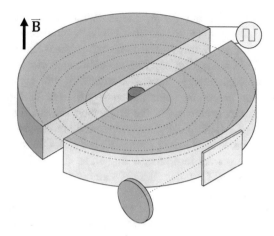

Fig. 4.14 Schematic of cyclotron used in the production of medical radioisotopes. Ions from source (red) are accelerated across a potential difference between the 2 dees (blue) but move in circular paths (dotted line) due to the Lorentz force provided by the static magnetic field ($\overline{B}$). A power supply (yellow) repeatedly switches the potential across the gap causing ion acceleration with each pass. Increasing ion speed increases the radius of circulation until near the edge of a dee, a deflector plate (orange) steers the energetic ion beam to the target (green)

the gap and obtain high energies. Higher energy ions approach the edge of the dees, where they can be steered with electrostatic deflector plates to hit a target (see Fig. 4.14). For example, protons accelerated against an oxygen-18 target (commonly in the form of water) can be used to produce ^{18}F via the $^{18}_{8}O(p^{+}, n)^{18}_{9}F$ reaction, where p^{+} is the proton and n is a neutron. Linear accelerators also use electrostatic fields to accelerate ions to energies sufficient to induce nuclear reactions in targets, and they too can be used to produce certain radiotracer isotopes, although linear accelerator configurations differ from that of a cyclotron.

 The energy used for nuclear medicine imaging is first stored in radioactive materials using devices that can enable nuclear reactions.

4.3   Data Acquisition

Because the high-energy photons emitted in NM come from diffuse distributions of a radiopharmaceutical, they must be collimated before detection in order to determine their ray paths for image production. Usually for single photon emissions, a lead collimator with many small parallel holes absorbs most of the photons approaching the detector, except for those along the paths of the holes. These high-energy photons are not naturally visible, so in NM devices they can be converted into visible light by means of a scintillating crystal. While losing energy in the scintillating crystal, an incoming photon excites atoms, which then emit visible light (scintillate) as the atoms return to the non-excited state. Photomultiplier tubes (PMTs), on the side of the crystal opposite the collimator, convert the visible light generated from the high-energy photons absorbed by the crystal into electrical signals available for processing. For the most part, the NM camera (called a gamma or Anger camera) consists of the collimator, the scintillating crystal, and the PMTs (see Fig. 4.24). Typically, the Anger camera can be rotated about the patient to acquire 3D data used to reconstruct 3D images of the radiopharmaceutical distribution inside the patient in a process known as single photon emission computed tomography (SPECT). Other SPECT camera configurations have become clinically available, such as multiple pinhole collimation and solid-state detection. Electronic collimation can be used in the case of positron-emitting radiotracers for positron emission tomography (PET) as well as in research devices such as Compton cameras [10]. These cases will be described below in further sections.

4.3.1   Collimation

Collimation establishes the γ-ray path. This is needed to get a projection of the signal along a path used for image creation. Without collimation, each point of radiotracer distribution within the patient could potentially cause scintillation in any part of the scintillating crystal and all positional information is lost. Currently, the most commonly applied collimation in gamma cameras is parallel hole collimation, defining ray paths that are all perpendicular to a large planar crystal. In parallel hole collimator design, for the same hole diameter, the longer the bore, the higher the spatial resolution (Fig. 4.15). This is because shorter holes allow more oblique ray paths to pass through the collimator. However, longer bore holes achieve greater spatial resolution by blocking these oblique rays, reducing detector efficiency (sensitivity): fewer γ-rays per decay are detected. Also, because ray paths accepted through the collimator holes include oblique rays, spatial resolution is superior closer to the collimator. Another consideration in collimator design is that the walls of the holes, the septa, must absorb gamma rays striking them. Collimators are therefore made of high-density materials such as lead (Pb). Although lead is economical, it is quite malleable and the collimator must be protected to avoid damage. For medium - and high-energy photons, thicker septa are needed to avoid septal penetration, making the collimator heavier and less sensitive. Septal penetration causes streaking as shown in Fig. 4.15e. Septal penetration gives rise to a six-pointed star artifact because the honeycomb packing of the holes align planes of parallel septa every 60° about any given hole. The figure was made by imaging 364 keV photons from a small ^{131}I source using a parallel hole collimator more suitable for the lower-energy 141 keV photons of ^{99m}Tc.

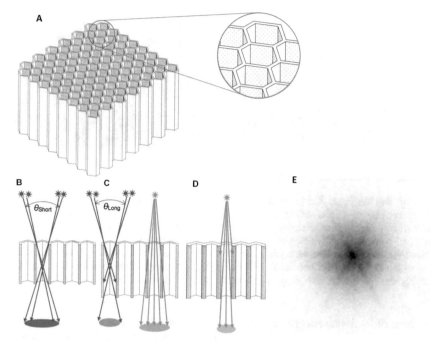

Fig. 4.15 Collimator choices. (**a**) Standard parallel hole collimators have holes arranged in a hexagonal pattern separated by thin lead (Pb) septa. (**b**) A shorter hole length has a wider acceptance angle (θ) allowing more photons to pass, increasing sensitivity. (**c**) For the same hole diameter, the longer the bore, the higher the spatial resolution. Also, spatial resolution is superior closer to the collimator. Higher-energy photons may penetrate thin septa (green arrows). (**d**) Thicker septa are needed to block high- and medium-energy photons, improving spatial resolution, but decreasing sensitivity. (**e**) Septal penetration produces a star artifact around a small source

 Without collimation in nuclear medicine imaging, there is no image. Collimators define the path of the high-energy photons used to create the image.

4.3.2   Scintillation Crystals

High-energy photons (γ-rays) cannot be seen by the human eye nor are they easily imaged. Anger cameras use scintillating crystals to convert γ-rays into visible light. The typical crystal used is sodium-iodide doped with thallium – NaI(Tl). The γ-ray interactions of Compton scattering and the photoelectric effect can produce a cascade of high-energy electrons within the scintillation crystal, leaving vacancies.

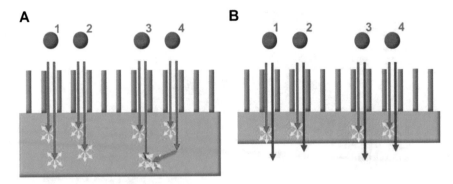

Fig. 4.16 Scintillating crystal (blue) thickness: (**a**) Thick crystals stop a larger percentage of the high-energy photons (green arrows) from the radiotracer sources (purple), producing visible light (yellow), providing greater efficiency and therefore higher sensitivity. Thicker crystals increase the likelihood of scatter within the crystal (orange), reducing spatial resolution by blurring source locations (3 and 4). (**b**) Thinner crystals have less internal scatter, maintaining better resolution (sources 3 and 4 resolved). The likelihood of stopping high-energy photons is less (red arrows), reducing sensitivity. Collimator septa (gray) ensure parallel ray paths of incoming photons

Electrons filling lower energy vacancies within the crystal can produce visible light. The precise details of the process can be found elsewhere [11]. The γ-ray photon produces energetic electron-hole pairs within the crystal which in turn cause much secondary excitation. The electron-hole pairs migrate to "luminescence centers" where they combine and release energy, in part, as visible light. Crystal purity and homogeneity are important as defects can delay or halt this migration by trapping. Scintillating crystals should have a high density and high atomic number to ensure a high number of γ-ray interactions via the photoelectric effect and Compton scattering. Scintillating crystals should be affordable, chemically inert, mechanically strong, and virtually transparent to visible light. Also such crystals should have light output that is proportional to γ-ray energy input, and an output color spectrum that is compatible with PMT input. Although NaI(Tl) crystals are hygroscopic (absorb moisture from the air) and relatively fragile, they are affordable, and established manufacturing techniques can produce large crystals suitable for Anger cameras: up to about 50 cm $\times$ 50 cm and 2.5 cm thick. Thick crystals stop a larger fraction of the high-energy photons providing greater efficiency and higher sensitivity (Fig. 4.16). However, scattering in thick crystals reduces resolution: as with collimator design, there is a trade-off between spatial resolution and sensitivity.

4.3.3   Photomultiplier Tubes

The visible light signal from a scintillating crystal is not strong: typically just several thousand photons for each γ-ray absorbed. In Anger cameras, photomultiplier tubes

Fig. 4.17 Photomultiplier
tube (PMT) schematic.
Several visible photons
(yellow) hit the
photocathode (blue) and
eject a photoelectron (single
orange arrow) which is
accelerated across a voltage
potential (V_C to V_1) to hit a
dynode (green). Several
secondary electrons are
accelerated through the next
potential difference to hit the
next dynode. Each dynode
amplifies the electron
cascade which is collected at
the anode (purple). The total
potential difference (V_C to
V_A) can be about 1 kV, with
each dynode at a higher
potential than the previous
($V_{i+1} > V_i$)

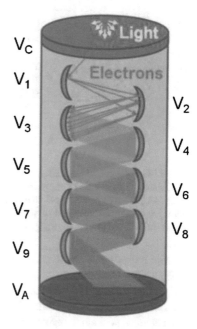

(PMTs) are used to convert the visible light into a useable electronic signal and to
boost this signal before further amplification. A PMT has an entrance window that is
flush either with the scintillating crystal or with a thin light guide between the crystal
and the PMT that mediates between the two. The window end of the PMT has a
photoemissive coating such as cesium antimony (with rubidium or potassium) that
ejects electrons when struck with visible light; therefore, this end is called the
photocathode. It can take up to ten visible light photons to eject one of these
electrons, called a photoelectron. The efficiency of photoelectron production varies
with the visible photon energy for each photocathode, and this spectrum of response
ideally matches the spectrum of the light output of the scintillator. A series of about
ten metal plates, called dynodes, are placed along the tube at progressively higher
voltage biases (Fig. 4.17). A photoelectron is thereby accelerated to the first dynode
where it hits and its kinetic energy is used to eject several secondary electrons. These
are accelerated to the next dynode where each electron ejects several electrons. The
process is repeated dynode to dynode producing a cascade of electrons which are
collected at the anode as an output electronic signal suitable for amplification. The
dynodes must be able to emit secondary electrons easily and a photoemissive coating
is often suitable. The tube is evacuated to keep a clear path for the electron cascade,
but such a design makes it susceptible to variations due external magnetic fields
because of Lorentz forces. The signal should be proportional to the energy of the γ-
ray detected and, ideally, the electronics are fast enough that each event (γ-ray) can
be detected separately. In an Anger camera, many dozen PMTs of about 5 cm in
diameter are typically close packed in a honeycomb arrangement against the back of
the crystal, although variations on this arrangement are common.

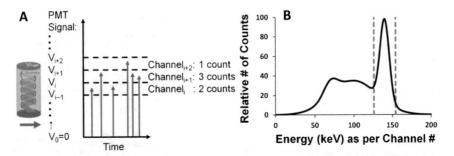

Fig. 4.18 Pulse height analysis. (**a**) A multichannel analyzer (MCA) bins the output pulses of the PMTs according to voltage range into many channels. (**b**) Usually, MCA channels scale linearly with energy (e.g., 1 or 2 keV per channel), such that a histogram of events per channel provides an energy spectrum. Events within a narrow energy window about the energy peak of interest (dotted lines) can be used to produce images with fewer, typically lower energy, scatter events

The total output of the PMTs is summed and ideally is proportional to the total energy deposited in the scintillating crystal for that γ-ray interaction. Each output pulse can be binned into a series of narrow contiguous ranges of pulse heights using a multichannel analyzer (Fig. 4.18a). Each bin, or channel, typically scales linearly with energy, so such a histogram of output pulse heights can be plotted as an energy spectrum (Fig. 4.18b). The spectrum is not monochromatic (i.e., one single energy, like 141 keV for ^{99m}Tc) because a number of processes broaden the energy peak, reducing the energy resolution. Compton scatter within the patient, off the collimator, and within the crystal broadens the peak in the lower energy direction in the spectrum. The scintillation process is stochastic so the number of visible light photons varies, also broadening the peak as will the electronic detection circuits since no design is ideal. Good energy resolution is desirable for distinguishing between two peaks in dual-isotope studies and aids in scatter rejection. The image can be constructed by using only photon events close to the peak of interest (e.g., 140.5 keV $\pm$ 10% for ^{99m}Tc) ensuring that these γ-rays have undergone little scatter. This is called energy windowing.

4.3.4   Positioning Circuits

In the 1950s, Hal Anger invented a gamma camera using the principle that light sharing among large photomultipliers can be weighted to determine the precise location of where the γ-ray struck the scintillating crystal. Light from the scintillating crystal is shared among several photomultipliers as depicted schematically in Fig. 4.19. In this example, the symmetrical nature of the light sharing enables one to intuitively determine the precise location of the photon location (precisely between PMTs 9 and 14 in Fig. 4.19). In practice, positioning circuits or algorithms weight the photomultiplier outputs to determine the position. Anger developed a resistor network and summing electronics which provided for a weighting of the

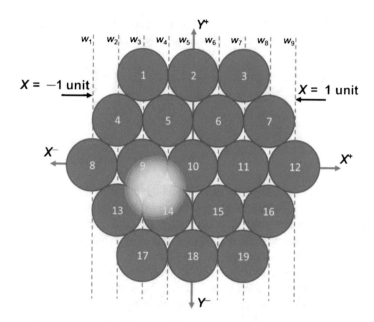

Fig. 4.19 Anger camera principle schematic. Here, the symmetrical nature of the light sharing among PMTs enables one to localize the γ-ray event as occurring precisely between PMTs 9 and 14

inputs from each photomultiplier tube enabling accurate positioning of the location of the gamma-ray scintillation. Currently, this is more typically done digitally, allowing for corrections of differences among PMTs or throughout the crystal. If the weighting of the ith column of n PMT columns is w_i, and the total light output of all the PMTs of that column is I_i, traditionally the summed outputs X^+ and X^- for the lateral direction (x-direction) are formulated as

$$X^+ = \sum_{i=1}^{n} w_i I_i \tag{4.8}$$

$$X^- = \sum_{i=1}^{n} (1 - w_i) I_i \tag{4.9}$$

$$X = \frac{(X^+ - X^-)}{(X^+ + X^-)} \tag{4.10}$$

where $w_i = (i-1)/(n-1)$, with the positions scaling linearly with signal output. The position X is scaled such that the first column is -1 unit and the last column is $+1$ unit laterally from the center position (Fig. 4.19). A similar calculation is done for the rows of PMTs for the axial (y-direction) position Y and summed outputs Y^+ and Y^-. Even with PMTs approximately 5 cm in diameter, gamma-ray localization can typically be determined to within about 5 mm close to the camera surface (close to the collimator). In the example shown, relative PMT outputs for PMTs 10 and

13 might be 1, and for PMTs 9 and 14, they might be 4. Summing left to right along the column (with most terms left out because they are zero):

$$X^+ = 0 \times 0 + \frac{1}{8} \times 1 + \frac{1}{4} \times 4 + \frac{3}{8} \times 4 + \frac{1}{2} \times 4 \qquad (4.11)$$

$$X^- = 1 \times 0 + \frac{7}{8} \times 1 + \frac{3}{4} \times 4 + \frac{5}{8} \times 4 + \frac{1}{2} \times 4 \qquad (4.12)$$

$$X = \frac{(3.125 - 6.875)}{(3.125 + 6.875)} \qquad (4.13)$$

giving $X = -0.375$. Visually, one can see that the center of the light distribution is shifted to the left 37.5% of the distance between the center of the center PMT 10 and the center of the far left PMT 8. Additionally, the total summed output of all the PMTs, traditionally called the Z output, provides a signal proportional to the total γ-ray energy deposited in the crystal for that event.

Plotting the scaled (X, Y) coordinates of each event as (x, y) forms a projection image of the radiotracer distribution, giving thousands or millions of recorded events. The process of localization of the radiotracer within the body is summarized as follows: radiotracer distribution ρ produces a proportional gamma-ray intensity $f(x, y)$, the gamma-ray intensity produces a proportional collimated gamma-ray signal at the crystal S, the crystal produces a proportional optical signal Q, the PMTs produce a proportional electrical signal I, and the position circuit or algorithm produces the scaled localization (x, y). Integration of the events at all positions (pixels) produces the image $\widehat{f}(x, y)$.

In practice, a phantom (a test object) enabling the excitation of the crystal in a precise grid of points can determine if there is any distortion in the (x, y) placement and tables can be constructed for the automatic digital correction of small placement errors for each pixel.

Given a more arbitrary arrangement of the scintillating crystal and PMTs, one could imagine performing a calibration of the weights by sequentially activating the crystal at many (n) known locations $(X_0 = [X_1, X_2, X_3, \ldots, X_n]'$ in the one-dimensional case) and recording the PMT readouts I ($[I]_{ij} = I_{ij}$, where i refers to the PMT readouts for location X_i and j is the jth PMT). Here, the notation A' means the transpose of A. Equations 4.8, 4.9, and 4.10 could be rewritten as $X = IW$, where the individual weights for the PMTs are arranged as a vector $(W = [w_1, w_2, w_3, \ldots, w_n]')$ and X contains the resultant positions of n γ-ray events. By knowing the n locations, X_0, the weights could conceivably be estimated $(\widehat{W})$ by a least squares fit:

$$\widehat{W} = (I'I)^{-1} I' X_0 \qquad (4.14)$$

where $[(I'I)^{-1} I']$ is the pseudoinverse of I. Demonstrating this is left as an exercise for the student.

4.3.5 Solid-State Detectors

It is conceivable to replace the PMTs in an Anger system with silicon photomultipliers or avalanche photodiodes, but the principles discussed above remain the same. The detection system is "solid-state" in that it uses no tubes. Such solid-state detection systems for preclinical scanners were the first to become commercially available (e.g., circa 2006, Flex Triumph LabPET, Gamma Medica-Ideas, Sherbrooke, Quebec, Canada), followed by such solid-state PET detection systems within clinical whole-body PET/MR (e.g., circa 2013, Biograph mMR, Siemens, Erlangen, Germany). The LabPET and mMR both employ APDs (avalanche photodiodes). Currently, four commercially available clinical systems employ SiPMs (SPECT Cardius, Digirad, Poway, USA; PET/CT Discovery MI, GE, Tirat Carmel, Israel; PET/CT Vereos, Philips, Haifa, Israel; Biograph Vision PET/CT, Siemens, Erlangen, Germany). Many of these systems are referred to as "digital" which is a misnomer to the extent that most of the systems are processing analog visible light outputs. Conversely, it can be argued that the traditional Anger detector systems have already been digital in that the electronic pulse from the PMT for each event is digitized after amplification or, at the latest, digitized after time integration. In clinical scanners, image formation, processing, and any needed reconstruction have been digital for decades, already.

A typical avalanche photodiode (APD) is a silicon-based semiconductor with a complex layered doping structure. Generally, visible light can generate electron-hole pairs with the semiconductor that are separated by a voltage bias. The electrons drift to a high electric field region where they are accelerated to produce more carrier electrons via impact ionization, which in turn produce more, amplifying the initial signal. The reverse bias applied within an APD is set to provide a linear response between the photocurrent produced by the visible light and the amplified signal. Since the visible light produced from the scintillating crystal is proportional to the gamma-ray energy absorbed, the APD can provide the energy information needed for an Anger camera. At even higher voltages, the APD can be set to operate in a Geiger mode: a large output signal can be produced from the detection of even a single visible light photon. However, like a Geiger, the output pulse height is constant. In Geiger mode, this is the case regardless of the number of electron-hole pairs produced by the photon and consequently can count individual photons, but not their relative energies. Silicon photomultipliers (SiPMs) commonly consist of a 2D array of thousands of individual APDs operating in Geiger mode. If designed such that the chance of any APD receiving more than 1 photon simultaneously is small, a sum of all the individual APD signals is then proportional to the amount of light produced by the scintillator. SiPMs can provide gains similar to that of PMTs, giving a solid-state solution to the detection of scintillated light.

A newer class of detectors involves the direct conversion of the gamma-ray photon into electron-hole pairs which are then separated by an electric field and the charge collected to provide a signal. Such detectors using high purity silicon or germanium/lithium crystal have been traditionally used in nuclear physics for

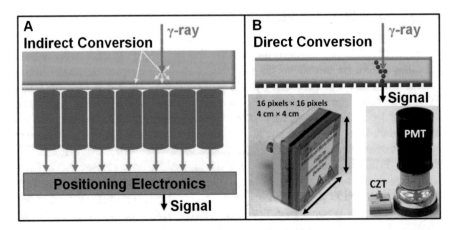

Fig. 4.20 Indirect vs. direct conversion. (**a**) Indirect conversion. The gamma-ray (green arrow) energy is converted to visible light (yellow arrows) within the crystal. The light guide (light yellow) channels this into PMTs (purple) where the energy is converted to an electron beam. Collection at the PMT anode provides a current (orange arrows) for positioning logic to localize the signal. (**b**) Direct conversion. Gamma-ray energy deposited in the semiconductor crystal (e.g., CZT) produces electron–hole pairs with the holes (red circles) collected by a cathode (gray). The pixelated anodes (magenta) collect the electron (blue circles) current virtually at the location of the gamma-ray interaction, directly providing a localized signal. Photo of module is shown in inset left and comparison to PMT size is shown in inset right

gamma-ray detection, but these are usually cooled by liquid nitrogen. In nuclear medicine, commercial scanners employing cadmium zinc telluride (CZT) have been available since circa 2008 since CZT has the advantage of room-temperature or near-room-temperature operation [12]. The first CZT modules employed in such scanners were 4 cm × 4 cm in area and 5 mm thick, with a pixelated charge collection on a 16 × 16 matrix. The compact size aids in the design of organ-specific cameras such as two popular solid-state dedicated cardiac SPECT cameras, D-SPECT (Spectrum Dynamics Medical, Caesarea, Israel), and the Discovery 530c (GE Healthcare, Tirat Carmel, Israel) or even a handheld gamma camera like the Crystal Cam (Crystal Photonics, Berlin, Germany). These modules were also used in the first commercial solid-state gamma camera suitable for whole-body SPECT (Discovery 670CZT, GE Healthcare), and this camera was redesigned with thicker crystals to improve detector efficiency.

Anger cameras use indirect conversion for detection in that the energy of the gamma ray is converted to visible light in the scintillator that is then channeled to PMTs via light guides and amplified to an electronic signal output. Positioning logic provides the localization. Each step in this process potentially degrades the energy and positioning information. In solid-state detection modules, like CZT, the energy of the gamma ray is directly converted to a proportional number of electron–hole pairs within the crystal, which provide the current for signal output. Pixelated charge collection enables a generation of current localized to the gamma-ray interaction, potentially increasing the spatial resolution of cameras made with such modules (Fig. 4.20). In commercial devices, CZT provides better energy resolution than NaI

(Tl) detectors, but periodic incomplete charge collection can underestimate the energy of some photon interactions.

4.3.6   SPECT Rotating Gantry Methods

Traditionally, Anger cameras employed disk-shaped NaI(Tl) crystals about 40 cm in diameter and about 1 cm or more thick. These single detector heads were often mounted on a gimbal enabling planar views to be acquired with the patient lying down, sitting, or standing. These evolved into cameras with rectangular crystals of about the same size or larger with the detector head or heads mounted such that they can rotate about the patient's longitudinal axis. Usually the gantries are circular and dual-headed (Fig. 4.21a), although Philips Healthcare's Precedence retained a gimballed system, Toshiba made a tri-headed camera, and single-headed systems are available. As with x-ray CT, multiple planar images from a camera rotating about the patient can be used to reconstruct a 3D map, but in this case it is a map of radiotracer distribution via single photon emission computed tomography (SPECT). An Anger SPECT acquisition might have 60–120 angular views, requiring 10–30 seconds for each view, giving acquisition times of 5–30 minutes for a dual-head camera. "Whole-body" SPECT covering eyes to thighs axially might require three such acquisitions. Practically, there is a limit to the duration a patient can lie still, so maximum SPECT acquisition times do not extend much beyond 45 minutes.

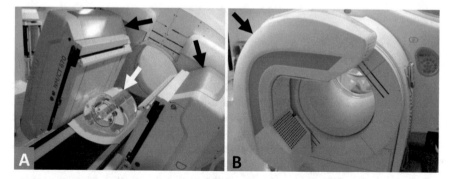

Fig. 4.21 Standard and dedicated cardiac SPECT. (**a**) Dual detector head (arrows) SPECT Anger camera with rotating gantry. A phantom (white arrow) rests on the bed. (**b**) Stationary gantry (arrow) containing 19 pinhole CZT gamma cameras for dedicated cardiac SPECT acquisitions. Note that the SPECT gantry in both cases is coaxial with a CT gantry in the background

4.3.7   SPECT Stationary Gantry Methods

Although computed tomography requires planar views from many angles about the patient, geometries that allow many views to be acquired simultaneously can provide sufficient data for tomographic reconstruction with no need to move or rotate the gantry during acquisition. For example, the D-SPECT dedicated cardiac camera uses nine detectors with an axial stack of four 4 cm CZT modules each arrayed about the patient to provide approximately 180 degrees of coverage centered on the left anterior oblique view of the heart. Experience shows that this is sufficient data for reconstructing myocardial perfusion images. Each detector "wiggles" about its own axis to provide views from additional angles, and the system is designed to focus on the heart after a short prescan. These detectors are mounted in an L-shaped gantry that shows no motion during data acquisition. Similarly, the Discovery 530c (see Fig. 4.21b in SPECT/CT configuration) dedicated cardiac camera (GE Healthcare, Tirat Carmel, Israel) uses a stationary gantry with 19 pinhole CZT gamma cameras arrayed in an L-shape about the patient and pointed at the heart. In both cases, CZT modules facilitate designs in which the detectors are brought very close to the patient, focused at an organ of interest, increasing camera sensitivity. Sensitivity is also increased because many views are acquired simultaneously or near simultaneously. These cameras can shorten the duration of data acquisition for myocardial perfusion imaging by a factor of about 5, compared to conventional methods, or they can be used to reduce the injected dose. Moreover, simultaneous acquisition facilitates dynamic scanning in which time series images can provide additional information, such as estimates of cardiac flow reserve.

4.3.8   Gating

Temporal resolution is a measure of the shortest acquisition duration in which a data set can be acquired. In diagnostic nuclear medicine, the requirements for dynamic imaging are not typically stringent. For example, renal functioning may require a time series of data sets (frames) for 2- to 4-second duration for a total of 1–2 minutes after the injection of ^{99m}Tc-mercaptoacetyltriglycine (^{99m}Tc-MAG3). A greater challenge is the imaging of the beating heart during a myocardial perfusion scan after the injection of ^{99m}Tc-sestamibi or the equivalent. Such a "motion picture" (cine) of the heart can be analyzed for pathological wall motion and ejection fraction (EF). The EF is the percent of the blood pool taken into the left ventricle during diastole that is ejected during systole, and it provides valuable diagnostic information regarding coronary health. Given that a human heart rate could be about 120 beats per minute under stress, making such a cine requires data acquisition durations sufficiently small to capture a heart cycle with a period of about 500 ms. This is typically done with 8–16 frames per cycle, setting the required acquisition times at about 60 ms apart. Clinical practice shows that these durations are too short

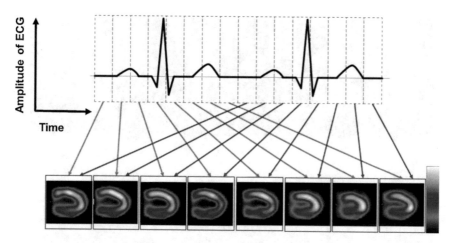

Fig. 4.22 Cardiac gating. During a heartbeat, image acquisition data from, for example, one-eighth of the period determined by an ECG is stored sequentially in one of eight frames (green arrows). During the next cycle the process is repeated (red arrows), matching systole with systole and diastole with diastole. Given hundreds of beats during an acquisition, sufficient data can be acquired in each of the eight frames to provide an image of sufficient quality to analyze ejection fraction and wall motion

to provide sufficient data for a left ventricle image of satisfactory quality to analyze for EF.

However, the gating principle for imaging can be used in such situations where motion is repeated in a fairly constant manner at fairly constant intervals. Electro-cardiography (ECG or EKG) provides the timing of the heart cycle via an electro-cardiograph connected to the patient during the nuclear medicine scan. The average period of the cycle is determined automatically from the ECG and divided into at least eight equal durations called bins or frames (Fig. 4.22). For each cycle, data acquired during the first 1/8 cycle, for example, is added to the first bin or frame, for the second 1/8 cycle it is added to the second frame, and so on. After data from the last duration is added to the last frame, the process is repeated for the next cycle. For a 16-minute total acquisition duration and eight frames, each frame has about 2 minutes worth of data acquisition which is sufficient to generate a useful "snap-shot" of the myocardium during that portion of the cycle. Displaying the eight frames sequentially provides a cine useful for the diagnosis of wall motion and further analysis can provide EF.

In nuclear medicine, respiratory motion is sometimes gated to produce sharper images of lesions in the lungs or near the diaphragm. The signal for the respiratory cycle can be provided by external physical devices that detect chest motion by belts or infrared cameras, although data-driven methods have also been successful since the motion is comparatively large and slow.

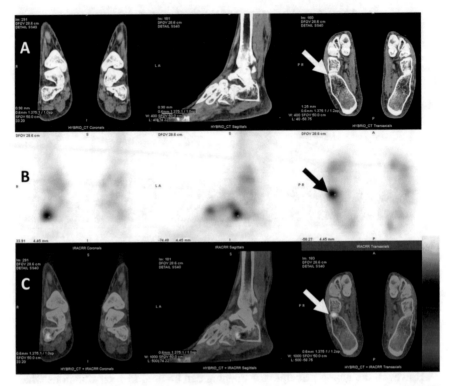

Fig. 4.23 SPECT/CT ^99m^Tc-MDP imaging. (**a**) CT acquisition. (**b**) SPECT. (**c**) Fused SPECT/CT images localizing focal uptake to the calcaneus (arrows)

4.3.9 Hybrid SPECT/CT Systems

Since the turn of the millennium, hybrid scanners have become more prevalent, making it possible to overlay NM images that show physiology onto CT or magnetic resonance (MR) images that show anatomy. This is typically accomplished using united coaxial scanners giving SPECT/CT, PET/CT, or PET/MR devices [13–15]. The CT or MR can provide localization for the radiopharmaceutical uptake. Additionally, the anatomical information can be used to correct image artifacts caused by the attenuation of emission photons in SPECT or PET [16]. In practice, this correction is often less critical in SPECT than in PET. Almost all the currently installed clinical PET devices are hybrid PET/CT scanners.

Because nuclear medicine images are relatively noisy and are limited by relatively poor spatial resolution, small lesions are difficult to detect, and the anatomical location of regions of radiotracer uptake can be difficult to determine in such images. The introduction of dual modality scanners has addressed the latter issue by providing nuclear medicine images showing physiology (functional imaging) registered with the anatomical information from CT. In these scanners, the patient lies still on a bed which is then translated through fixed, mechanically aligned coaxial CT and SPECT (or PET) gantries (Fig. 4.21) so that the data acquired are precisely co-registered in space. The acquisitions are closely sequential to minimize the effects of patient motion. After image reconstruction, the high-resolution anatomical images (from CT) are overlaid with the functional images (from SPECT or PET) to provide precise localization of radiotracer uptake. For example, in a SPECT/CT of feet, the ^{99m}Tc-MDP focal uptake is obvious but hard to localize precisely in the SPECT image (Fig. 4.23b), whereas in the CT image (Fig. 4.23a), the pathology is not obvious to a naïve observer. In the fused SPECT/CT image, the focal uptake is readily localized to the calcaneus (heel bone), even though it is relatively distal. The ability to localize pathology to soft tissue, bone, or a specific organ can greatly impact diagnosis and patient care.

4.4 Image Formation

4.4.1 Planar Imaging

One can imagine a radiation detector enclosed in shielding with a small long hole through the shielding to the detector. Such a device would be very directional: giving a reading only if it happened to be pointed directly at a radiation source so that a ray path was aligned through the hole into the detector. This concept is used for a handheld gamma probe in surgery: the surgeon can locate suspicious lymph nodes that have been previously labeled with typically ^{99m}Tc-sulfur colloid. One can image using such a device to make a plot of radiotracer distribution in a patient. The patient would lie on a table with the probe accepting γ-rays at the perpendicular. A precise x-y Cartesian location could be recorded, and if the duration at each location were identical, a plot of the counts in the detector would give the relative planar distribution of the radiotracer in the patient. This is the principle of a rectilinear scanner, employed before the invention of the Anger camera. The use of the collimator and the Anger principle enables an Anger camera to acquire the relative intensity of all planar points of the distribution simultaneously (Fig. 4.24).

 The high-energy photons in nuclear medicine imaging cannot be seen: their energy must be converted to a visible form by a detection system.

Planar images are usually not quantitative and pathology can become obvious by comparing relative uptake. For example, axially asymmetric uptake in a planar bone scan is not usually expected in a normal subject. Figure 4.24b shows a focal uptake in the right foot in a planar image of a ^{99m}Tc-MDP bone scan, but the precise localization is not apparent in a planar scan. The geometry of a planar scan is similar in concept to a standard radiograph (e.g. a chest x-ray) because of overlap, and

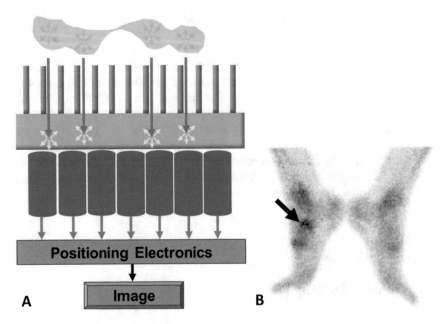

Fig. 4.24 Planar imaging. (**a**) Gamma camera. The radiotracer distribution (light green) emits gamma rays, some (dark green arrows) of which pass through the collimator (gray) to be absorbed in the scintillating crystal (blue). Scintillated visible light (yellow arrows) is converted to an electrical signal in PMTs (purple), and positioning electronics determine the planar x–y location of the events. (**b**) Medial planar image of a ^{99m}Tc-MDP bone scan showing focal uptake in right foot (black arrow)

knowledge of anatomy is needed to understand that the ribs are anterior and posterior to the lungs. An x-ray CT scan is needed to provide 3D data. Similarly, all the events recorded in Fig. 4.24b are overlapped in a 2D image, whereas SPECT and SPECT/CT can provide the 3D localization (Fig. 4.23). As with x-ray radiographs, planar NM images have the advantages of being simpler and faster than tomographic imaging. For example, a whole-body planar bone scan could be performed within 12 minutes, whereas a whole-body SPECT image of the same could take over 45 minutes and is not possible on some cameras that have a limited axial FOV for SPECT.

4.4.2   Dynamic Imaging

Dynamic imaging is typically a time series of planar images designed to track the movement of the tracer within the body. For example, in a three-phase bone scan, the patient is injected with the radiotracer (e.g., ^{99m}Tc-MDP), while the region of interest is within the camera's FOV. Immediately after IV injection, a time series of 15–30 planar acquisitions of a duration 2–4 seconds each can be used to evaluate blood flow. Similar blood flow images can also be used in renal scans, but with other tracers (e.g., ^{99m}Tc-mecaptoacetyltriglycine or MAG3 to assess kidney function and ^{99m}Tc-dimercaptosuccinic acid or DSMA for renal cortical imaging). Dynamic scans of 1–2 minutes for 20–30 minutes total of a MAG3 renal study can be used to measure the washout of the tracer in the kidney and thus determine if a renal obstruction is present. Dynamic scanning is essential in evaluating the arterial blood flow to a transplanted kidney. Another application is the measure of coronary flow reserve (CFR) in myocardial imaging. CFR is the ratio between resting and maximal possible coronary blood flow and can be determined by dynamic SPECT or PET. Using about 60 serial frames of about 5 seconds each, the calculation of CFR is accommodated by stationary gantry emission tomography devices, such as the dedicated CZT cardiac cameras or PET, because projections from multiple angles are acquired simultaneously. Standard SPECT cameras requiring the heads to move during the dynamic acquisition lead to inconsistencies in the acquired data, complicating image reconstruction and analysis.

4.4.3 Image Digitization, Storage, and Display

In nuclear medicine, the standard for image storage has been digital, with film production and storage being phased out around the turn of the millennium. The signal from the PMTs described in Sect. 4.3.3 is digitized with analog to digital convertors, giving a discrete representation of the energy bins for each pulse. The events within each window at a certain x-y Cartesian location are summed for the duration of the scan. Typically, for planar studies, y describes the axial direction and x the lateral. Positioning electronics digitize the location signals by ADC and establish pixilation. The camera designers decide on the effective pixel grid, or matrix, overlaid on the projection data, typically giving the user a choice of 64×64, 128×128, 256×256, 512×512, and 1024×1024 matrix sizes, although other choices are offered with different cameras. Whole-body planar images can be acquired on most cameras by translating the patient in the axial direction slowly (e.g., 10–14 cm/minute) between two detector heads to give anterior and posterior views. By synchronizing bed position with the effective pixel position for the detectors, events can be binned into the proper location. Such a planar scan is typically up to 256×1024 pixels with two views, like an anterior and posterior view. Each pixel might have up to 32,767 gray levels (dynamic range of 16 bits, signed), although cameras vary. Consequently, NM data sets have traditionally not been large in the amount of information stored: under 1 megabyte (MB) for a planar data set. Emission tomographic data is often organized as several hundred transverse images or "slices" each with a matrix size as mentioned above. In tomographic data, coordinates x and y usually define the transverse plane and z defines the axial direction. A SPECT MPI data set might have 64 slices with each image having a 64×64 matrix size, requiring about 0.3 MB. Alternatively, a whole-body SPECT requiring perhaps five bed positions to cover the patient's axial length could give up

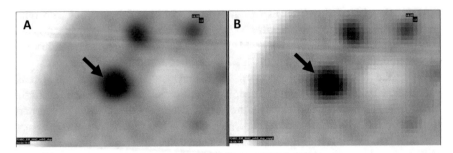

Fig. 4.25 Image digitization and display. ^{123}I-MIBG SPECT image of a NEMA body phantom. (**a**) As displayed, interpolated pixels on a fine mesh render the edges of this 37-mm-diameter hot sphere smooth (arrow). (**b**) The true underlying image matrix has a courser matrix with a 4.42 mm pixel width, and edges are visibly stepped (arrow)

to $5 \times 256 = 1280$ slices with each image having a 256×256 matrix size, giving >80 MB in data (5×256^3). Regardless, these memory requirements are generally smaller than those of MRI or CT imaging covering the equivalent anatomy since MRI and CT typically have better resolution using small pixel sizes and therefore larger matrix sizes (a finer "mesh") to represent them. Also, NM data is photon poor compared to x-ray radiographs or CT, so if the pixel size is too small, there are an insufficient number of events per pixel, giving poor count statistics and too much image noise.

Because of the relative course mesh, NM images are rarely displayed according to the discrete pixel representation of the image matrix. It is more common to interpolate these pixels to a finer matrix suitable for the screen used for viewing. This has the effect of rounding out the visible "steps" due to pixilation within edges or curved features of the object being imaged (Fig. 4.25). Most users prefer this interpolation, but when measuring relative values within the image, it is useful to understand the underlying course matrix structure. For example, one can measure the intensity across a gradient without recording a change because a visual gradient exists due to interpolation, but in the underlying data the measurement is made within one pixel. If display and measurement software return the interpolated values, it is wiser to select a software mode in which the actual pixel values are used for medical reporting.

Nuclear medicine images, like other medical images, are normally stored or transferred in formats standard for clinical images. These formats have a section of the image file called a "header" that contains clinically relevant information such as the patient's name, identification number, date of birth, and other identifying characteristics. For NM images, the header also typically contains the pixel size, pixel scaling, start time of the scan, duration of the scan, and slice thickness for emission tomography. For absolute quantitative SPECT (and PET), additional relevant information is included, such as activity of injected dose, time of injection, dose remaining in the syringe after injection, half-life of the radiotracer used, camera sensitivity, patient weight, and patient height. For planar scans, the camera sensitivity is taken from a calibration measurement that relates the number of events detected by the camera to the radiotracer activity within the FOV of the camera (e.g., in kcounts/s per MBq). If available, this information can then be read from the header and used to relate the intensity of the image within an ROI to an absolute radiotracer concentration. Common formats used in NM or PET are DICOM (digital imaging and communications in medicine) and ECAT (emission computed aided tomography).

4.4.4 SPECT Image Reconstruction

Raw data for SPECT is typically organized as sinograms. This organization comes about quite naturally as the detector heads rotate about the patient. This is illustrated in Fig. 4.26. A sinogram $P(\theta, t)$ is a plot of pixel intensities (counts) along the lateral direction t of the camera head when it is at an angle θ on the gantry. Since the total count in each pixel is ideally proportional to the sum of the radiotracer distribution along the ray path hitting the detector at that location, the sinogram or set of sinograms is sometimes called the projection data. If an $n \times n$ matrix is used for acquisition, the direction t will have n discrete counts, and there will be a total of n sinograms: one for each slice in the axial direction. Each line of the sinogram represents a viewing angle θ relative to some fixed direction. In Fig. 4.26, $\theta = 0$ corresponds to the horizontal or x-direction. The series of angles θ will have m discrete components where $m = 180°/\Delta\theta$ or $m = 360°/\Delta\theta$ depending on whether the sinogram spans half a rotation or a full rotation, respectively. Here $\Delta\theta$ is the step between stops in the camera detector head rotation about the patient and is typically between $2°$ and $6°$. The data in Fig. 4.26 is based on a dual-headed camera with the detectors set at $90°$ to each other (L-mode) stopping every $6°$ for 15 stops, giving a total of 30 64×64 planar projections spanning $180°$. This standard acquisition for MPI spans from the patient's right anterior oblique angle to left posterior oblique angle. Experience shows that for heart scans, fewer photons exit the back right-hand side of the patient, because many are absorbed by the bones (such as the spine) and dense tissue. Therefore, only half a sinogram is typically acquired for MPI, although a few sites scan the whole $360°$. Consequently, the planar images in Fig. 4.26 are 64×64 and the sinogram is 30×64, and there would be 64 sinograms: one for each slice. For comparison, a bone SPECT of a hip might have 128 sinograms of 60 angles and 128 pixels per plot across the detector, spanning $360°$.

Each sinogram is reconstructed by methods described in Chap. 2. Iterative reconstructive techniques like OSEM (Sect. 2.3.3.3) were used in NM before CT because the data sets are noisier (photon poor). Traditional FBP (Sect. 2.2.8) does not handle image noise well because the ramp filter required emphasizes the high spatial frequency where NM noise resides. Rolling off the ramp filter at high frequencies, by using a Hann filter on the ramp, for example, can reduce this problem. Pre-filtering the planar data in image space to reduce statistical noise is another approach. Iterative techniques provide the opportunity to employ

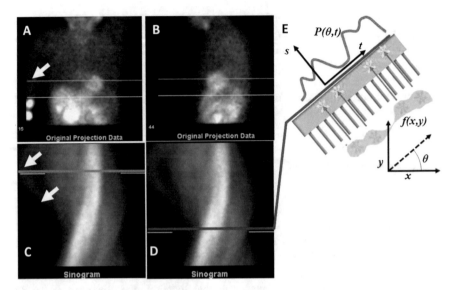

Fig. 4.26 SPECT sinograms of ^{99m}Tc-MIBI MPI. (**a**) Anterior planar image on a 64 × 64 matrix – 1 of 60 views spanning 180° around and 64 slices axially. Fine green line marks a slice through the heart and arrow marks the right arm. (**b**) Lateral planar view. (**c**) Sinogram for slice marked in (**a**). Right arm traces ½ a sine wave (arrows). Orange line maps the same count intensity as the green line in (**a**). (**d**) Same sinogram with the red line marking the count intensity plot of the green line in (**b**). (**e**) For one slice, one line of the sinogram $P(\theta, t)$ maps the count intensity (dark green line) at camera angle θ, along lateral pixel direction t, for the 2D radiotracer distribution of that slice, $f(x, y)$

regularization (smoothing) as part of the iterative process. Similarly, corrections due to inconsistent data sets can be introduced in the iterations. Common corrections employed account for γ-ray photon attenuation and scatter. Iterative reconstruction algorithms also lend themselves to detector response modeling within the system matrix (Sect. 2.2.6), leading to superior image quality. A schematic representation of iterative reconstruction in SPECT is given in Fig. 4.27, which is a specific example of a similar schematic given in Fig. 2.23.

4.4.5 Attenuation Correction

During the emission process, photons traveling along a ray path a have certain probability of being absorbed along that path as described in Chap. 3 and Sect. 4.1.2. For a given voxel, the intensity I_0 along a ray path into the detector is reduced to I as described by Eq. 4.3. The attenuation is largely dependent on the electron density of the material penetrated, as dense material is more likely to reduce beam intensity via Compton scattering and the photoelectric effect, and is characterized by the attenuation coefficient μ (Eq. 4.3). Attenuation artifacts appear as a relative reduction in the radiotracer intensity compared to the true distribution. In planar

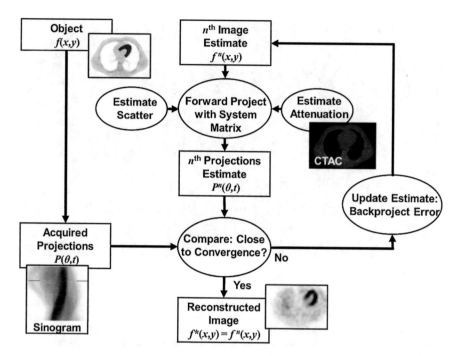

Fig. 4.27 Iterative SPECT reconstruction. Each slice of true radiotracer distribution $f(x, y)$ (simulated object) is represented by projection data (sinogram) within the acquisition system, $P(\theta, t)$. With iterative reconstruction, an image estimate $f^n(x, y)$ is forward projected using a system matrix to model the acquisition system to produce an estimate of the projections $P^n(\theta, t)$ for that guess. During forward projection, the attenuation and scatter of each ray source can be modelled for correction. Attenuation estimates are often based on CT (CT attenuation correction, CTAC). The estimated projection data is compared to the true projection data and if they are close enough in value, the image estimate is deemed to be a sufficient reconstruction $f^*(x, y)$ of the object $f(x, y)$. If not, the error is back projected to update and correct the previous estimate, and the cycle is repeated

imaging, the effects of attenuation are somewhat mitigated by typically acquiring two opposing views simultaneously, such as anterior and posterior views in a planar bone scan: the sternum is prominent in the anterior view and the spine is prominent in the anterior view. A geometric mean of these two views is sometimes used to reduce the attenuation effects in the relative uptake presented by a planar image. In SPECT imaging, attenuation is often ignored in the image reconstruction, especially if the organ is near the body's surface and attenuation artifacts are well understood. This is the case for MPI, with clinical images commonly read without attenuation correction (AC). However, the American Society of Nuclear Cardiology recommends using AC methods, if available, to increase the confidence of the reading. Figure 4.28 shows an attenuation artifact caused by photon absorption in the breast tissue that is then corrected using anatomical information from a spatially registered CT.

As mentioned in Sect. 4.4.4, attenuation correction in modern clinical scanners is typically done as part of iterative reconstruction. In this process, the image is

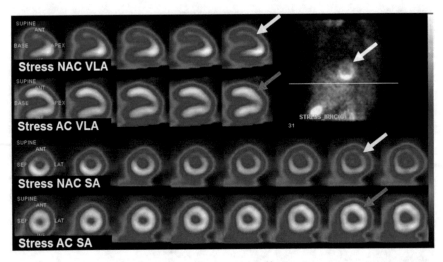

Fig. 4.28 SPECT MPI breast attenuation artifact. In the ^{99m}Tc-mibi projection data shown, the "shadow" presented by photon attenuation of the breast overhangs the anterior myocardial wall (white arrow), reducing the apparent radiotracer intensity. In the reconstructed vertical long-axis (VLA) and short-axis (SA) images without attenuation correction (NAC), there is an artifactual decrease in apparent uptake in the anterior wall (yellow arrows). Attenuation correction (AC) using anatomical information from a spatially registered CT restores this uptake to normal intensity (red arrows)

estimated, perhaps by FBP. In the iterative model, as the image is forward projected into the detectors by means of the system matrix, the path of the ray from each pixel is calculated. This path is compared to the attenuation map. The attenuation map contains an assigned attenuation coefficient for each pixel. The number of counts from each pixel being projected is reduced by the amount of attenuation given along each path (Fig. 4.29). The estimated attenuated sinogram from these projections is compared to the true sinogram (which has been attenuated by the patient). The error in the sinogram comparison is back projected onto the estimated image (using the system matrix). This gives a new estimated image and the process can be repeated (iterated).

An attenuation map containing the attenuation coefficients of each pixel in a tomographic slice can be constructed a number of ways. Initially, for brain SPECT, the outlines of the head could be estimated from non-attenuation corrected (NAC) images and a single value of an attenuation coefficient assigned to the interior. This value would be close to the attenuation coefficient of water for photons of the isotope being used. This method does not work well in the thorax where there are sharp differences in attenuation along most path lengths since the lungs provide little attenuation and the soft tissue and bone provide more, in a non-homogeneous arrangement. An improved solution was to use a long-lived isotope, like gadolinium-153 (^{153}Gd), to provide gamma photons for a transmission scan through the patient. The source was rotated about the patient and provided a measure of attenuation for each pixel after tomographic reconstruction. The map needed to be

Fig. 4.29 Attenuation estimation of the forward projection $P(\theta, t)$ of an image f. There are $N = n \times n$ pixels of width Δx to forward project at each discrete angle θ, into detectors at t with a physical detector bin width of Δt. For any pixel f_i, the attenuation along the path length l to the edge of the patient (object) needs to be estimated and the signal from f_i be reduced by exp $(-\mu l)$ where μ is the average attenuation coefficient along that path length

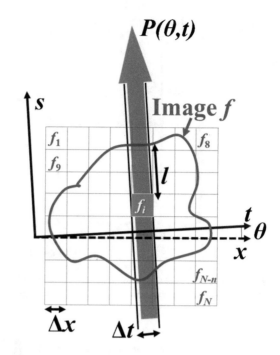

scaled since, for example, the 141 keV photons of ^{99m}Tc see less attenuation than the approximately 100 keV photons from ^{153}Gd. Additionally, such maps are time-consuming and count poor, so the attenuation correction process adds noise to the final image. The advent of SPECT/CT hybrid imagining provided a ready solution in that CTs naturally give an attenuation map that is not noisy, has high resolution, and is acquired significantly faster than the SPECT itself. Again, because of the difference in the average energy of the CT imaging photons (perhaps 80 keV) compared to that for γ-rays used for SPECT (e.g., 141 keV for ^{99m}Tc), the map needs to be appropriately scaled (Fig. 4.30). In practice, look-up tables converting CT Hounsfield units to attenuation coefficients can be constructed for γ-ray energies that might be used in SPECT. Additionally, the attenuation map is also typically blurred (filtered) to match the spatial resolution of the SPECT to avoid introducing artifacts in the AC SPECT image from the sharper features of the original CT.

4.4.6 Scatter Correction

SPECT image reconstruction has an assumed model of the acquisition process. For example, simple FBP reconstruction is based on the forward projection of the radiotracer emission along parallel ray paths into detectors that are arrayed or moved around the patient. In Sect. 4.4.5, there is a description of how a model of attenuation can be added to the forward projection process during iterative

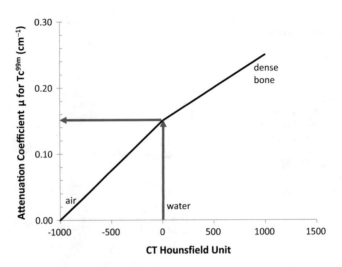

Fig. 4.30 CT attenuation correction (CTAC) for ^{99m}Tc SPECT. The CT Hounsfield unit (HU) measurement is scaled to an attenuation coefficient using a bilinear mapping

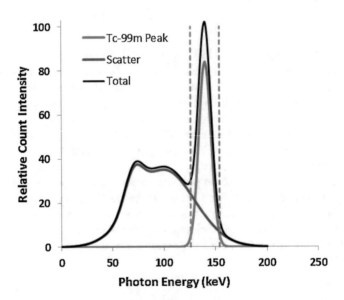

Fig. 4.31 The total energy spectrum for a SPECT acquisition (black) consists of "true" counts (blue) and scattered counts (red) from photons which have lost energy during scattering. Limits on the energy window accepted (dotted lines) reject most of the scatter on the lower energy end of the spectrum. Scatter correction aims to estimate and account for the scattered counts that remain within the peak energy window (here 140.5 keV ± 10%)

reconstruction. Essentially, such models try to emulate the physical truth of the emission/detection process so that one can iteratively reduce the difference between what projections our estimate of the radiotracer distribution would give and actual projections from the true acquisition from the true radiotracer distribution. Scatter complicates that model.

Instead of the assumed straight ray path, a photon may be scattered within the patient, in the collimator, or in the scintillating crystal, for standard Anger cameras. With Compton scattering, since the energy of the scattered photon decreases monotonically with the angle of deviation from the original ray path, photons lose more energy the more they are scattered. For monochromatic gamma radiation, the sharp peak of the emitted photon energy spectrum will be blurred to lower energies as photons are scattered (Fig. 4.31). Setting an energy window such that these lower energies are rejected from counts used for image formation greatly decreases the amount of scatter in the image. Since scattered photons are not following straight ray paths, this greatly decreases the blur in the image. Unfortunately, because of the limited energy resolution of clinical Anger cameras (about 10–15% FWHM), one cannot make the energy window too narrow, or the image becomes count deficient as unscattered photons are rejected too. Therefore, there are typically scattered photon counts within a reasonable energy window (e.g., 140.5 keV $\pm$ 10% for ^{99m}Tc). The scatter can be estimated by setting a narrow contiguous energy window just below the peak energy window in a method called dual energy window (DEW) correction. In triple energy window (TEW) correction, a similar window is set just above the peak energy window, and an average of the lower and upper energy window counts gives an estimate of the amount of scatter within the peak energy window. Of course, the number of counts has to be scaled according to the energy widths of each window since a narrower window will give proportionally fewer counts. The simplest method of scatter correction is to subtract the scatter estimate from the projection data and then reconstruct (Fig. 4.32). Because of sparse data, this can introduce noise and negative counts which must be dealt with. Alternatively, the scatter estimate can be added to the forward projection during iterative reconstruction since the projection data is assumed to consist of "true" counts and "scattered" counts (Fig. 4.27). Further improvement can be achieved by using the anatomical information, as is often available from CT, to model the scatter for photons generated from any pixel into any detector position, similar to what is done for attenuation correction. One manufacturer (Philips, Healthcare, Amsterdam) employs such modeling for SPECT reconstruction using a technique called effective-source scatter estimate (ESSE). For PET reconstruction, such modeling is more common, tracking up to three scattering events per photon in some systems.

4.4.7 Resolution Recovery

As is evident from the geometry shown in Fig. 4.15, the spatial resolution provided by a parallel hole collimator degrades for sources more distant from the collimator surface. The figure exaggerates the effect because the hole width to length is not to

Fig. 4.32 Simple scatter
correction by counts
subtraction. In triple energy
window scatter correction
(TEW), a weighted average
of the scatter estimate counts
(blue) is removed from the
counts within the peak (red).
(**a**) Scatter is estimated from
adjacent energy windows on
either side of the peak. (**b**)
The estimated scatter counts
within the peaks can be
subtracted. (**c**) There are
fewer events recorded in the
cold sphere of the scatter
corrected phantom image,
improving contrast

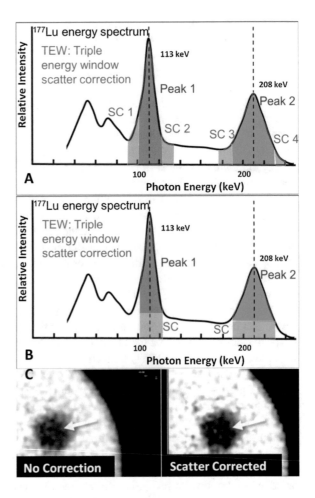

scale. Typically, parallel hole collimator designs might have a ratio of 20:1–50:1 for
the collimator hole length compared to the collimator hole diameter. This greatly
narrows the acceptance angle of ray paths into any particular hole, providing better
spatial resolution than might be indicated by the schematic in Fig. 4.15. Regardless,
the spatial resolution of a parallel hole collimated Anger camera might vary from
5 mm FWHM at the detector head surface to 15 mm FWHM at 25 cm away from the
surface. Effectively, the ray paths are not parallel beams aligned into each hole, but
are cones of acceptance angles pointed into each hole. Reconstruction by FBP
assumes parallel beam paths, but iterative construction affords the opportunity to
introduce more accurate collimator response. Since the system matrix describes the
physical forward projection process, these wider acceptance angles can be modeled
within the forward projection step of iterative reconstruction (Fig. 4.27). Effectively,

a point source far from the collimator will have ray paths that enter more holes than the same source close to the collimator: the source far away is blurred. Some of the first commercial iterative reconstruction software that modeled this used the term "wide-beam reconstruction" (UltraSPECT, Ra'anana, Israel), and the terms "resolution recovery" (especially for SPECT) and "point spread function modeling" (especially for PET) are also used. Resolution recovery has enabled clinics to reconstruct to the same image quality with half the number of counts compared to reconstruction methods without. Therefore, protocols using resolution recovery for SPECT reconstruction can reduce the acquisition duration or the injected dose by half.

4.4.8 Post-Processing

Traditionally, planar NM images are displayed with little or no post-processing. Some systems may display interpolated pixels as mentioned in Sect. 4.4.3. Nuclear medicine physicians, after much experience, use the noise in the image to intuitively estimate count density and the lack of post-processing image filtering retains the spatial resolution of the camera to aid in the detection of small features indicating pathology. The projections used for emission tomography reconstruction, like for SPECT, are acquired usually with much shorter durations than planar studies and therefore suffer from noise. As described in Chap. 2, FBP reconstruction of tomographic images from projections accentuates this image noise due to the ramp filter employed in Fourier space. Therefore, SPECT images have usually been filtered as a post-processing step, even after most clinics began to use iterative reconstruction techniques. This filtering suppresses noise by smoothing the pixel-to-pixel differences in the image, usually by a weighted averaging process. Since the emission/ detection processes is stochastic, following Poisson statistics, image noise in nuclear medicine is typically grainy with some large random differences between adjacent pixels that are expected to have similar radiotracer uptake. Some of this noise can be regularized during construction, such as using Bayesian reconstruction techniques (Sect. 2.3.3.4) that have parameters to penalize the variation between adjacent pixels during reconstruction or impose some filtering during reconstruction. For example, image reconstruction for the dedicated cardiac camera depicted in Fig. 4.21 uses such a penalized likelihood reconstruction algorithm. Regardless, tomographic images are usually smoothed after reconstruction. The degree of filtering is usually decided in consultation with nuclear medicine physicians. A good starting point is to attempt to smooth features that are smaller than the spatial resolution of the camera system and adjust from there. Figure 4.33 shows that without image smoothing for MPI ^{99m}Tc-mibi SPECT, there appears to be sharp variations of uptake in the myocardium which is artefactual and due to noise, and these are removed with image filtering.

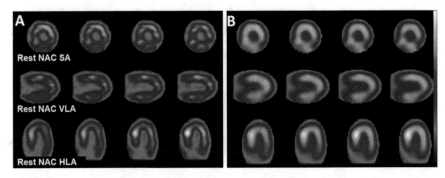

Fig. 4.33 Post-processing: image filtering for MPI ^{99m}Tc-mibi SPECT at rest. (**a**) These non-attenuation corrected (NAC) short-axis (SA), vertical long-axis (VLA), and horizontal long-axis (HLA) views of the myocardium have not been filtered (smoothed), showing apparent non-homogeneous uptake due to image noise. (**b**) The same data and reconstruction has been smoothed with an 8.5 mm FWHM Gaussian filter, showing more homogeneous uptake compared to the unfiltered images

Other post-processing actions, usually provided as software tools, include automatic scaling for absolute radiotracer concentration measurements. A sensitivity calibration is performed with a phantom to relate the true concentration of a radiotracer within the FOV to the counts per pixel or counts per voxel in the displayed images of a known acquisition duration. This sensitivity calibration (e.g., counts per second per Bq) usually varies between cameras and varies with time, so it is usually determined regularly (e.g., quarterly) if absolute quantitation is needed. Additionally, interest in quantitative or semi-quantitative measurement usually focuses on an ROI that might be a lesion or an entire organ. Consequently, software segmentation tools are used as a post-processing step to delineate the ROI before measurement of the radiotracer concentration within. With regard to MPI, tools have existed for over a decade to segment and find the contours of the left ventricle in a semi-automatic manner. By performing this on the stages of a gated cardiac study, the end systolic and end diastolic volumes can be estimated and an ejection fraction calculated. This aids in the diagnosis of abnormal heart function. Post-processing might also involve generating time-activity curves for ROIs from dynamic studies, dosimetry calculations, and other reasons.

4.4.9   Hybrid Image Fusion

Hybrid scanning lends itself to displaying CT and NM separately but on the same screen or distributed over two screens. There are usually software tools to facilitate triangulation onto ROIs in coronal, sagittal, and transverse views, sometimes with options for oblique viewing. Fused SPECT/CT images typically overlay artificially colored SPECT images on top of grayscale CT images. This can be done, for example, if the dynamic range of each data set is reduced by half, they are aligned and interpolated to the same display pixel size, and then the two images are added, taking care with the appropriate color representation. A color map is the assignment of which color to which grayscale level, the simplest being assigning white to the highest value pixel and black to the lowest or the reverse. Other examples of color map graduation are given in Figs. 4.10, 4.22, and 4.23 as bars of color at the edge of the images.

4.4.10   Quality Control

Regular daily quality control (QC) of an NM camera includes physical inspection and any start or restart of the controlling software. Generally, daily QC should be performed about an hour before the start of the clinical day to enable basic tests to be done and to troubleshoot problems that do not require intensive service. An early start on QC inspections can also determine if a camera cannot be used clinically and this may mean rescheduling patients. Since many radiotracers require uptake time, it is important to determine the camera's readiness to avoid unnecessarily injecting patients who might have to be rescheduled. This avoids unnecessary radiation exposure to patients and staff. QC schedules can be established by national regulation or from recommendations by professional organizations (e.g., American College of Radiology, ACR) and international bodies (e.g., International Atomic Energy Agency, IAEA) [17]. Tests to characterize camera performance, often used for acceptance testing of a new camera, are based on protocols to which manufacturers agree (e.g., National Electrical Manufacturer Association, NEMA).

Primarily, imaging quality tests are based on test objects called phantoms. A known configuration of radiotracer is placed in the FOV of the camera, and the images after acquisition can be compared to a known expectation. As test objects for imaging, phantoms have two competing characteristics. First, phantoms should have a known, measurable configuration against which the image can be compared. However, phantom acquisitions should emulate typical imaging conditions, including effects of anatomy. The first drives the phantom design to be simple. For example, daily tests of an Anger camera include a uniform flood source to check detector uniformity: the planar images should look homogenous. These sources are usually flat sheets (Fig. 4.34a) of a longer-lived isotope (typically ^{57}Co). However, no patient is thin and flat: it does not give a volumetric uniformity. Therefore, the second drives the phantom to be complicated. For example, the Hoffman brain phantom (Fig. 4.34d) mimics anatomy, but it is hard to measure quantifiable features because of its more complicated structure. Consequently, phantom design often attempts to balance these two goals. The Jaszczak phantom used to check SPECT image quality is a cylindrical phantom of about 5 L that contains cold features such as spheres and rods. The phantom is mostly water that can be made to generate a signal by an injection of, for example, 600 MBq of ^{99m}Tc-pertechnetate. Because of the known geometry, simple calculations like contrast can be made, and an estimate

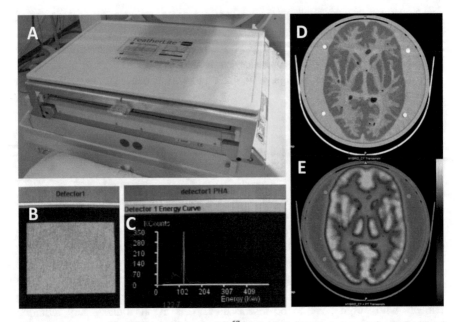

Fig. 4.34 Phantoms: test objects. (**a**) Daily QC [57]Co flood phantom on detector head. (**b**) Flood phantom image. (**c**) Flood phantom energy spectrum. (**d**) Hoffman brain phantom CT. (**e**) Hoffman brain phantom fused image display with the emission tomography image (color) overlaid onto CT (grayscale)

of the effective spatial resolution can be made by noting which size rods can be resolved (Fig. 4.35).

Measurable quantities on the phantom images are usually compared to standards to determine if the camera is operable. For example, the daily QC flood test for an Anger camera from one manufacturer checks the following:

- Total detector counts of 4000 kcounts.
- No default visual problems in the uniformity.
- Background noise in the absence of flood source is less than 26 kcounts/minute.
- Energy peak for [57] Co is 122 keV to within 3 keV.
- Energy resolution is within 12.0% FWHM.
- Uniformity (pixel variation) is within 5.5%.

Should any specifications be outside the norm, the camera is typically removed from clinical use until acceptable performance can be restored. Practically, it is usually worthwhile to rerun a test since erroneous results might be caused by, for example, the misplacement of the phantom or a new very active phantom that might require special placement.

There are a variety of test objects used for daily, quarterly, or yearly QC. The spectrum of their intricacy, according to the competing principles of phantom design mentioned above, ranges from a point source to estimate spatial resolution to

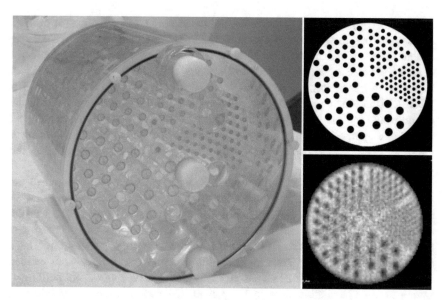

Fig. 4.35 Image quality Jaszczak phantom. (**left**) Solid acrylic rods and spheres are surrounded by ~5 L of radiotracer solution. (**upper right**) Ideal transaxial image through the acrylic rods of 9.5–31.8 mm in diameter. (**lower right**) Reconstructed ^{99m}Tc SPECT image of the same under clinical conditions

anthropomorphic (human-like) phantoms with separate compartments mimicking organs of varying radiotracer concentrations which are used to evaluate dosimetry software. Additionally, custom (bespoke) phantoms are regularly designed and built for research purposes. Generally, the goal of such tests is to optimize scanners and their protocols to provide consistent diagnostic quality images to minimize acquisition time and ionizing radiation dose to the patient.

4.5 Clinical Example

Myocardial perfusion imaging (MPI) SPECT has been the mainstay of nuclear cardiology for a couple of decades, providing a means for assessing infarct or perfusion defects of the myocardium for the left ventricle. Radionuclide MPI aids in the assessment of obstructive coronary artery disease (CAD) in terms of whether or not it is present and its location and severity if it exists. The left ventricle is responsible for pumping oxygenated blood to the whole body, as opposed to the right ventricle which supplies venous blood to the lungs for oxygenation. Given the closeness of the lungs to the heart and their pliable, spongy nature, it is not difficult to reason that the work done by the left ventricle is substantially larger than the right. Consequently, pathology in the left ventricle can be felt as acute pain during exertion. Figure 4.36 shows an MPI SPECT of a healthy heart compared to one with a myocardial infarction, as evident by a drastically reduced ^{99m}Tc-mibi uptake in the apex (tip) of the left ventricle toward the septum (wall separating it from the right ventricle). In this case, if tissue is viable but not perfused, the patient may benefit from surgical intervention by means of inserting a stent into the left anterior descending coronary artery, thereby opening up the vessel's lumen allowing blood to perfuse to this region. Viability can be tested with thalium-201 (^{201}Tl) SPECT or ^{18}F-FDG (fluorodeoxyglucose) PET imaging. Uptake of either radiotracer in this region would indicate viability. In the former, ^{201}Tl acts as a potassium analog and myocardial uptake (1 day post-injection) indicates that those cells are still alive. In the latter, FDG is a glucose analog showing sugar metabolism at regions of uptake, again indicating viability.

In standard MPI, a SPECT performed on a resting heart is compared to one performed after the heart has been stressed. The rest and stress scans may be performed on consecutive days, but it is not uncommon to perform them on the same day. In the latter case, the second scan is performed with an IV injection of three times that of the first scan (e.g. 900 MBq vs. 300 MBq of ^{99m}Tc-mibi). This is so that counts due to the second injection dominate the image. Some sites perform the stress study first, and if it is assessed as normal, the rest study is not performed, saving time and radiation dose to the patient. Otherwise, a rest study follows a minimum of 3 hours after the stress study. Same-day rest/stress protocols are common. An approximately 300 MBq bolus of ^{99m}Tc-mibi is injected and scanning performed after about 45 minutes. The stress study can be performed immediately afterward. The patient is exercised until the heart is stressed or pharmacological stress is applied (e.g. adenosine). About 900 MBq of the radiotracer is injected at peak exercise. The patient is then imaged at least 15 minutes after exercise or 30–45 minutes after pharmacological stress. The difference in the relative radiotracer

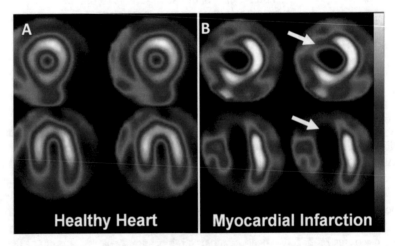

Fig. 4.36 SPECT MPI. (**a**) In a healthy heart the myocardium is well perfused as evident by the relatively homogeneous distribution of the ^{99m}Tc-mibi radiotracer throughout the muscle in the short-axis (SA) view at the top and the horizontal long-axis (HLA) view at the bottom. (**b**) In this infarcted heart, there is a reduction in the radiotracer uptake in the apical-septal region (arrows)

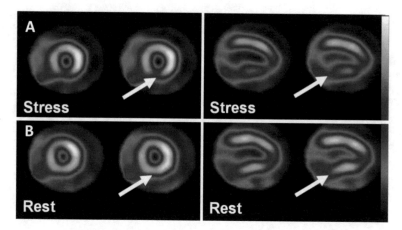

Fig. 4.37 SPECT MPI rest/stress study. (**a**) During stress, the ^{99m}Tc-mibi has a reduction in uptake in the anterior wall of the myocardium (arrows) in the SA views on the left and the vertical long-axis (VLA) views on the right. (**b**) During rest, the radiotracer is well perfused throughout the anterior wall (arrows)

uptake between stress and rest images can indicate CAD (see Fig. 4.37). In obstructive CAD, one expects that the heart may be well perfused during rest, but that regions of the myocardium supplied by an obstructed artery will be lack sufficient perfusion during exercise, and this will be revealed by a reduction in radiotracer uptake. Such reversible perfusion defects may be ameliorated by surgically placing stents in the regions of obstruction within the appropriate coronary artery. The most likely artery with obstruction giving a reversible perfusion defect in the anterior wall shown in Fig. 4.37 is the right coronary artery. The wait times after injection are slightly different for ^{99m}Tc-tetrafosmin, a radiotracer similar to ^{99m}Tc-mibi.

4.6   Challenges and Emerging Directions

Standard NM imaging has been remarkably successful in providing functional information relevant to diagnosis for over 60 years. It remains a key method in the assessment of CAD, bone pathology, pulmonary embolism, and oncology. The radiotracer concept and ease of use of gamma cameras also enable techniques to be developed for special cases like tracking gastric emptying or gastrointestinal bleeding. However, gamma cameras in general and SPECT acquisitions in particular suffer from a low detector sensitivity: few of the photons emitted are registered in the detector as suitable for imaging. In principle this is due to three factors. The first is that high-energy photons, being very penetrating, do not readily interact with matter to give a signal. As discussed in Sect. 4.3.2, the bulk of the scintillating crystal provides volume for this interaction, but making crystals too thick degrades image spatial resolution due to increased scatter. To some extent, new solid-state camera

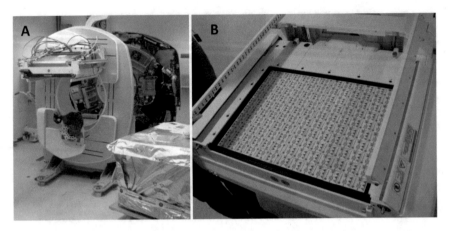

Fig. 4.38 First clinical install of a whole-body solid-state SPECT/CT (Discovery 670 CZT, GE Healthcare, Tirat Carmel, Israel) at Rambam Health Care Campus, Haifa, Israel, in 2015. (**a**) In the foreground, one of two detector heads is mounted on the NM gantry. The CT ring is in the background. (**b**) Each head contains 130 CZT modules, visible here before the collimator is placed in front

designs employing materials such as CZT for direct energy conversion, as described above, provide avenues for mitigating this issue. Secondly, the collimator principle establishes the ray paths necessary for image formation by means of absorbing and excluding photons that are not along the path, consequently removing most of the signal (Sect. 4.3.1). One advantage to PET, described in Chap. 5, is the removal of the collimator in its design and relying on electronic collimation. Thirdly, standard camera designs have one or two heads mounted on a gantry to rotate about the patient, meaning that at any given time, most of the angles of the required projections are not being acquired. Again, this problem is reduced in PET where most designs employ a full ring of detectors to acquire all projections simultaneously, and the dedicated cardiac cameras described in Sect. 4.3.7 acquire multiple projections simultaneously. Current SPECT camera development aims to increase camera sensitivity while maintaining or improving spatial resolution.

Recently, whole-body solid-state CZT SPECT cameras have been designed and made commercially available. In 2015, the first clinical install of a general-purpose whole-body SPECT (GE Discovery 670 CZT) provided the same flexibility and similar sensitivity as a standard dual-headed gamma camera with moderately improved spatial resolution (Fig. 4.38). Each detector head consisted of 10 rows of 13 columns of 4 cm square CZT modules of the same design as used in the 2 CZT dedicated cardiac cameras described in Sect. 4.3.7 (Discovery 530c, GE Healthcare, and the D-SPECT, Spectrum Dynamics). Current commercial versions employ modules of similar design but with a 7.2-mm-thick crystal instead of 5 mm, increasing sensitivity by about 20%. Although meeting the goals of increasing sensitivity with improved spatial resolution, the traditional geometry of this scanner requires 260 modules, compared to 76 modules used in the Discovery 530c or 36 modules in

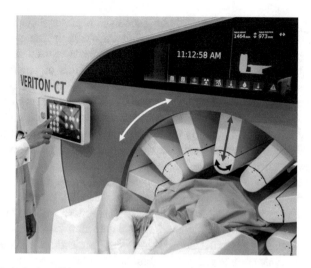

Fig. 4.39 Whole-body solid-state SPECT/CT (Veriton image courtesy of Spectrum Dynamics Medical, Caesarea, Israel). Twelve detector heads can be rotated to position (white arrows). CZT modules are mounted on gimbals at the end of cantilevers that can be moved radially (red arrows) to conform to the patient's shape. Swivel motion of the detectors (black arrows) inside the housing provides the required angular projections. This flexibility in detector geometry enables organ specific scanning

the D-SPECT. Since the detectors in emission tomography systems are generally the most expensive component, there is also motivation to reduce the number of detectors needed or their cost, without compromising camera performance.

Two commercial SPECT designs provide full-ring or near-full-ring detector coverage, enabling the acquisition of projections at all required angles simultaneously or nearly simultaneously. The G-SPECT (MILabs, Utrecht, the Netherlands) uses stationary multi-pinhole collimation and a NaI(Tl)/PMT scintillation detector system to provide simultaneous full 360° coverage with 3 mm spatial resolution, which is at least twice as good as typical SPECT systems. The Veriton (Spectrum Dynamics, Caesarea, Israel) extends the concept of the D-SPECT to cover the whole-body case (Fig. 4.39). The Veriton and D-SPECT were both invented and developed by Yoel Zilberstien and Spectrum Dynamics. In the Veriton, twelve detector heads each have 7 CZT modules aligned axially to give 7×4 cm $= 28$ cm axial coverage. The 12 detectors are mounted at the end of cantilevers that surround the patient in a star-like pattern in the transverse plane. Each cantilever sits at an angle of $360°/12 = 30°$ from the adjacent one. The cantilevers can be moved radially in and out so that the detectors can be placed very close to the patient to improve spatial resolution. Additionally, each detector is on gimbals that can be rotated back and forth in the transverse plane to provide projections at many different angles as needed. Consequently, many angular projections can be acquired at once, and a full data set can be acquired much faster than with a standard SPECT design. Additionally, the entire gantry holding the cantilevers can rotate. Because of the flexibility in the direction that the detectors are aiming, acquisition protocols can be written to acquire counts from a specific organ, like the heart for MPI, or from an specific ROI. By optimizing the detector directionality, the detectors can be used more effectively since they spend less time pointing at background counts or empty space. This design makes good use of a limited amount of detector material (only 84 CZT modules), provides projections from multiple angles, has good sensitivity, and has good spatial resolution. Compared to standard designs, this geometry can reduce scan times by a factor of 2 or 3. By providing multiple projections simultaneously, both the G-SPECT and Veriton facilitate dynamic SPECT acquisitions.

Future developments likely include organ-specific cameras such as needed for breast imaging or brain imaging. A commercial solid-state breast scanner exists (GE Healthcare Discovery NM750b), and efforts are being made to reduce the patient radiation exposure required in order to make these studies appropriate for screening in mammography. Modular solid-state detection facilitates organ-specific design, but the number of studies done globally per year is a factor in determining if these designs become commercially available. Solid-state detection may advance the development of a Compton camera that does not use a collimator. Although the details lie beyond the scope of this chapter, a Compton camera makes use of the detection of a γ-ray scatter event and a second detection of an absorption event to determine ray paths. By eliminating the collimator, the goal is to greatly increase camera sensitivity. The primary constraints on clever gamma camera designs for the clinic are affordability and diagnostic need.

References

1. Carrasquillo JA, O'Donoghue JA, Pandit-Taskar N, et al. Phase I pharmacokinetic and biodistribution study with escalating doses of ^{223}Ra-dichloride in men with castration-resistant metastatic prostate cancer. Eur J Nucl Med Mol Imaging. 2013;40:1384–93.
2. Selwyn RG, Nickles RJ, Thomadsen BR, DeWerd LA, Micka JA. A new internal pair production branching ratio of ^{90}Y: the development of a non-destructive assay for ^{90}Y and ^{90}Sr. Appl Radiat Isot. 2007;65:318–27.
3. Lhommel R, Pierre Goffette P, Van den Eynde M, Jamar F, Pauwels S, Bilbao JI, Walrand S. Yttrium-90 TOF PET scan demonstrates high-resolution biodistribution after liver SIRT. Eur J Nucl Med Mol Imaging. 2009;36:1696.
4. Abikhzer G, Kennedy J, Israel O. ^{18}F NaF PET/CT and Conventional bone scanning in routine clinical practice: Comparative analysis of tracers, clinical acquisitions protocols, and performance indices. In: Beheshti M (ed) PET Clin. 2012;7(3):315–28.
5. US Nuclear Regulatory Commission. Regulatory guide 8.39 release of patients administered radioactive materials. USNRC. Apr 1997. Available at: http://www.nrc.gov/reading-rm/doc-collections/.
6. Stabin MG, Breitz HB. Breast milk excretion of radiopharmaceuticals: mechanisms, findings, and radiation dosimetry. J Nucl Med. 2000;41:863–73.
7. Delacroix D, Guerre JP, Leblanc P, Hickman C. Radionuclide and radiation protection data handbook. Radiat Prot Dosim. 2002;98(1):5–168. Available at https://www.nuc.berkeley.edu/sites/default/files/resources/safety-information/Radionuclide_Data_Handbook.pdf. Accessed 6 Mar 2018.

8. Israel O, Front D, Lam M, Ben-Haim S, Kolodny GM. Gallium 67 imaging in monitoring lymphoma response to treatment. Cancer. 1988;61:2439–43.

9. Walrand S, Flux GD, Konijnenberg MW, et al. Dosimetry of yttrium-labelled radiopharmaceuticals for internal therapy: ^{86}Y or ^{90}Y imaging? Eur J Nucl Med Mol Imaging. 2011;38(Suppl 1):S57–68.

10. Kim Y, Lee T, Lee W. Double-layered CZT compton imager. IEEE Trans Nucl Sci. 2017;64 (7):1769–73.

11. Dahlbom M, King MA. Principles of PET and SPECT imaging. In: Dahlbom M, editor. Physics of PET and SPECT imaging. New York: CRC Press/Taylor & Francis Group; 2017. p. 413–38.

12. Ben-Haim S, Kennedy J, Keidar Z. Novel cadmium zinc telluride devices for myocardial perfusion imaging—technological aspects and clinical applications. Semin Nucl Med. 2016;46(4):273–85.

13. Hasegawa BH, Gingold EL, Reilly SM, Liew SC, Cann CE. Description of a simultaneous emission-transmission CT system. Proc SPIE. 1990;1231:50–60.

14. Townsend DW, Beyer T, Kinahan PE, Brun T, Roddy R, Nutt R, et al. The SMART scanner: a combined PET/CT tomograph for clinical oncology. Conference record of the 1998 IEEE nuclear science symposium. Vol 2, pp. 1170–1174, 1998.

15. Catana C, Wu Y, Judenhofer MS, Qi J, Pichler BJ, Cherry SR. Simultaneous acquisition of multislice PET and MR images: initial results with a MR compatible PET scanner. J Nucl Med. 2006;47:1968–76.

16. Bailey D. Transmission scanning in emission tomography. Eur J Nucl Med. 1998;25:774–86.

17. IAEA. IAEA Human Health Series No. 6: quality assurance for SPECT systems. Vienna: International Atomic Energy Agency; 2009.

Chapter 5
Positron Emission Tomography (PET)

Synopsis: In this chapter the reader is introduced to the physics of clinical imaging using positron emitters as radiotracers.

The learning outcomes are: The reader will be able to define the function of positron emission tomography (PET) scanning, including PET radiotracers, and explain the technical reasons why virtual all clinical PET scanners are hybrid PET/CT or PET/MR scanners. Students shall be able to analyze the design constraints provided by conventional, time-of-flight (TOF), and solid-state PET scanners.

Positron emission tomography (PET) is a nuclear medicine imaging technique that uses radiotracers for diagnostic imaging in a manner similar to the techniques used by gamma camera and SPECT imaging. Consequently, this is a sister chapter to Chap. 4, applying the principles described there to the case of PET. As mentioned in Sect. 4.6, gamma cameras suffer from a low sensitivity (photon detection efficiency) in part because of the physical collimation used: photon ray paths are determined by recording only the small fraction of emitted photons that happen to travel paths defined by small holes in a strong absorber. This limitation is somewhat mitigated in PET. Electronic collimation can be used in the case of positron-emitting radiotracers. Within 1 or 2 mm of the location of emission within tissues, emitted positrons meet and mutually annihilate their anti-particles (electrons). Two high-energy annihilation photons are generated and travel in virtually opposite directions. A coincidence detection of both photons by two detectors determines the ray path of the photons, which is then used in image reconstruction. The detectors typically employ scintillating crystals and PMTs, and the absence of a physical collimator enables the detection of proportionally more photons, thus increasing sensitivity. In practice, arrays of detectors placed in rings around the patient allow for the acquisition of 3D data used for tomographic image reconstruction. Solid-state methods of detecting

© Springer Nature Switzerland AG 2020
H. Azhari et al., *From Signals to Image*,
https://doi.org/10.1007/978-3-030-35326-1_5

scintillations have become clinically available in newer PET designs. Although there are a large variety of studies that can be performed, PET is primarily used in oncological studies.

5.1 Physical Phenomena

5.1.1 Radioactive Decay and Positron Emission

The decay of a parent nuclide by positron emission is described above in Sect. 4.1.1. Table 4.1 lists nine positron (β^+) emitters that are used in diagnostic imaging, but this is not an exhaustive list. In the clinic, the most commonly used positron emitters have half-lives considerably shorter than the ^{99m}Tc "workhorse" NM isotope. Most of these are made in a cyclotron (Fig. 4.14). For radioisotopes like ^{11}C and ^{13}N, the cyclotron must effectively be on-site with the clinic, because the short half-lives do not accommodate transportation. Fluorine-18 remains the most common radioisotope in clinical PET, and with a half-life of almost 2 hours, radiotracers made from ^{18}F can be transported reasonable distances from the cyclotron to the clinic. Gallium-68 can be produced from germanium-68 (half-life of 270.8 days) in a generator similar in idea to the ^{99}Mo/^{99m}Tc generator discussed in Sect. 4.2.5. The reaction is by EC: $^{68}_{32}$Ge$(e^-,\gamma)^{68}_{31}$Ga. With a half-life of about an hour, ^{68}Ga radiotracers can also be transported from production to clinical sites, but this is not ideal. For PET cardiac studies, rubidium-82, a potassium analog, can be produced in a generator from strontium-82 (e.g., CardioGen-82, Bracco Imaging, Monroe Township, NJ), via EC. The reaction is $^{82}_{38}$Sr$(e^-,\gamma)^{82}_{37}$Rb with a half-life of 25.55 days. With a half-life of 1.273 minutes, the ^{82}Rb generation must be in the imaging room for this type of MPI PET study. Unlike using ^{201}Tl, also a potassium analog, a scan the next day cannot give cardiac viability information since the ^{82}Rb has virtually all decayed. Usually ^{18}F-FDG (fluorodeoxyglucose) is used for PET cardiac viability scans. These examples show some of the considerations with respect to decay rates when using PET isotopes and are hardly an exhaustive list.

5.1.2 Particle/Anti-particle Annihilation

As described in Sect. 4.1.1, a positron emitted from within a body travels several millimeters before it slows and undergoes annihilation with an electron in the tissue. The average distance travelled depends on the energy of the emitted positron and the electron density of the tissue or matter surrounding the emitter. A second particle, an anti-neutrino, is also emitted in positron decay, and the potential energy available to the decay process is shared between the anti-neutrino and the positron, but not in a constant ratio. The positron energy is thus a spectrum with the maximum energy

Table 5.1 Effective positron range in water for some PET radionuclides

Radionuclide	Element	Maximum β^+ energy (MeV)	Effective positron range (mm)
^{11}C	Carbon	0.97	0.39
^{18}F	Fluorine	0.64	0.23
^{68}Ga	Gallium	1.90	1.2
^{82}Rb	Rubidium	3.39	2.6

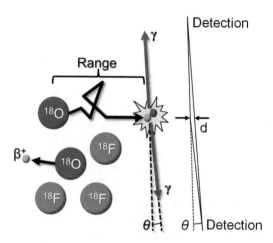

Fig. 5.1 Positron annihilation in tissue. Here the $^{18}_{8}F(,\beta^+)^{18}_{8}O$ decay generates positrons (green) that each annihilate with an electron (purple) generating two 511 keV annihilation photons (red arrows) travelling virtually in opposite directions. Positrons usually lose energy in matter (black arrow) before annihilation, and the total displacement from emission (range) adds to the uncertainty in the emission location. Because the positron carries momentum, the photon paths deviate from 180° by a small angle (θ). This deviation also adds a small uncertainty to the emission location (d) diagrammed at the right for a case in which coincidence detection is about equidistant from the annihilation

equal to the energy available for the nuclear reaction of decay (called the Q-value). The average positron energy is about 40% of this maximum value. The effective range of positrons from an emitter is usually given as the root mean square of the distances from the emission at which annihilation occurs for a particular medium (Table 5.1) [1]. Theoretically, positron range adds to image blur since the annihilation events will be displaced from the emission events [2]. Practically speaking, given that the intrinsic spatial resolution of clinical PET cameras is about 4 mm at best, this degradation due to positron range is not large, especially for PET's "workhorse" radiotracer, ^{18}F-FDG.

Occasionally, the subsequent matter/anti-matter annihilation can give rise to one photon emission (if the electron being annihilated is tightly bound to a nucleus and can transfer momentum to it) or to three or more photons sharing energy and momentum, but generally a photon pair is produced (Fig. 5.1).

PET scanners detect the high-energy photon pairs generated from positron annihilation within a body, not the positrons themselves.

5.1.3 Coincident Photon Pairs

Generally, annihilation produces two 511 keV that travel in virtually opposite directions, with each photon having energy equal to the rest mass of an electron, as mentioned above. Near-simultaneous, or coincidence, detection of these two photons establishes a line of response (LOR) between two detectors, and ideally the emission event falls close to this line. Coincidence detection obviates the need for physical collimation. However, since the positron carries momentum, the photon paths deviate from 180° by a small angle of up to 0.25° for ^{18}F (Fig. 5.1), and, as with the positron range, this deviation also adds a small uncertainty to the emission location. This deviation could be as much as 2 mm, for clinical PET scanners. Note that the positrons do not interact with the camera itself, but rather undergo annihilation within the object being imaged. Like gamma cameras, PET cameras detect high-energy photons that are emitted from the body.

5.2 Signal Sources

5.2.1 Emission Versus Transmission Scanning

As mentioned in Sect. 4.2.1, in a transmission image, internal structure is revealed by the transmission and absorption of photons that originate from outside the object being imaged, whereas in emission imaging, the internal structure is revealed by the emission of photons originating from within the object (Fig. 4.7). Historically, whole-body PET transmission images were generated from positron emitters that could transmit 511 keV photons through the patient to build an attenuation map. One method was to use pin sources of ^{68}Ge. The $^{68}_{32}$Ge $(e^-,\gamma)^{68}_{31}$Ga reaction produces ^{68}Ga, which is a positron emitter. Alternatively, ^{137}Cs sources have been used, and the map has been adjusted to scale the 662 keV photons emitted to the 511 keV attenuation map required. Typically, the sources were rotated about the patient to activate detectors on the other side of the patient, along LORs passing through the patient. Such transmission scans suffered from noise typical of NM images and could take 20 minutes to acquire. Clinics could perform a transmission scan, inject the radiotracer with the patient on the bed, wait for uptake, and then scan, giving a lengthy acquisition process. Alternatively, if the patient underwent transmission and emission scanning after uptake, there would be cross-talk between transmission and emission data, since the energy windows were the same or nearly the same. This diminished image quality. With the advent of commercial hybrid PET/CT scanners in 2001, these methods were replaced with the CT for transmission scanning: within 2 years virtually all PET scanners sold were hybrid. The CT provides a fast (<1 minute) transmission scan for attenuation maps and localization giving low noise images with good resolution and no cross-talk with PET [3].

5.2.2   F-18 FDG: [^{18}F]-2-Fluoro-2-deoxy-D-glucose

The most common radiotracer used for PET is fluorine-18 FDG ([^{18}F]-2-fluoro-2-deoxy-D-glucose) [4]. This is a radioactive sugar analog that is incorporated into cells as if it were glucose, but becomes only partially metabolized within, thereby marking the cells with ^{18}F. Figure 5.2 shows that its molecular structure is similar to glucose but a hydroxyl group has been replaced by ^{18}F. Patients are requested to fast for about 4 hours before the scan, enabling the skeletal muscles to switch to the creatine cycle, minimizing their absorption of glucose and ^{18}F-FDG. This reduces the background signal for the regions of interest: many types of cancerous lesions are metabolically very active and will uptake glucose and ^{18}F-FDG more than the surrounding tissue. Also, the brain always remains with glucose metabolism, so an ^{18}F-FDG PET scan of the brain can show the distribution of its metabolic activity which can be diagnostically useful (Fig. 5.3). Under certain conditions, the myocardium preferential uptakes glucose and ^{18}F-FDG, showing the metabolic activity throughout the heart muscle, which is useful for myocardial viability scanning.

5.2.3   Other Positron Emission Radiotracers

Biologically active radiotracers are often organic molecules. As such, replacing a stable carbon atom (^{12}C) with ^{11}C generates a potential PET radiotracer, with the simplicity that the two isotopes of carbon are virtually chemically identical so the

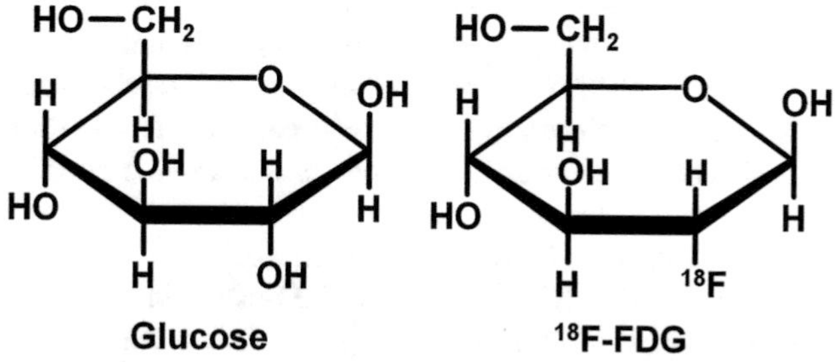

Fig. 5.2 F-18 FDG: [^{18}F]-2-fluoro-2-deoxy-D-glucose (right) is similar in molecular structure to glucose (left). In this schematic, unlabeled vertices represent carbon

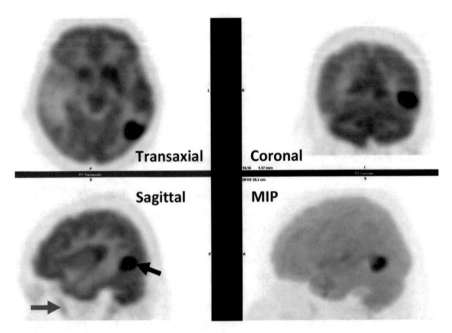

Fig. 5.3 F-18 FDG brain PET: Lung cancer metastasized to the brain. Both the tumour (black arrow) and the brain uptake FDG at a higher concentration than the other tissues (gray arrow). MIP is a maximum intensity projection plotting here the maximum voxel value along ray paths perpendicular to the sagittal plane

radiotracer is also virtually chemically identical to the non-radioactive molecule. An example of such a radiotracer is ^{11}C-choline which targets neoplasms and can be used to image prostate cancer and brain tumors. With a short half-life (~20 minutes), such ^{11}C tracers are usually used at the same site where they are produced. The production of ^{18}F-fluorocholine (fluoromethyl-dimethyl-2-hydroxyethyl-ammonium [FCH]) gave a choline analog that mimics natural choline uptake in the body due to its structural similarity to choline. The longer half-life of ^{18}F (~2 hours) better accommodates transportation and use in the clinic.

As mentioned previously, ionic forms of ^{18}F (NaF) and ^{82}Rb (RbCl) in solution, without ligands, can be used for PET bone studies and PET MPI, respectively. Also without a ligand, ^{124}I (NaI) has potential for the diagnosis and treatment of differentiated thyroid cancer. For MPI, ^{13}N-ammonia can be used but special arrangements need to be made for delivery to the patient because of the very short half-life of ^{13}N (~10 minutes). There is a family of compounds that target somatostatin receptors that are sometimes overexpressed in neuroendocrine tumors: ^{68}Ga-DOTATATE (^{68}Ga-1,4,7,10-tetraazacyclododecane-1,4,7,10-tetraacetic   acid   [DOTA]– octreotate), ^{68}Ga-DOTANOC (Ga^{68}-DOTA–1-NaI3-octreotide [NOC]), and similar radiotracers. Prostate-specific membrane antigen (PSMA) has been labeled with either ^{18}F or ^{68}Ga for imaging of metastatic prostate cancer (as well as with ^{177}Lu for treatment and SPECT imaging). ^{18}F-DOPA   (6-[^{18}F]-L-fluoro-L-3,

4-dihydroxyphenylalanine) can be used, for example, to image neuroendocrine tumors which are not rich in somatostatin receptors or as a brain imaging agent to assist in the diagnosis of Parkinson's disease.

This is not an exhaustive list, but many of them have been used routinely in the clinic. Because of the higher spatial resolution and system sensitivity of PET compared to SPECT, there is much interest in further developing PET imaging agents.

5.3   Data Acquisition

Clinical PET scanners use the Anger principle for detection and event location as discussed in Sect. 4.3.4, with some variations to the design. Dual-headed Anger cameras with thicker NaI(Tl) crystals (~2.5 cm instead of ~1 cm) have been used for coincidence detection (e.g., a version of the Infinia Hawkeye Nuclear Gamma Camera, GE Healthcare, Tirat Carmel, Israel): a slot collimator accepted coincident photons only in the transaxial (transverse) planes, and the opposing heads could record coincidence events. The dual heads needed to be rotated about the patient to acquire projections across 180°, similar to a standard SPECT acquisition. The thicker crystals were needed to stop a larger fraction of the 511 keV photons which otherwise penetrate the thinner crystals much more effectively than the 141 keV photons of ^{99m}Tc, for example.

In clinical scanners, it is common to find a modular structure to the Anger system. Figure 5.4 shows two modules in which small bismuth germanate (BGO) scintillating crystals are stacked flush to PMTs with a light guide. In this case, each PMT is subdivided to act as four PMTs, and each module acts as a mini-Anger camera. Other designs use different sizes and arrangements of PMTs, but the division of the scintillating material into small crystals is common. The modules are typically arrayed to provide a cylinder (Fig. 5.5) around the axis of the patient with the longer dimension of the crystal roughly parallel to annihilation photons which arrive approximately along the radial direction. This 20 or 30 mm of crystal depth is needed to stop a sufficient number of photons. The smallest crystal dimension (~3 to ~6 mm) is laid out tangentially in the transaxial plane to optimize the in-plane transverse spatial resolution. The dimension of the crystal along the axis of the patient (~3 to ~8.5 mm) limits the axial spatial resolution between slices. The elongated nature of the crystal also contributes to the total internal reflection of the visible scintillated light, helping to channel light into the PMTs.

For scintillating crystal material, other than BGO, clinical scanners may also use lutetium oxyorthosilicate (LSO) crystals and variations thereof, such as lutetium-yttrium oxyorthosilicate (LYSO). Other materials have been used in preclinical and research devices but are not commonly found in the clinic.

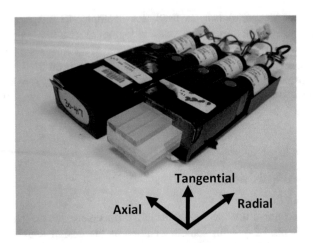

Fig. 5.4 Two discarded PET detector modules. Four PMTs with square faces 24 mm × 24 mm comprise two modules here. In the foreground, with covering removed, the faces of 2 PMTs are partially visible with stacks of BGO scintillating crystals in front. Originally an array of 6 × 6 crystals covered these two faces, with each crystal 2 mm × 8.5 mm × 30 mm in the tangential, axial, and radial directions. Each PMT is subdivided to act as four PMTs. Between the crystals and the PMTs is a light guide. This scanner design (GE LS PET/CT, GE Healthcare) has 3 modules axially and 112 tangentially in a ring, giving 12,096 crystals

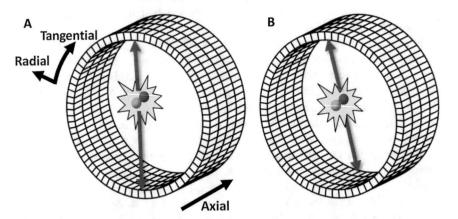

Fig. 5.5 PET detector blocks or modules arranged for a cylindrical FOV. (**a**) Dozens of detector blocks are arranged tangentially to define a circle about 50–80 cm in diameter. Axially, about three to five blocks define an axial FOV about 15–26 cm. (**a**) Legacy PET designs used septal rings to separate the photons received by crystals to a transaxial plane (2D mode). Schematically here, an emitted positron (green) annihilates with an electron (purple) to produce two 511 keV photons travelling in opposite directions (red arrows) and is recorded by detectors at the top and bottom at the same time and in the same transaxial plane. (**b**) In current clinical practice, septa are not used (3D mode). Annihilation photon pairs travelling at angles oblique to the transaxial plane can be recorded by detectors which are not in the same transaxial plane

5.3.1 Physical Collimation: Septa

Older PET designs used septal rings to limit the pairs of photons received by crystals to a transaxial plane (2D mode). This had the advantage of reducing scatter by excluding photons scattered out of plane. The 2D mode also aided in tomographic reconstruction since it is amenable to the FBP algorithms initially available. Since the detector modules are arranged in a ring or cylinder, the septa are annular rather than the simpler slot collimation described above. In 2D mode (Fig. 5.4), typically the data from crystals directly across from each other was accepted, defining a transaxial slice. The septa are typically designed to also accept LORs that cross one crystal width in the axial direction, and this defines a "cross-plane" slice. Since there is a plane and cross slice for each axial crystal width, the PET slice thickness is usually taken as ½ the crystal width. In the example shown in Fig. 5.5, the crystal width in the axial direction is 8.5 mm, so 4.25 mm is the stated PET slice thickness. The PET ring is 3 modules of 6 crystals set axially per module defining 18 "PET rings" with 18 planes in direct opposition and 17 cross-planes giving 35 slices of 4.25 mm. This is an axial FOV of about 15 cm. Axial FOVs range up to 26 cm for commercial scanners. Newer designs have no physical collimation and acquire data in "3D mode" [5].

5.3.2 Electronic Collimation: Coincident Photon Detection

An annihilation event is recorded as having occurred along an LOR if two scintillations are counted as coincident (Fig. 5.6). As with SPECT, scintillations are only counted if they fall within an acceptable energy window to reduce acceptance of scattered photons and other erroneous events. Typical windows could be 300–650 keV for a scanner in 2D mode using BGO crystals or a more restrictive 425–650 keV window for clinical scanners in 3D mode using LSO-type crystals [6]. An accepted scintillation starts a timing window of perhaps 15 ns or less. If a second scintillation occurs within that timing window, it is considered to be a coincidence event. At count rates of up to 1000 kcounts/s, the timing between annihilation events detected is on the order of microseconds, so ideally this window is small enough to track annihilation events as being sequential. Figure 5.6 depicts a 2D mode in which annular septa restrict the acceptance of photon pairs to detectors that are either directly across from each other defining the same transaxial plane or shifted by one crystal size axially, defining a cross-plane as described in Sect. 5.3.1. For example, the modules displayed in Fig. 5.4 were from a scanner with 672 crystals arrayed around a ring which enables projections every $180°/(672/2) = 0.5357°$. Geometric correction was applied to give detector spacing of 1.97 mm across 553.6 mm, so the 2D sinograms were 281×336, suitable for FBP reconstruction. The PET sinogram structure is tied to the geometry of the detectors.

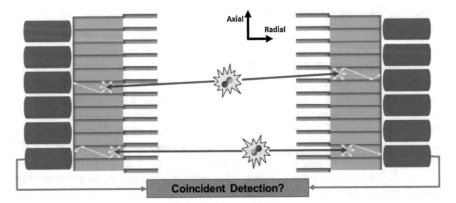

Fig. 5.6 PET 2D mode. In the lower annihilation event, a positron (green dot) annihilates with an electron (purple dot) to produce two photons travelling in opposite directions (red arrows) that scintillate crystals (blue) simultaneously and the visible light (yellow arrows) is detected by the PMTs (purple cylinder). Electronic circuitry determines if the detected events are coincident and determines the LOR if so. In the upper annihilation event, the septa (gray) accept an LOR at cross-planes, axially. LORs at more oblique angles are not detected in 2D mode since at least one of the photons will be stopped by the septa. Generally, annihilation events occur sequentially with respect to the timing speed of electronic detection, so that the upper and lower events are seen as separate

Electronic collimation provides a method of obtaining projection data without using hole collimation as described in Sect. 4.3.1 for SPECT. Figure 5.7 shows that by organizing data collection such that LORs are parallel to each other, parallel ray projections can be collected, and sinograms can then be constructed for the 2D mode case.

A geometric correction to this projection data is usually applied to account for the fact that the spacing between LORs decreases for LORs that pass further from the center of the ring. The data is typically rebinned to even out the spacing of the detectors. Since any one detector crystal can have multiple valid LORs with other crystals across the transaxial FOV, many projection angles can be established. Because LORs depend on two coincident photons travelling in opposite directions, a full PET sinogram covers only 180°, in 2D mode.

Also shown in Fig. 5.7 are two other main confounding factors in recording true events [7]. Coincidence detection of a scattered photon establishes an erroneous LOR that passes close to, but not through, the annihilation event. As with SPECT, scatter adds to blur and diminishes contrast. Secondly, two annihilation events occurring within one timing window can establish an erroneous LOR if their companion photons may remain undetected. These are called random events and must be corrected. One method to determine the rate of detection R_{random} of random events between two detectors i and j is to displace the timing window (e.g., 15 ns wide) to a time when there cannot be a coincidence (e.g., 1 μs later). Consequently, any "coincident" events must be random under these conditions. The probability of

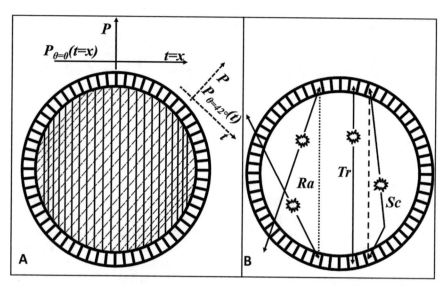

Fig. 5.7 PET electronic collimation. (**a**) Detector crystals are arranged in a ring in this schematic and coincidence detection of annihilation photon pairs establish LORs (solid parallel lines) which determine a projection $P_{\theta=0}(t=x)$ onto the x-axis. Note that the spacing between LORs decreases for LORs that pass further from the center of the ring. Therefore, unlike traditional SPECT, the spacing (t) of the projection is not constant. The "geometric correction" of PET projections rebins the data by interpolation to even out the spacing. Any one detector crystal can have multiple valid LORs with other crystals across the transaxial FOV. Shown here is an example second projection $P_{\theta=42°}(t)$ (dashed parallel lines). In 2D mode, a sinogram of such projections covering 180° provides a full data set for the reconstruction of one tomographic slice. (**b**) A true event establishes an LOR (marked Tr). Coincidence detection of a scattered photon establishes an erroneous LOR (Sc) that passes close to, but not through, the annihilation event (star). This scatter adds to blur and diminishes contrast. Two annihilation events randomly occurring within one timing window can establish an erroneous LOR (Ra). Their companion photons may remain undetected

random detection increases with the rate of detection at each detector and can be calculated as

$$R_{\text{random}} = 2\tau R_i R_j \tag{5.1}$$

where τ is the width of the timing window and the 2 accounts for either detector i or j detecting an event first. For a small source in the center of the FOV, the count rates among the detectors are about equal ($R_i \approx R_j$), so the random rate increases as the square of the count rate. At the extreme, one can imagine that if there was sufficient activity within the FOV to activate all the detectors virtually all the time, there would be no true coincidence information and therefore no data available to reconstruct the image.

Collimation in PET is electronic: the path of the high-energy
annihilation photon pair is determined by coincidence detection.

5.3.3 3D Acquisition

As described in Sect. 5.3.1, in 2D mode septa restrict LORs to a transaxial plane
(or cross-plane). In 2D mode, the axial direction is at right angles (or nearly so) to the
LORs since they fall within transaxial planes that are normal to this axis. Removing
the septa allow LORs to fall at oblique angles to the axial direction (Fig. 5.5b). Some
older PET designs allowed for both 2D and 3D scanning as the septa could be
extended or withdrawn. Because of wider acceptance angles, many more LORs are
legitimate. Count rates in 3D mode can be more than five times higher compared to
the count rates for the same object being scanned in 2D mode. Effectively, these
scanners acquire a set of sinograms at oblique angles to the axis in the axial direction,
which are typically rebinned to equivalent transaxial sinograms using a Fourier
rebinning technique [8] to enable slice-by-slice image reconstruction by traditional
means like FBP. More modern clinical scanners use 3D mode only, with fully 3D
image reconstruction techniques.

Whole-body PET scanners designed in the 1990s had axial FOVs of about 15 cm.
This is suitable for brain or cardiac imaging since these organs can usually fit into
one FOV of this size. However, oncologic ^{18}F-FDG scans are typically performed
"eyes to thighs" covering about 1 m of the patient length. Some studies, such as for
the diagnosis of melanoma, often require the entire body to be imaged, up to about
2 m in axial length. Consequently, whole-body PET scans can require 6–15 "bed
positions": a step-and-shoot method translates the patient axially by about 15 cm
every few minutes. Total scan durations for whole-body imaging could range from
about 15 to 50 minutes, depending on the clinical need. In 2D mode, most protocols
include a 1- or 2-slice overlap between bed positions to reduce discontinuities in the
image at the overlap (Fig. 5.8). In 3D mode, this overlap is typically 25–50% of the
slices, due to lower count statistics near both ends of the PET bore. This is because
sources in slices near the beginning or the end of the bore have fewer LORs at
oblique angles that will intersect the PET ring for both of the annihilation photon
pairs. Consequently, 3D mode for whole-body imaging usually requires more bed
positions, compared to 2D mode, to cover the same axial distance, but the higher
detector efficiency in 3D mode leads to much shorter scan durations at each bed
position. Discussed below in Sect. 5.6, newer clinical PET scanners may come with
axial FOVs of up to 26 cm or higher, and this geometry can be used to substantially
reduce scan duration or injected radiotracer activity.

Fig. 5.8 PET bed positions. (a) Multiple bed positions (red) are required to collect data "eyes-to-thighs" for a whole body scan of a patient (blue). There is little overlap (dark red) between bed positions in 2D mode. (b) In 3D mode there is typically 25–50% overlap between bed positions, requiring more stops, but the scan durations are shorter at each bed position

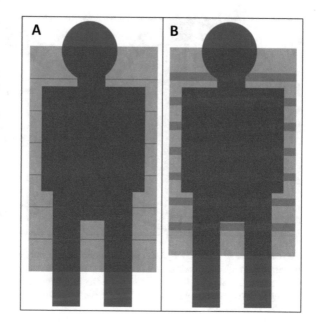

5.3.4 Time-of-Flight Acquisitions

Although "coincidence" detection has been defined above as two events detected within about 15 ns of each other, one can imagine that if the timing of the two events were even more precise, one could plot the location of that event along an LOR rather than merely establishing an LOR. This is the time-of-flight (TOF) principle for PET [9]. For example, a point source in the center of the FOV would generate annihilation photon pairs that would reach any pair of detectors along an LOR at exactly the same time (Fig. 5.9). If the source were moved upward (anterior direction for a supine patient), the detectors in the upper half of the scanner would record the event before the lower detectors because that photon wouldn't have to travel as far. Precise timing could theoretically exactly locate the event along the LOR. In Fig. 5.9, the annihilation event is located at a distance r from the midpoint of an LOR of length D. The upper detector is a distance d_1 from the event and the photon arrives at time t_1 after positron/electron annihilation. Similarly, the lower detector records an event at time t_2 and is a distance d_2 from the annihilation. If c is the speed of light, the distances d_1 and d_2 are just ct_1 and ct_2. Therefore:

$$ct_1 = \frac{D}{2} - r$$

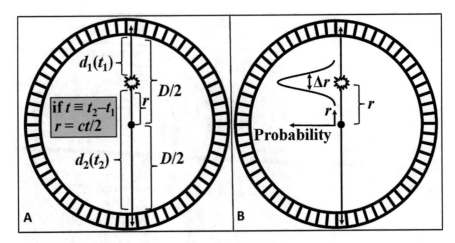

Fig. 5.9 Time-of-flight (TOF) PET. (**a**) An annihilation event (star) displaced a distance r up from the center of an LOR (black dot) will travel a shorter distance d_1 to the upper detector and take a shorter time t_1 compared to the photon travelling to the lower detector. (**b**) The uncertainty Δr in placing the event along the LOR is ~10 cm for the first clinical TOF scanners and as good as ~3 cm for later TOF models

$$ct_2 = \frac{D}{2} + r$$

$$c(t_2 - t_1) = \left(\frac{D}{2} + r\right) - \left(\frac{D}{2} - r\right)$$

$$r = ct/2$$

where t is the difference in time between t_2 and t_1. It follows that the temporal resolution Δt (error in timing) causes an error Δr in locating the event along the LOR:

$$\boxed{\Delta r = c\Delta t/2} \tag{5.2}$$

To determine the location to a typical voxel size, perhaps 2 mm, would require a timing resolution of $2 \times (2 \times 10^{-3}$ m$)/(3 \times 10^8$ m/s$) = 13$ ps. A scanner with such a timing resolution could plot the events in 3D and have no need for image reconstruction.

With the time-of-flight principle, a more precise timing of the detection of each photon's arrival helps localize the original annihilation event.

However, the first clinical TOF scanners introduced have temporal resolutions of around 540–650 ps, giving an uncertainty in position along the LOR of about 8–10 cm. This is insufficient to locate the event to within a voxel, but this added information can improve image contrast when introduced to an image reconstruction algorithm. Because a slender patient might be just 20 cm thick in the anterior/posterior direction, using TOF is generally more visually evident in larger patients for these scanners. The so-called "digital" TOF PET systems employing solid-state techniques have temporal resolutions ranging from about 220 to 390 ps, reducing this uncertainty in location to about 3–6 cm, further improving contrast.

5.3.5 Calibration

As previously mentioned, the projection data must be corrected for random, scatter, and geometric effects. Additionally, the data is corrected for dead time, which is usually a small percentage of the time a detector is unavailable to record an additional event after a previous photon is detected. Normalization calibration refers to corrections necessary for varying detector efficiencies. Ideally this is done by irradiating all detector pairs with the same known source of annihilation photons. Scanners that have a pin source available to rotate about the FOV, as described in Sect. 5.2.1, can produce a so-called blank scan of the FOV without a patient or object within. Detector pairs giving slightly higher or lower than expected count rates can be scaled accordingly. Blank scans take a long time to acquire, and their noise also contributes to the noise of the acquisition, so more sophisticated methods of accounting for detector efficiency variations are currently more common.

In the clinic, PET is typically used in a quantitative or semi-quantitative manner. Consequently, a scaling is required between the relative intensities represented in the reconstructed image (proportional counts) and the absolute radiotracer concentration in Bq/mL. This can be done volumetrically using a featureless phantom: usually a ~ 20cm diameter cylinder with a precisely known radiotracer concentration centered in the FOV (Fig. 5.10). This phantom is long enough to extend activity throughout the axial FOV. After a long acquisition to reduce variations due to random noise, the image is reconstructed with all corrections including attenuation correction. The proportional counts are then scaled to the known radiotracer concentration. The term "well counter calibration" or WCC refers to this process.

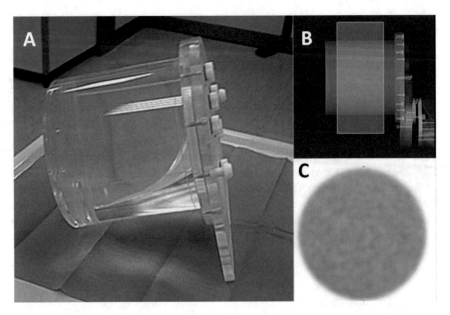

Fig. 5.10 Well counter calibration. (**a**) This ~5 L WCC phantom is loaded with a precisely known activity of radiotracer solution. (**b**) Scout x-ray scan of phantom mounted within the PET/CT FOV. Green locator marks segment to be scanned. (**c**) PET image intensity is scaled to the known radiotracer concentration

5.3.6   Noise Equivalent Count Rate

Correcting for scatter and random detection shown in Fig. 5.7b adds noise to the true events that are used to reconstruct a PET image. This noise can be substantial at high count rates. The noise equivalent counting rate (NECR) is a measure of to what degree the number of true events has been degraded. As mentioned in Sect. 5.3.2, the random count rate R_{RAND} increases as the square of the count rate, which would be proportional approximately to the square of the activity in the FOV. However, the true count rate R_{TRUE} would be expected to be proportional approximately linearly to the activity in the FOV. Consequently, there is a point at high count rates where the random counts far surpass the true count rate to the point that little or no information is available to construct an image. At high count rates, correcting for randoms adds too much noise to the image. Similarly, the count rate due to scatter R_{SCAT} adds to noise, although it remains roughly at about 30–40% of the true count rate. NECR is defined as

$$\mathrm{NECR} = \frac{(R_{TRUE})^2}{R_{TRUE} + nR_{RAND} + R_{SCAT}} \tag{5.3}$$

where $n = 2$ for the delayed window method of estimating randoms described in Sect. 5.3.2. NECR has the property that for a small source with small activity, the

Fig. 5.11 Noise equivalent count rate (NECR) output from a clinically installed PET using the manufacturer's software (edited). Note that while the scatter count rate tracks the trues count rate, the random count rate increases quadratically with the activity concentration. Here the NECR peaks at 125 kcounts/s for an activity concentration of 30 kBq/mL

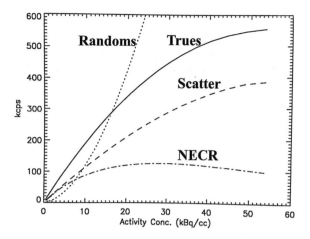

random and scatter rates would be small, adding little noise, so the noise equivalent count rate is roughly the same as the true count rate. However, for large objects with much scatter and high activity, the NECR is reduced and in fact peaks at a certain activity concentration. Peak NECR might vary from 100 kcounts/s at 20 kBq/mL to 350 kcounts/s at 50 kBq/mL for clinical scanners (Fig. 5.11). Older scanners in 2D mode had higher activity concentrations for the peak, as might be expected since the septa drastically reduce the number of possible events accepted.

5.3.7   Solid-State Methods

Section 4.3.5 explains that avalanche photodiodes (APDs) and silicon photomultipliers (SiPMs) can take the place of photomultiplier tubes (PMTs) within detection systems, and it lists several commercial PET scanners using such solid-state detectors. However, PET detection systems retain the use of scintillating crystals since there is currently no material that can provide an efficient solution for the direct conversion of annihilation photons to electron/hole pairs for charge collection, because of the relatively high energy of these photons compared to γ-rays used in SPECT imaging. Solid-state PET systems enabled the design of hybrid systems integrating PET with magnetic resonance imaging (PET/MR). Photomultiplier tubes rely on a cascade of electrons through an evacuated tube for the amplification of the signal, and because of Lorenz forces, such cascades of charged particles are sensitive to the high magnetic fields presented by MR gantries. PMTs fail in the 1–7 Tesla magnetic fields used in MR.

Since SiPMs commonly consist of a 2D array of thousands of individual APDs operating in Geiger mode (SPADs or single photon avalanche diodes), there is the design option of analyzing the integrated output from all of these or counting these individually. The latter method is called a digital photon counter (DPC), since a signal is counted for every photon detected from the scintillator. One commercial system uses DPCs, which can give reduced temperature sensitivity and aid in improving timing resolution when compared to SiPMs used in the integrative mode. Regardless of the method employed, these solid-state devices are often referred to as "digital" PET scanners. The major PET manufacturers offer "digital" PET/CT systems giving improved image quality when compared to their analog systems.

5.3.8 Hybrid PET/CT Systems

Transmission scanning in PET can provide an attenuation map for the object for attenuation correction. However, as discussed below, transmission scanning using radioisotope sources can cause cross-talk with the emission scan, generate relatively noisy transmission scan attenuation maps, and require lengthy acquisition times. The hybridization of CT with PET [10] provides a solution to these issues as well as provides a high-quality anatomical map to for the localization of PET radiotracer uptake [11, 12]. Introduced to routine clinical use in 2001, the first commercial PET/CT scanners were literally an pre-existing design of a whole-body PET wedded to a pre-existing design of CT by coaxially mounting their gantries and integrating their hardware and acquisition software. The CT bore diameter was about 10 cm wider than the PET bore diameter, and this is evident in Fig. 5.12 in which there is a narrowing of the PET/CT bore toward the PET ring at the far end. All major PET manufacturers developed integrated PET/CT systems giving fast CTAC and good anatomical localization which improves patient management [13]. Within 2 years of their introduction, virtually all clinical PET installs were hybrid PET/CT systems.

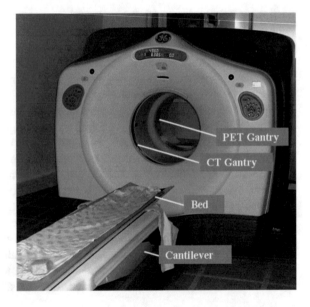

Fig. 5.12 An early commercial hybrid PET/CT scanner (Discovery LS, GE Healthcare, Tirat Carmel, Israel), installed at Rambam Health Care Campus, Haifa Israel, July 2001. Note that the PET gantry bore diameter is noticeably narrower than that for the CT

In PET/CT, the anatomical information of the CT complements the physiological information of the PET, aiding attenuation correction, localization, and clinical diagnosis.

5.4   Image Formation

5.4.1   Tomographic Principle

As shown in Fig. 5.7a, the LORs can be taken as parallel sets. Accounting for geometric corrections, these sets are projection data of the radiotracer distribution within the object being scanned. One could display the projection from the vertical anterior/posterior LORs (solid parallel lines in Fig 5.7a) to form a planar image, as with a gamma camera. This is not done routinely in the clinic since all the projections covering 180° are acquired simultaneously and available for tomographic reconstruction. As discussed in Chap. 2, with tomography the imaging data is represented in 3D, usually as cross-sectional slices. The relationship between the projection data and the PET tomographic reconstruction is similar to the relationship between a chest x-ray radiograph and a chest CT. In the former, the sternum, mediastinum, lungs, and spine may all be visible, but they overlap, and only our knowledge of anatomy is used to place the sternum anterior to the spine. In a transaxial CT slice, this placement is visually evident. Similarly, PET image reconstruction locates the PET radiotracer distribution in 3D as a series of transaxial slices.

5.4.2   Image Reconstruction

In the absence of computing power to apply more sophisticated image reconstruction methods, filtered back projection (FBP) was traditionally used (see Sect. 2.2.8). PET relies on physical phenomena which are based on discrete events that follow Poisson counting statistics, due to the nature of the radiotracer decay and the detection process. An expectation maximization (EM) method of image reconstruction (see Sect. 2.3.3.2) was shown by L.A. Shepp and Y. Vardi (1982) to be based on a proper stochastic model for such emission scans. Compared to FBP, EM provides images with reduced noise without excessive smoothing. Given as set of exact projections, EM image reconstruction results converge. Ordered subsets expectation maximization (OSEM) was developed about a decade later by H. Malcolm Hudson and Richard S. Larkin to implement EM with an order of magnitude acceleration in the processing time (Sect. 2.3.3.3). OSEM and its variations have become standard in nuclear medicine reconstruction, especially for PET. Initially, OSEM was applied only on sinograms for each individual transaxial slice, as appropriate for PET scanners operating in 2D mode. Sinograms from projection slices at oblique angles to the transaxial plane, from acquisitions in 3D mode, needed to be interpolated to equivalent sinograms in 2D mode. By the 1990s this was performed using an algorithm called Fourier rebinning or FORE [8]. Affordable computing power has enabled 3D OSEM algorithms to be developed that do not require such rebinning. Extensive 3D system matrices can be stored and used. These can model the point spread function [14] of the detector response (PSF modelling), which is somewhat equivalent to the collimator modelling (resolution recovery) described in Sect. 4.4.7 for SPECT. Typically, a scatter model is used in PET, with some algorithms tracking up to three scattering events for per photon. The schematic of iterative reconstruction shown in Fig. 4.27 remains relevant for PET.

Modern reconstruction methods for PET also incorporate TOF information when available. Besides OSEM, one manufacturer has also implemented RAMLA reconstruction (Sect. 2.3.4) and fashions reconstruction algorithms to be "blob"-based rather than voxel-based, since few biological structures follow a Cartesian layout. Bayesian reconstruction methods (Sect. 2.3.3.4) have also become available for clinical PET. These algorithms typically have an adjustable parameter or parameters that can be used to "tune" the amount of variation allowed from voxel to voxel. When used correctly, Bayesian reconstruction techniques can render sharp edges and small localized uptake, as can be done with many iterations of unmodified OSEM algorithms, but without amplifying the image noise associated with many EM iterations. This increases the signal-to-noise ratio with the potential of improving lesion detectability.

5.4.3   Attenuation Correction

Correcting for attenuation in PET is similar to that for SPECT in that a transmission scan can provide an attenuation map for the object, as discussed in Sect. 4.4.5. If the transmission scan is a CT, the Hounsfield units of the CT must be scaled to attenuation coefficients appropriate for 511 keV photons, as shown in Fig. 4.30 but with different scaling (e.g., 0 HU scales to 0.093 cm^{-1} for PET). Scaling also needs to be done if a γ-ray source is used. For example, ^{137}Cs transmission sources for PET have the primary photon peak at 662 keV, not 511 keV. However, if a ^{68}Ge/^{68}Ga pin source is used for the transmission scan, no such scaling is needed since these sources produce annihilation photons at 511 keV. As mentioned in Sect. 5.2.1, radionuclide pin sources provide noisier attenuation maps than CT and take much longer to acquire than CT [3]. Consequently, PET attenuation correction is virtually all based on CT (CTAC) in the clinic.

Figure 5.13 shows how attenuation correction can be applied in PET in an object with a homogeneous radiotracer distribution $\rho(x,y)$ and homogeneous attenuation coefficient throughout $\mu(x,y)$. Along a line of response in one direction from the annihilation event, the photon traverses a path length L_1 through the object before

Fig. 5.13 PET attenuation correction. For an object (gray) with a homogeneous radiotracer concentration ρ, and homogeneous attenuation coefficient μ, any annihilation event (star) along the LOR (straight black line) has the same probability of being not counted due to attenuation regardless of where it occurs along the LOR

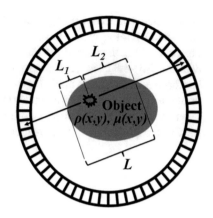

passing through the air to the detector ring. Comparatively, there is no attenuation of the 511 keV photons in air. In the other direction, the companion photon travels a distance L_2 through the object. If the probability P_1 of detecting a 511 keV photon entering a detector is ε, the probability of detecting the first photon is this efficiency ε multiplied by the probability that the photon will exit the object without attenuation. Using Eq. 4.3, this probability can be written as I/I_0, where I_0 is the intensity (photon per unit time) emitted from that annihilation location and I is the attenuated intensity:

$$P_1 = \varepsilon \frac{I}{I_0}$$

$$P_1 = \varepsilon e^{-\mu L_1}$$

A similar equation holds for the probability P_2 that the second photon will be detected. However, PET requires that both photons are detected in coincidence, so the probability of a coincidence detection P in this case is the product of P_1 and P_2:

$$P = P_1 P_2$$
$$P = \left(\varepsilon e^{-\mu L_1}\right)\left(\varepsilon e^{-\mu L_2}\right)$$
$$P = \varepsilon^2 e^{-\mu(L_1 + L_2)}$$
$$P = \varepsilon^2 e^{-\mu L} \tag{5.4}$$

The probability of coincidence detection, with respect to attenuation, depends only on the total path length L of the LOR through the object for the detector pair and the attenuation coefficient along that path. This holds true for any annihilation event along the LOR, even if it occurs outside the object. For example, if a ^{68}Ge/^{68}Ga pin source were placed by the first detector on the left of Fig. 5.13, the probability of a photon being detected along that LOR in that detector would be $P_1 = \varepsilon$. The companion photon would have to traverse the entire length L through the object before reaching the second detector with a probability of $P_2 = \varepsilon\cdot\exp.(-\mu L)$ of being detected. So with an external pin source, the probability of detection ($P = P_1 P_2$) also conforms to Eq. 5.4. It's unlikely that the detector efficiencies ε are identical, but the principle that the degree of attenuation depends on the total path length through the object remains [15]. The general argument can be extended to smaller sections along the LOR, and one can define an attenuation factor f along the LOR as

$$f = \exp\left(-\int_{\text{LOR}} \mu(x,y)ds\right)$$

where s is the distance along the LOR and $\mu(x, y)$ is the attenuation map defined on the x-y plane. Figure 5.7 shows that every LOR relates to a data point on the acquired sinogram $P_\theta(t)$, so an attenuation correction factor ACF can be define to "undo" the effects of attenuation:

$$\text{ACF}_\theta(t) = \exp\left(+\int_{\text{LOR}_\theta(t)} \mu(x,y)ds\right) \tag{5.5}$$

Since the attenuation map $\mu(x, y)$ can be determined from a transmission source outside the object, PET provides the potential to calculate an exact attenuation corrected sinogram P' by multiplying each point on the sinogram by the appropriate attenuation correction factor:

$$P'_\theta(t) = P_\theta(t) \cdot \text{ACF}_\theta(t) \tag{5.6}$$

The attenuation corrected sinogram can then be reconstructed by FBP or iterative reconstruction methods such as OSEM. However, the ACFs apply the largest scaling on LORs that have the most attenuation and therefore likely poor count statistics and more noise. Attenuation weighted OSEM (AW-OSEM) uses the statistics of the non-attenuation corrected sinogram and accounts for the attenuation by weighting it in a forward projection step and avoids this amplification of such noise.

 In PET, the attenuation along the path of a photon pair does not depend on the depth of the radioactive decay generating that annihilation pair.

Figure 5.14a shows a transaxial PET image with and without attenuation correction. An NAC (non-attenuation corrected) PET image is usually recognizable in that the outer contours of the body show artifactually higher radiotracer uptake than the expected background uptake in the body. LORs that are tangential to the body carry

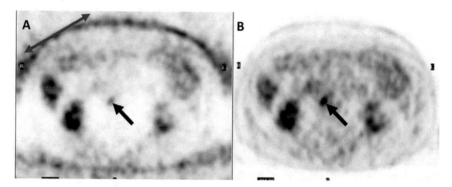

Fig. 5.14 PET CT attenuation correction. (**a**) In this non-attenuation corrected transaxial PET image, LORs tangential to the patient's body (e.g., gray double arrow) have the least attenuation and therefore a relatively stronger signal. A pathological uptake of ^{18}F-FDG in an abdominal lymph node is not visually obvious (black arrow). (**b**) With CTAC, the body contour shows typical background uptake of ^{18}F-FDG, and the pathological uptake of the radiotracer in an abdominal lymph node is visually evident (arrow)

the signal of the uptake near the surface, but have very little attenuation because they pass mostly just through air. Consequently, upon image reconstruction without attenuation correction, these LORs provide a strong signal localized to the surface of the body. In the CTAC PET image (Fig. 5.14b), the background uptake is more balanced and features deep within the body are more evident.

5.4.4   Scatter Correction

As in SPECT, scattered photons add to the blur in the PET projection data because the LORs of scattered events are displaced from the annihilation location (Fig. 5.7b). Energy discrimination can be used to remove some of these scattered events, since scattered photons will be less than 511 keV in energy. In PET detectors, the energy resolution is worse, and the peak energy window for acquisition is much wider, when compared with SPECT. As a percentage of the peak energy, the energy window can easily be twice as wide for PET (>40%) as those used for SPECT (often 20%). Consequently, using the DEW and TEW scatter correction techniques described in Sect. 4.4.6 for SPECT don't work well with PET.

With the ubiquitous use of PET/CT scanners, the CT readily provides a high-resolution map of, fundamentally, electron density of the anatomical structures of the patient. Since for 511 keV photons scatter is predominantly due to the Compton effect, the CT provides information as to the likely location and degree of scattering. Scatter modelling in clinical PET reconstruction can track up to three scattering events for any particular ray path of an annihilation photon. Conceptually, the blur due to scatter is subtracted from the emission projection data. Practically, it is modelled in the forward projection step of iterative reconstruction as diagrammed schematically in Fig. 4.27.

5.4.5   Point Spread Function Modelling

Because of geometric effects, the point spread function (PSF) is not typically the same throughout the FOV of the PET ring. For uptake near the center of the FOV, the relevant LORs also pass near the center and are therefore more or less normal to the detector faces at the PET ring circumference. The point at which the annihilation photon creates scintillation within the crystal, either near the point of entry or farther into the crystal toward the PMT, makes little difference in the positioning of these LORs. However, for LORs that pass farther away from the center, this is not the case. As evident from Fig. 5.7a, these LORs can be at very oblique angles to the detector face. Consequently, there is an uncertainty in the placement of the LOR that is dependent on the depth of interaction of the annihilation photon with the crystal of the detector. This uncertainty translates to a progressively poorer spatial resolution, that is, a broader PSF, for events placed more and more radially from the center of the FOV. This effect can be modelled within the system matrix used for iterative image reconstruction, analogous to the resolution recovery method described in Sect. 4.4.7 for SPECT. The geometry for PET is more complex than for SPECT, so in some cases a heuristic (experimental) mapping of the PSF variation throughout the FOV of a PET scanner has been performed, exploiting some symmetry to reduce the number of points to be determined by measurement.

5.4.6   Post-Processing

The most common post-processing performed on PET tomographic data is smoothing (filtering) with a Gaussian filter. Since the voxels are not isotropic, the smoothing in the axial direction (often called the z-axis) can be different from the smoothing in the transaxial plane. Typically, the filter is chosen such that the FWHM of the Gaussian matches or is close to the value of the spatial resolution: in the range of 3–8 mm FWHM, depending on the type of scanner and the requests of the physician reading the study. Since the purpose of filtering in post-processing is to suppress the image noise, high count images with good statistics might receive less smoothing (smaller FWHM for the Gaussian filter) than lower count images. Regardless, the purpose of image filtering is to aid the physician in reading the study and reaching a diagnosis, similar to the goals in post-processing SPECT as illustrated in Fig. 4.33.

The higher spatial resolution, straightforward attenuation correction methods, and greater count rates provided by clinical PET, compared to SPECT, have facilitated the practical application of quantitation and semi-quantitative methods routinely in the clinic. Dynamic PET might be further analyzed via time-activity curves to establish compartmental models for radiotracer uptake. For example, the ^{18}F-FDG uptake of a lesion might be modelled as an FDG-avid metabolically active lesion residing in surrounding tissue that uptakes FDG more slowly. Such radiotracer-kinetic models have the potential to aid diagnosis since differing pathologies are likely to have different rates of radiotracer uptake. Although commercial software to aid in this dynamic analysis has been available for several years, it is becoming more common to see this software as an option in the consoles of newer PET/CT scanners.

5.4.7 Quantitation

As mentioned in the previous paragraph, quantitative and semi-quantitative methods in PET have been used routinely in the clinic [16]. As mentioned in Sect. 5.3.5, quantitation requires that there is a calibration performed that scales the proportional counts represented in the reconstructed image and the absolute radiotracer concentration in Bq/mL. This is usually done periodically with a phantom containing a precisely known activity of the radiotracer of interest.

However useful the absolute radiotracer concentration might be, it will vary from scan to scan depending on the activity of radiotracer injected, the uptake time before the scan, and the weight of the patient. Usually, these three factors are corrected or normalized to remove their effect resulting in a quantity known as standardized uptake value (SUV). Figure 5.15 illustrates the normalization process. Conceptually, the SUV is an attempt to assign the same value to two lesions that are physiologically similar but exist within two different patients and scanned under different conditions. SUV is defined as

$$\boxed{\text{SUV} = \frac{C}{A/m}} \qquad (5.7)$$

where C is the voxel concentration in Bq/mL, A is the injected activity decay corrected to the start time of the acquisition, and m is the patient weight in grams (g). Here, the patient weight can be thought of as a volume surrogate, with $1\ \text{g} \approx 1\ \text{mL}$ for the human body, since our average density is close to the density of water. Considering this surrogacy, the denominator represents the average radiotracer concentration throughout the body, ignoring uptake eliminated in the urine right before scanning. Consequently, the SUV of the background in many tissues throughout the body is about 1, for an ^{18}F-FDG PET study. Ideally, this number would be unit-less, but because of the patient weight substitution for volume, the units are [g/mL]. Some organs typically uptake the radiotracer more readily, with the SUV of a liver being about 2–4 g/mL, and the brain can be >4 g/mL, for an ^{18}F-FDG PET study. For these studies, patients are asked to fast for at least 4 hours before the exam so that the skeletal muscles stop using glucose. Otherwise, the background uptake in the skeletal muscles decreases the contrast of any existing lesions.

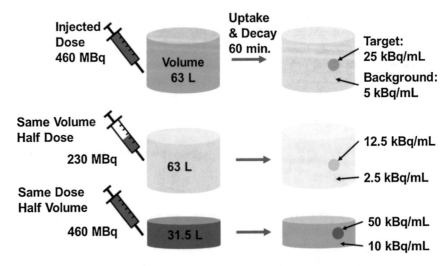

Fig. 5.15 PET SUV quantitation. A target-to-background ratio of 5:1 after uptake is shown for all three cases, but the higher uptake in the "lesion" ranges from 12.5 to 50 kBq/mL. SUV normalizes the uptake for injected dose activity at scan time and for patient volume, using patient weight as a surrogate. In all three cases, the SUV of the "lesion" is 5 g/mL and the SUV of the background is 1 g/mL

The SUV may be reported as an average (SUV$_{mean}$) within an ROI, but this tends to reduce the value reported as voxels with intense uptake are averaged with less intense voxels. Also, the placement of the ROI will affect this value. It is more common to use the maximum value (SUV$_{max}$) of the voxels within the ROI in order to reduce these effects. However, the value of SUV$_{max}$ is very sensitive to the amount of smoothing applied to the image either during or after reconstruction and is also sensitive to noise. To increase consistency, another measure of SUV is to translate a 1 mL spherical volume throughout the volume of the lesion of interest and report the maximum average of this process. This is called SUV$_{peak}$, although other definitions of SUV$_{peak}$ exist. While being less sensitive to noise because of the averaging, the SUV$_{peak}$ ROI is not user-dependent, and the small volume ensures that it is sensitive to the values of individual high-intensity voxels. Because it is more computationally intensive, SUV$_{peak}$ is not available as a measurement on all PET workstations. Additionally, some investigators have noted that ^{18}F-FDG uptake in body fat is minimal, so only lean body mass should be used in the calculation of Eq. 5.7 (SUV$_{lbm}$), or the SUVs will be higher than expected (e.g., a background tissue uptake much greater than 1 g/mL). Others have recommended corrected SUVs for blood glucose levels. Again, the goal of all these adjustments is to produce a measure that can relate the physiology of similar lesions even under different study conditions. Since each adjustment can introduce variability, SUV$_{max}$ has remained a popular in reporting because of its simplicity.

In addition to the variability found in SUV discussed above, accurate quantitation in PET is reliant on many other factors [17] which may each contribute some uncertainty, but all together can render variations in SUV of ±50%. In the clinic, quantitative measurement is used, for example, to track response to treatment for some oncology patients. At times, the knowledge of whether or not tumors are responding to treatment is important to the management of patient care and for prognostics. Being able to report that a lesion is reduced in volume and in metabolic activity can be valuable information. However, reporting that the SUV has decreased by 20% in, let's say, the past 3 months, is meaningless if variations are ±50% due to the measurement process. Some factors that must be controlled for accurate quantitation are the calibration of the dose calibrator used to measure radiotracer activity, calibration of the PET, measurement of the residual activity in the syringe after injection, the synchronization of all clocks involved, acquisition parameters, reconstruction parameters, patient wait time for uptake, and patient motion. For example, it can be shown from Eq. 4.2 that the percent change in the activity is equal to the decay rate constant (as a percent) with respect to a change in the unit time. The rate constant λ for ^{18}F is $\ln(2)/\tau$, where τ is the half-life. This gives a change of 0.6% per minute. For ^{68}Ga, this is 1.0% per minute, and for ^{11}C this is 3.4% per minute. Therefore, synchronization to within 1 minute of all clocks used is important for PET quantitation. This is easy to do, but one needs to be meticulous about this and the other factors, or the cumulative error is too large for the quantitation to be useful. The European Association of Nuclear Medicine and the American College of Radiologists have established accreditation programs to help harmonize the quantitative results among different scanners at different sites. Regardless, for tracking response to treatment, it is recommended that, where possible, to repeat follow-up scans on the same machine under the same conditions.

5.4.8 Hybrid Image Fusion

As mentioned in Sect. 4.4.9 concerning SPECT/CT image fusion, hybrid scans are usually displayed with CT and NM separately on one or more viewing screens. Fused images typically overlay artificially colored PET images on top of grayscale CT images. Figure 5.16 shows different color maps applied to the PET portion of the image, rendering the lesion with varying degrees of visual contrast. The color map chosen is usually determined by the diagnosing physician, and workstations designed to display hybrid PET/CT images allow color map selection on the fly. The color map cold "hot iron" is used frequently in publications of fused PET/CT images, giving a color scale tracking radiotracer intensity from dark, to red hot, to white in appearance. The "perfusion" color map has the odd characteristic of assigning "black" to very low value pixels, as expected, but also to medium-intensity pixels. This property assigns dark values to pixels on the intensity gradient surrounding an intense uptake, greatly increasing the contrast in the immediate neighborhood of intense uptake. While the artificially enhanced contrast of this and some

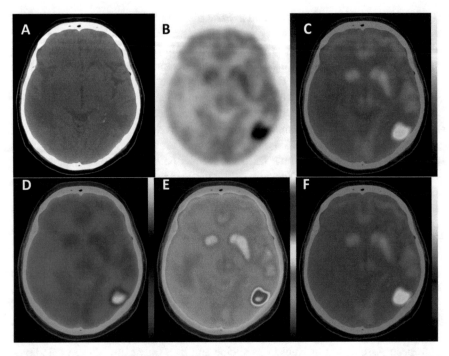

Fig. 5.16 Hybrid image fusion PET color maps. (**a**) Transaxial CT image through the brain. CT remains a grayscale in the fused images. (**b**) ^{18}F-FDG PET grayscale showing a lesion (darkest uptake) of lung cancer metastasized to the brain. (**c**) Fused PET/CT image with PET as a "hot iron" color map. (**d**) PET "perfusion" color map. (**e**) PET "French" color map. (**f**) PET "warm" color map

other color maps is desirable for some readers, other physicians prefer the more linear representation of radiotracer intensities provided by the "hot iron" color map.

5.4.9   Quality Control

Daily quality control for a PET scanner includes, at minimum, a check of the detector efficiencies, equivalent to a blank scan mentioned in Sect. 5.3.5. A long-lived positron emitter usually provides the source of annihilation photons. This is commonly either a ^{68}Ga/^{68}Ge source, as mentioned above, or ^{22}Na. The sources can be pin sources rotated close to the detector ring about the FOV, cylindrical phantom sources placed in the center of the FOV, or pin sources placed in the center of the detector ring. Ideally, all the LORs (detector pairs) would receive counts during daily QC (quality control), but only the first configuration achieves this. Regardless, daily QC provides a count rate to all detectors that can be tracked day-to-day for any unexpected variations. Primarily, this is to check if there has been a significant change in detector efficiencies. Other parameters that can be measured during daily

QC include tracking unexpected changes in the coincidence timing and the energy peak.

QC tests performed on a monthly or quarterly basis include WCC or the equivalent. Image quality and radiotracer quantitation can be checked monthly or quarterly, for example, by using a Jaszczak phantom (Fig. 4.35) modified by removing the cold sphere features and using an Esser lid. This lid provides cold features to check scatter correction and hot features of varying sizes to check SUV (Fig. 5.17) [18]. Manufacturers often provide phantoms to check PET/CT registration that can consist of long-lived sources used to provide a PET signal, embedded in small dense objects easily visible on the CT. Generally, PET and CT should be registered to within one voxel. This means that the location information encoded into the images of the PET and the CT allow them to be displayed relative to each other to a positioning error of no more than a voxel.

Generally, sensitivity, spatial resolution, and NECR are determined at the time of acceptance of a newly installed scanner and after any major change to the PET detectors. Some of these tests require special phantoms. Acceptance testing is usually performed at the PET site with the cooperation of the manufacturer who may temporarily supply special resources for the testing. As mentioned in Sect. 4. 4.10 for gamma cameras, these tests often are based on NEMA protocols [19]. Usually a NEMA body phantom (e.g., see Fig. 4.21a) or the equivalent is performed

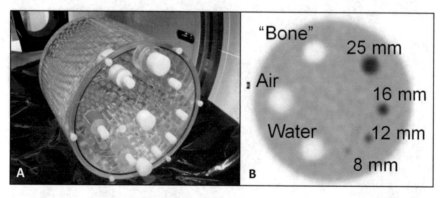

Fig. 5.17 Quality control: image quality test using a modified Jaszczak phantom with an Esser lid. (**a**) Phantom on PET bed. Blue solution within the phantom indicates a higher ^{18}F-FDG concentration. The red lasers from the CT gantry aid in positioning. (**b**) PET transaxial image through a slice with 4 "hot" regions (dark circles) loaded to have a concentration 2.5 times that of the background (gray). The hot regions range from 8 to 25 mm in diameter. "Cold" regions excluding radiotracer are used to estimate scatter into a bone surrogate (Teflon® - polytetrafluoroethylene), air, and water

yearly as an image quality check. Comprehensive QC schedules can be established by national regulation or from recommendations by the ACR (American College of Radiology), EANM (European Association of Nuclear Medicine), SNMMI (Society of Nuclear Medicine and Molecular Imaging), or IAEA (International Atomic Energy Agency) [20].

5.5   Clinical Example

As mentioned above, the ability of ^{18}F-FDG PET to provide functional images of glucose metabolism is a powerful tool for diagnosis and tracking response to treatment. Figure 5.18 shows a case that had been tracked over a number of years. The patient had prostate cancer metastasized to the colon 2 years later, both of which were dealt with by resection. The PET/CT scan in Fig. 5.18 was performed 5 years after the initial diagnosis showing further metastasis to the stomach and liver, with a fairly small 1.8 cm lesion in the lung that becomes very evident in the ^{18}F-FDG scan. Fused hybrid PET/CT images aid in the precise localization of the lesions.

Some of the first applications of whole-body ^{18}F-FDG PET included the diagnosis of cases with a single pulmonary nodules (SPN). In the absence of further metastasis, which is more clearly evident on PET than CT, SPN typically leads to resection giving a good prognosis. However, if there is metastasis as evident on PET, other treatments may be started immediately, obviating unneeded surgery.

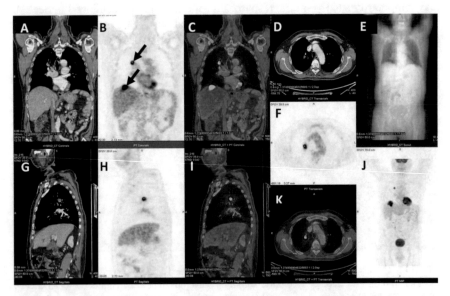

Fig. 5.18 Clinical example of ^{18}F-FDG PET/CT showing prostate cancer metastasized to the liver and lung (arrows) as well as other sites. (**a**) CT coronal. (**b**) PET coronal. (**c**) Fused PET/CT coronal. (**d**) CT transaxial. (**e**) CT scout scan. (**f**) PET transaxial. (**g**) CT sagittal. (**h**) PET sagittal. (**i**) PET/CT fused sagittal. (**j**) PET MIP. (**k**) PET/CT fused transaxial

5.6   Challenges and Emerging Directions

As described in the previous section, ^{18}F-FDG PET has been instrumental in oncology for diagnosis of disease. Whereas the PET scanners of the 1980s were primarily research devices suitable for metabolic brain studies, the 1990s saw the introduction of commercial whole-body PET scanners for the clinic. However, transmission scanning was done with radioisotopes leading to issues with cross-talk to the emission scan, relatively noisy transmission scan attenuation maps, and lengthy acquisition times. The advent of clinical PET/CT around 2001 provided a solution to these issues by means of the CT. About a decade later, the next innovation, TOF PET, became routinely available in the clinic. Additionally, afford-able computing power enabled the clinical application of more advanced reconstruc-tion algorithms. These last two developments lead to substantially improved image quality with reduced acquisition times and/or injected radiotracer activity. However, as with SPECT, PET remains photon-poor when compared with x-ray CT, giving noisier images and longer acquisition times. Also, although the CT of PET/CT systems provides anatomical localization of radiotracer uptake, CT soft tissue contrast is inherently poor sometimes making it difficult to localize uptake precisely within the brain or within poorly differentiated soft tissue.

The introduction commercial clinical systems that are a modality hybrid of PET and magnetic resonance imaging (MR) has begun to address the issue of localizing radiotracer uptake within soft tissue. MR provides excellent soft tissue contrast and is very flexible in the types of contrast it can provide. For example, in brain imaging, MR images typically clearly delineate gray matter, white matter, and cerebral spinal fluid, enabling more precise localization. Three vendors provide PET/MR designs with two of the designs incorporating the PET ring within the confines of the MR gantry. Because PMTs will not function correctly within high magnetic fields, these PET rings incorporate solid-state detectors. Figure 5.19 shows a clinical example from a PET/MR scanner. Since there is not the one-to-one mapping of x-ray photon attenuation to annihilation photon attenuation found in PET/CT scanners, special acquisitions of the MR must be performed to provide a reasonable attenuation map for PET. These special acquisitions continue to be developed and improved.

So-called "digital" PET systems with solid-state detectors have not just enabled the development of PET/MR scanners. As digital PET/CT hybrids, these scanners generally offer better spatial resolution and better TOF temporal resolution while maintaining or improving PET detector sensitivity [21]. Figure 5.20 shows an example of a PET/CT solid-state scanner. Some makes of these scanners may be offered with options such as continual bed motion [22]. The bed moves slowly through the scanner during the acquisition. This enables the user to choose an axial FOV of arbitrary length rather than being "locked-in" to discrete steps set by multiple bed positions. This can reduce scan duration by scanning only the ROI as well as reducing radiation exposure from the CT transmission scan by limiting it to this ROI. Other advances include projection-driven respiratory gating [23]: by monitoring the data in real time for motion due to respiration, the scanner can extend

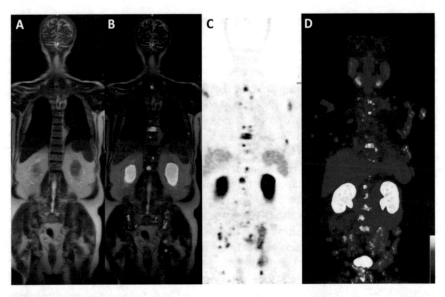

Fig. 5.19 Clinical PET/MR example of a 74-year-old man with metastatic prostate cancer. (**a**) Coronal HASTE MR (half-Fourier acquisition single-shot turbo spin-echo magnetic resonance imaging). (**b**) ^{68}Ga PSMA (prostate membrane antigen) fused PET/MR. (**c**) Coronal AC PET image. (**d**) PET MIP image. Demonstrating multiple PSMA-avid bone metastases. (Images were obtained on a MR-PET Scanner (Siemens, Erlangen, Germany) and are courtesy of Prof. David Groshar, Assuta Medical Center, Tel Aviv, Israel)

Fig. 5.20 Solid-state
PET/CT: Vereos (courtesy
of Philips, Haifa, Israel).
First commercial PET/CT to
have no PMTs

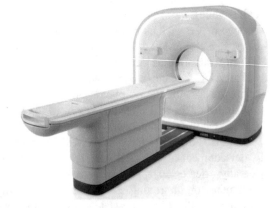

the scan time in the region of the motion (lungs or diaphragm) and retain only those counts with minimal motion (the quiescent breathing cycle [24]). Such motion correction can reduce blur in the lesion image [25]. Improved image reconstruction algorithms such as Bayesian reconstruction methods (Sect. 2.3.3.4) enable the iterative reconstruction to converge at a sharp image without making the image

Fig. 5.21 Extension of the axial FOV of the PET ring. For a point source (star) in the center of a PET ring (blue), many lines of response (red) will pass obliquely out of the PET ring over a large solid angle (Ω_1). Doubling the axial FOV reduces this solid angle ($\Omega_2 < \Omega_1$), so about twice as many LORs intersect the PET ring for potential detection, for this central point source

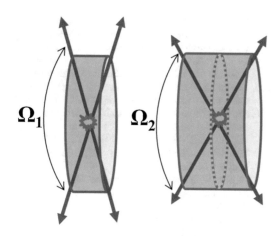

overly noisy. While some of these improvements are offered on their more traditional PET designs, most the major manufacturers of PET/CT systems offer a "digital" model with improved image quality when compared to their analog systems.

For a couple of decades, the standard PET ring for whole-body scanners had an axial FOV of about 15 cm. With a ring diameter ranging from 55 cm to 80 cm, this means that most of the potential LORs of photon pairs "escape" the ring at oblique angles and are lost to detection. Figure 5.21 illustrates that doubling the number of detector rings and thereby doubling the axial FOV greatly reduces the solid angle of the LORs that do not intersect the rings and cannot be detected. A "total-body" PET scanner has been built with an axial FOV of 2 m, giving potentially a 40-fold increase in PET sensitivity compared to the standard 15 cm FOV [26]. PET scans can then be done with greatly reduced injected activity or time. Also, total-body dynamic studies are feasible since the bed does not need to change position to cover all ROIs. These "total-body" PET scanners are being commercialized, but may remain uncommon at first because of the cost. Routine clinical scanners are available with axial FOVs of 25 or 26 cm, almost tripling the PET ring sensitivity compared to 15 cm FOVs, due to this improved geometry alone. Routine clinical scanners of the future may have axial bore lengths of 50 or 60 cm, but it remains to be seen if the benefits of such scanners would outweigh the costs.

References

1. Derenzo SE. Mathematical removal of positron range blurring in high-resolution tomography. IEEE Trans Nucl Sci. 1986;33(1):565–9.
2. Cherry SR, Sorenson JA, Phelps ME. Physics in nuclear medicine. 4th ed. Philadelphia: Saunders; 2012.

3. Kamel E, Hany TF, Burger C, Treyer V, Lonn AHR, von Schulthess GK, Buck A. CT vs [68]Ge attenuation correction in a combined PET/CT system: evaluation of the effect of lowering the CT tube current. Eur J Nucl Med. 2002;29:246–350.

4. Gambhir SS, Czernin J, Schwimmer J, Silverman DHS, Coleman RE, Phelps ME. A tabulated summary of the FDG PET literature. J Nucl Med. 2001;42:1S–93S.

5. Cherry SR, Dahlbom M, Hoffman EJ. Three-dimensional PET using a conventional multislice tomograph without septa. J Comput Assist Tomogr. 1991;14:655–68.

6. Schmand M, Eriksson L, Casey ME, Andreaco MS, Melcher C, Wiebard K, et al. Performance results of a new DOI detector block for a high resolution PET-LSO research tomography HRRT. IEEE Trans Nucl Sci. 1998;45(6):3000–6.

7. Tarantola G, Zito F, Gerundini P. PET instrumentation and reconstruction algorithms in whole-body applications. J Nucl Med. 2003;44(5):756–69.

8. Defrise M, Kinahan PE, Townsend DW, Michel C, Sibomana M, Newport DF. Exact and approximate rebinning algorithms for 3-D PET data. IEEE Trans Med Imag. 1997;16 (2):145–58.

9. Conti M. Focus on time-of-flight PET: the benefits of improved time resolution. Eur J Nucl Med Mol Imaging. 2011;38(6):1171–81.

10. Beyer T, Townsend DW, Brun T, Kinahan PE, Charron M, Roddy R, et al. A combined PET/CT scanner for clinical oncology. J Nucl Med. 2000;41(8):1369–79.

11. Kinahan PE, Townsend DW, Beyer T, Sashin D. Attenuation correction for a combined PET/CT scanner. Med Phys. 1998;25:2046–53.

12. Watson CC, Rappoport V, Faul D, Townsend DW, Carney JP. A method for calibrating the CT-based attenuation correction of PET in human tissue. IEEE Trans Nucl Sci. 2006;53 (1):102–7.

13. Bar-Shalom R, Yefremov N, Guralnik L, Gaitini D, Frenkel A, Kuten A, et al. Clinical performance of PET/CT in evaluation of cancer: additional value for diagnostic imaging and patient management. J Nucl Med. 2003;44(8):1200–9.

14. Arbizu J, Vigil C, Caicedo C, Penuelas I, Richter JA. Contribution of time of flight and point spread function modeling to the performance characteristics of the PET/CT biograph mCT scanner. Rev Esp Med Nucl Imagen Mol. 2013;32(1):13–21.

15. Ollinger JM, Fessler JA. Positron-emission tomography. IEEE Trans Sig Proc. 1997;14:43–55.

16. Coleman RE. Is quantitation necessary for oncological PET studies? For. Eur J Nucl Med Mol Imaging. 2002;29(1):133–5.

17. Boellaard R. Standards for PET image acquisition and quantitative data analysis. J Nucl Med. 2009;50:11S–20S.

18. American College of Radiology. PET phantom instructions for evaluation of PET image quality. Reston: ACR; 2013.

19. National Electrical Manufacturers Association. Performance measurements of positron emission tomographs: NEMA standards publication NU 2–2012. Rosslyn: NEMA; 2012.

20. IAEA. IAEA Human Health Series No. 1: quality assurance for PET and PET/CT systems. Vienna: International Atomic Energy Agency; 2009.

21. Nguyen NC, Vercher-Conejero JL, Sattar A, Miller MA, Maniawski PJ, Jordan DW, Muzic RF Jr, Su KH, O'Donnell JK, Faulhaber PF. Image quality and diagnostic performance of a digital PET prototype in patients with oncologic diseases: initial experience and comparison with analog PET. J Nucl Med. 2015;56(9):1378–85.

22. Dahlbom M, Reed J, Young J. Implementation of true continuous bed motion in 2-D and 3-D whole-body PET scanning. IEEE Trans Nucl Sci. 2001;48(4):1465–9.

23. Kesnera AL, Kuntner C. A new fast and fully automated software based algorithm for extracting respiratory signal from raw PET data and its comparison to other methods. Med Phys. 2010;37 (10):5550–8.

24. Liu C, Alessio A, Pierce L, Thielemans K, Wollenweber S, Ganin A, Kinahan P. Quiescent period respiratory gating for PET/CT. Med Phys. 2010;37(9):5037–43.
25. Xu Q, Xie K, Yuan K, Yu L, Wang W, Ye D. A statistical study of the factors influencing the extent of respiratory motion blur in PET imaging. Comput Biol Med. 2012;42:8–18.
26. Cherry SR, Jones T, Karp JS, Qi J, Moses WW, Badawi RD. Total-body PET: maximizing sensitivity to create new opportunities for clinical research and patient care. J Nucl Med. 2018;59:3–12.

Chapter 6
Magnetic Resonance Imaging (MRI)

Synopsis: In this chapter the reader is introduced to the phenomenon of magnetic resonance, with its associated physical aspects (i.e., the magnetic moment, susceptibility, magnetization precession, etc.). The reader will then learn about principles of MRI pulse sequencing and signal generation, the gradient fields, and the methodologies implemented for image formation and for flow imaging and magnetic resonance angiography (MRA).

The learning outcomes are: The reader will understand how the different magnetic fields interact with the magnetic moments and how the hydrogen proton magnetization can be manipulated, will know how an MRI signal is generated, will comprehend the different mechanisms of contrast in MRI, will know how spatial mapping is achieved in MRI, and will be able to follow a basic pulse sequence.

6.1 Introduction

Among the imaging modalities described in this book, MRI is apparently the most complicated one. It relies on a fine structured combination of physics, electrical engineering, computer science, mathematics, and of course medicinal knowledge. The current technology was achieved by numerous contributions from many people from different fields. Some may seem more fundamental, and some may seem relatively minor, but as in any puzzle board, every piece is essential to obtain the final picture. Consequently, it is not easy to comprehend how it works, and some

© Springer Nature Switzerland AG 2020
H. Azhari et al., *From Signals to Image*,
https://doi.org/10.1007/978-3-030-35326-1_6

parts are not obvious. I[1] recall the lecture of Paul Lauterbur (who received later a Nobel Prize for his contribution) given in the ISMRM meeting in Honolulu in 2002, where he described the development of the MRI scanner from his historical perspective. He told us that his group was given a large grant to build the "first" (see debate in Chap. 1) human scanner. The process was slow and complicated. And then one night at about 11 pm he received a panic phone call from his post doc. The post doc said: "I have gone all over the design and computations – it is never going to work!" Fortunately, for the world it did eventually work.

MRI is currently used all over the globe with various types and strengths of magnetic fields and configuration designs. It provides superb contrast in soft tissues. It can provide images with a 1 mm and even finer spatial resolution. It can be used to display the anatomy. It can display functionality of the heart and the brain (this field is known as fMRI). It can display blood flow and blood vessels and can be used to provide different contrast sources, such as diffusion, temperature, elasticity, and many more. MRI can also be combined with spectroscopy to study fine chemical aspects of tissue properties. And in recent years, it has been more and more used for image-guided minimal intervention procedures. At a certain time, it was thought that this modality would be an ultimate clinical "one-stop shop," meaning that it will provide all the radiological information needed in diagnosis and treatment. Although MRI is indeed very versatile and informative, nonetheless, it is still a very expensive imaging modality. It is cumbersome and slow to operate. The strong magnetic field and need to function under a "Faraday cage" configuration impose very limiting restrictions at and around the scanner in terms of metallic objects and RF-emitting equipment. And finally, the tight bore space highly limits access to the patient and creates a claustrophobic feeling in many people. Thus, until all the above drawbacks will be adequately solved, MRI will apparently remain a leading hazardless imaging modality but with limited availability to the public.

6.2 The Nuclear Magnetic Resonance (NMR) Phenomenon

In order to understand how MR images are produced, we have to go through many steps and understand several different physical phenomena as well as apply some mathematical derivations. To begin, let us start with the nuclear magnetic resonance (NMR) phenomenon. As outlined in the historical introduction in Chap. 1, the discovery of the phenomenon is attributed to Isidor Rabi, but the work of Felix Bloch and Edward Purcell brought it closer to become an applicable tool.

The basic setup for producing NMR signals is comprised of a strong magnet in the center of which the scanned object is placed (see Fig. 6.1). One antenna which is placed close to the object transmits radiofrequency (RF) waves in a direction perpendicular to that of the magnetic field. Another antenna which is placed

---

[1]HA

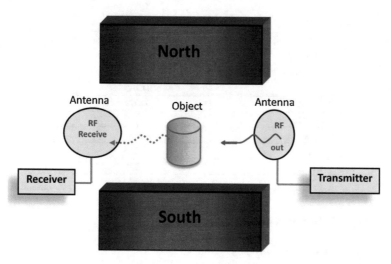

Fig. 6.1 The basic NMR setup. An object is placed within a strong magnetic field and is irradiated by RF waves. At the resonance frequency, the object responses by emitting waves with the same frequency

similarly to the first is used for receiving signals from the object. When transmitting bursts of RF waves with various frequencies toward the object, most of the time, the waves will pass through, and nothing will happen. However, at a certain frequency which is called the "resonance" frequency, the object responses by transmitting much weaker waves with the same frequency. The waves from the object decay fast. The phenomenon is very narrow-banded, i.e., slightly off-resonance transmissions will not yield any signal. Another important feature is that when the strength of the magnetic field is changed so does the resonance frequency. The relation is linear!

In addition to the above, it was found that on a relatively "macro" scale, the phenomenon is attributed to atoms and not to molecules. That is why the same frequency can be used to scan different tissues. On a fine frequency scale however, different materials will have different spectral signatures around the main frequency. This fact is used in an NMR subfield called magnetic resonance spectroscopy (MRS) to analyze molecular structures.

It is important to note that the NMR phenomenon is associated only with a certain group of atoms. A short list of atoms which are relevant to medical imaging is shown in Table 6.1. The most important atom in this context is hydrogen, which is used for medical imaging. Finally, it may be useful to know that electrons also have magnetic resonance effect, which is referred to as electron spin resonance (ESR).

Table 6.1 NMR frequencies for several atoms of interest [1]

Atom	Natural abundance [%]	Resonance frequency at 1 Tesla [MHz]
^{1}H	99.989	42.576
^{3}He	$1.34 \cdot 10^{-4}$	32.434
^{13}C	1.07	10.705
^{14}N	99.636	3.077
^{17}O	0.038	5.772
^{19}F	100	40.05
^{23}Na	100	11.262
^{31}P	100	17.235
^{129}Xe	26.4	11.777

6.3 Associated Physical Phenomena

6.3.1 Magnetic Fields

Magnets are considered attractive "creatures" which seem to operate in a mysterious way. They are used daily in numerous applications: from toys to electricity generators. Every child thinks that he/she knows how a magnet looks like with its famous North and South polarity. However, the truth is that magnets are not objects at all, but rather a manifestation of a physical phenomenon called the electromagnetic field. To be described in the most simplistic manner, it can be stated that similar to the fact that *static* electrical charges apply forces on each other, consequently their surrounding is regarded as an "electrostatic field." *Moving* electric charges, on the other hand, generate different forces. Hence, their surrounding is conceived as a "magnetic field." The simplest demonstration of this fact was done by André-Marie Ampère about two centuries ago. He put two loose conducting wires close to each other and observed how they attract or repel each other when electric currents ran through them. Similarly, by placing a compass next to an electric wire, one can observe the rotation of its needle to align with invisible rings surrounding the wire. It is therefore not surprising that the definition of the basic unit of the magnetic field strength (in CGS units), which is referred to as 1 Gauss, is the magnetic field which occurs at a distance of 1 cm from an electric wire running a current of 5 A (see Fig. 6.2). For comparison, the strength of the Earth's magnetic field varies from about 0.32 Gauss (at the equator) to about 0.65 Gauss (at the magnetic poles).

If, however, instead of using a straight conducting wire, a circular conductive ring is used and the ring is connected to an electric power source, then, in accordance with the "right-hand thumb rule," a magnetic field will be generated along a direction which is perpendicular to the plane in which the ring is located, as shown schematically in Fig. 6.3. The strength of this electromagnet is proportional to the current strength and inversely proportional to the area of the ring.

In MRI scanners the magnetic fields are much stronger than the Earth's magnetic field. Hence, they are measured in "Tesla" units where 1 Tesla $= 10^4$ Gauss. (Tesla is the standard SI unit.) Commercially available MRI scanners with permanent

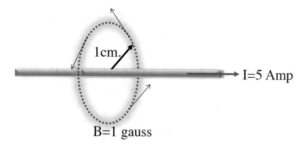

Fig. 6.2 The definition of the magnetic field strength unit of 1 Gauss is the magnetic field generated at a distance of 1 cm from a conducting wire running a current of 5 A. In accordance with the "right-hand thumb rule," the direction of the field is tangential to a circle surrounding the wire as indicated by the arrows

Fig. 6.3 A current loop will produce a magnetic field for which the central lobe is directed perpendicular to its plane. Its strength is proportional to the current and inversely proportional to the loop area

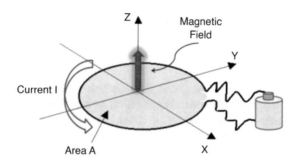

magnets commonly start with fields of about 0.1 T at the low end, while those using electromagnets have field strengths of up to 9 T and even higher. Currently, the most commonly used fields in the clinics are of 1.5 T, 3 T, or 7 T. This means that they are tens of thousands of times stronger than the magnetic field of Earth. The implication is that an enormous force will be applied onto any iron object which is placed near the magnet. In fact, this is one of the most common safety hazards. Many accidents have occurred at MRI sites because some maintenance equipment containing iron was mistakenly brought into the scanner room and was "grabbed" by the magnet (see pictures in the Internet). Apart from the danger to people who could be hurt during the process, the need to shut down the scanner for several days in order to release the trapped objects may also induce financial damage. In order to prevent such accidents, modern scanners have strong magnetic shields that substantially reduce the magnetic field outside the scanner. In addition, a red line marking of the 5 Gauss level (a standard safety limit) is commonly drawn on the floor around the scanner to warn people.

6.3.2 Magnetic Susceptibility: The Response of Matter to Magnetic Fields

Everyone knows that iron objects are strongly attracted to magnets. Moreover, they can be also magnetized by attaching them to a magnet. A simple experiment to demonstrate this phenomenon can be done by taking two paperclips and a magnet. When attaching one of the paperclips to the magnet, it will be magnetized and will be able to pull the other paperclip. The concept is schematically depicted in Fig. 6.4.

Less familiar is the fact that similar effect is also produced when other materials such as water, aluminum, diamonds, or even table salt are put inside a magnetic field. This stems from the fact that the effect is substantially much weaker. In general, the magnetization phenomenon is related to the strength of the magnetic field into which the material is inserted through a property which is called the *magnetic susceptibility*. The relation is given by

$$\overline{M} = \chi \cdot \overline{H} \tag{6.1}$$

where $\overline{H}$ is the magnetic field strength (its units are Ampere per meter: [A/m]), $\overline{M}$ is the magnetization (also in [A/m] units), and χ is the dimensionless susceptibility constant. Nonetheless, it should be noted that this relation is valid only for an isotropic material. In the case of medical imaging, the relation given in Eq. 6.1 is sufficient in most cases. In a more general case, however, for an anisotropic material, the relation may vary according to the spatial direction; hence, the susceptibility is actually defined by a susceptibility tensor.

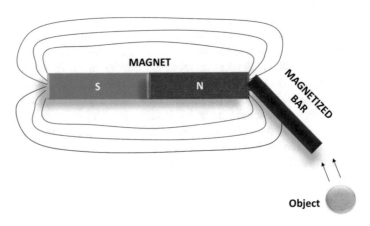

Fig. 6.4 A non-magnetic iron bar will be magnetized when attached to a magnet and will pull other iron objects

As a result of the magnetization phenomenon, the field in and around the magnetized material changes. The overall magnetic field which is also termed the magnetic induction is given by

$$\overline{B} = \mu_0\left(\overline{H} + \overline{M}\right) = \mu_0\overline{H} \cdot (1 + \chi) \tag{6.2}$$

where $\overline{B}$ is the resulting overall magnetic field strength and μ_0 is a dimensionless coefficient called the permeability coefficient.

Basically, in the context of magnetic susceptibility, there are three different types of materials:

I. *Diamagnetic* materials – When magnetized, these materials generate a field which *opposes* the external magnetic field. Accordingly, their susceptibility is negative, i.e., $\chi < 0$.

II. *Paramagnetic* materials – When magnetized, these materials generate a field which *enhances* the magnetic field. Accordingly, their susceptibility is positive, i.e., $\chi > 0$.

III. *Ferromagnetic* materials – These are actually paramagnetic materials with extremely strong susceptibility coefficients, i.e., $\chi \gg 0$. As the name implies, these materials contain iron atoms for which the Latin name is *Ferrum* (the corresponding symbol is *Fe*). Except for iron itself which has the strongest susceptibility, the second natural ferromagnetic material is called magnetite (Fe_2O_4).

Water and most of our body materials have diamagnetic susceptibility. Interestingly, although our red blood cells are made of a material called hemoglobin which contains four iron atoms, it is also diamagnetic in nature. Importantly, hemoglobin changes its susceptibility between oxygenated and non-oxygenated states. And this fact is used in functional MRI through an imaging method called BOLD (blood oxygenation level-dependent).

6.3.3 Magnetization of the Hydrogen Nucleus

MRI is mainly focused on the hydrogen nucleus which actually contains only one proton. Hydrogen is found almost everywhere in our body. In addition to water which constitutes about 60% of our body weight, hydrogen atoms constitute about half of the atoms in proteins [2], and fatty tissues are also very rich in hydrogen atoms. Thus, it is quite natural to choose it as a substance for medical imaging.

The response of hydrogen protons to magnetic fields may be viewed using two different complementary approaches: (i) classical physics and (ii) quantum physics.

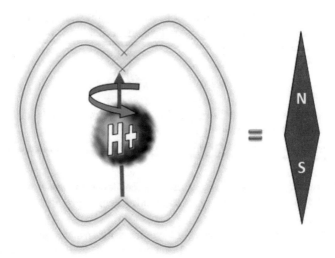

Fig. 6.5 Using a naïve approach, the proton is viewed as a positively charged spinning sphere. The rotating electric charge creates a magnetic field along the proton axis

A. *The Classical Physics Approach*

This extremely naïve and yet very practical approach considers the hydrogen proton to be a small spinning sphere with a single positive charge located on its surface as depicted in Fig. 6.5. The rotation of the electric charge around the proton axis is equivalent to a current loop. Consequently, a magnetic field is generated along the proton axis (as in Fig. 6.3). Hence, the proton can be considered as an equivalent to a small magnetic dipole as depicted.

Next, we recall that magnetic fields are directional. Consequently, when combining the magnetic fields from two or more hydrogen protons, vector summation should be applied. This implies that if the axes of the protons are spatially aligned, then the overall field will be larger than the individual fields. However, if the protons point at opposing directions, their fields will cancel each other, and the overall field will be small or even nulled.

As recalled, one mole of water weighs about 18 gr and contains two Avogadro's number of hydrogen atoms (about $1.2 \cdot 10^{23}$ atoms). Accordingly, a voxel of water with a size of 1 mm^3 will contain about $6.66 \cdot 10^{18}$ hydrogen atoms (quite a large number). If this voxel is not subjected to any external magnetic field, then this multitude of hydrogen atoms will be randomly oriented in space. As a result, the vector summation of all their magnetic fields will be nulled on the average. On the other hand, when inserted into the strong magnetic field of the MRI scanner (marked as B_0), the protons will tend to align with the field (similar to compass needles). Consequently, the vector summation of all their magnetic fields will add up to an average magnetization (magnetic moment per unit volume) of magnitude M_0. To slightly complicate the model and make it more realistic, it should be pointed out that the hydrogen protons – which are commonly referred to as "spins" – may reorient along two opposite directions. One direction is called "parallel," and the other is called "anti-parallel." The distribution between the two orientations is uneven. Hence, they do not entirely cancel each other, and the excessive magnetization is actually what we refer to as M_0, as depicted schematically in Fig. 6.6.

B. *The Quantum Physics Approach*

Using this approach we first relate to the Zeeman effect. When an atom with a spin number I is exposed to an external magnetic field, $(2I + 1)$, energy levels are generated. This means that the subatomic particles can be located in only one of these $(2I + 1)$ levels. Therefore, in order to change position from one level to another, the particle has to either absorb (going up) or emit (going down) a photon with a frequency f which equals $f = \Delta E/h$, where ΔE is the energy gap between the levels and h is Planck's constant ($h = 6.622 \cdot 10^{-34}[J \cdot sec]$), as depicted schematically in Fig. 6.7.

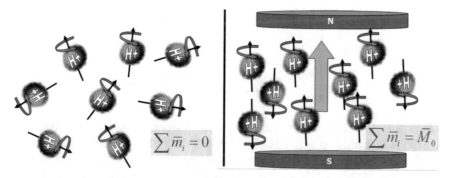

Fig. 6.6 (Left) Without any external magnetic field, the spins (hydrogen protons) in a voxel are randomly oriented in space. As a result, the overall magnetization of the voxel will be zero. (Right) When put inside the magnetic field of the MRI scanner, the spins will reorient either parallel or antiparallel to the external field. The net magnetization will be Mo. (Note: the arrows directions do not represent the spinning orientation)

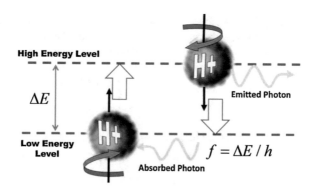

Fig. 6.7 According to Boltzmann's effect, under a magnetic field, the hydrogen protons have two energy levels (spin up, low; spin down, high). In order to shift from one level to the other, a photon has to be absorbed or released

In the case of a hydrogen proton, the spin number is $I = 1/2$. Consequently, there will be only two energy levels at which the proton can be positioned:

$$
\begin{cases}
E_1 = -\dfrac{1}{2}\gamma \cdot \hbar \cdot B_0 & \text{Low Energy Level} \\[2mm]
E_2 = +\dfrac{1}{2}\gamma \cdot \hbar \cdot B_0 & \text{High Energy Level}
\end{cases}
\tag{6.3}
$$

where B_0 is the magnetic field of the MRI scanner, h is Planck's constant, $\hbar = h/2\pi$, and γ is called the gyromagnetic ratio which value differs for different atoms. In this case it has a value applicable for the hydrogen proton ($\gamma = 2.675 \cdot 10^8$ [rad/s $\cdot$ T]).

As can be noted, the energy gap between the two levels is linearly proportional to the scanner's magnetic field B_0. The population of hydrogen protons is distributed between the two levels according to Boltzmann's distribution. The probability of finding a proton in each energy level is thus given by

$$
\begin{cases}
P_1 = \dfrac{\left[e^{+\frac{mB_0}{2kT}}\right]}{\left[e^{+\frac{mB_0}{2kT}} + e^{-\frac{mB_0}{2kT}}\right]} & \text{High Energy} \\[6mm]
P_2 = \dfrac{\left[e^{-\frac{mB_0}{2kT}}\right]}{\left[e^{+\frac{mB_0}{2kT}} + e^{-\frac{mB_0}{2kT}}\right]} & \text{Low Energy}
\end{cases}
\tag{6.4}
$$

where $m = \gamma \cdot \hbar \cdot I$ [Joule per Tesla : J/T] is the magnetic moment of the proton (see the following), k is Boltzmann's coefficient, and T is the temperature in Kelvin units.

The net magnetization (magnetic moment per unit volume) for the hydrogen proton population M_0 [J/T $\cdot$ m^3] is thus determined by the gap between the two energy level distributions, i.e.,

$$
M_0 = \rho \cdot m \cdot (P_1 - P_2) = \rho \cdot \gamma \cdot \hbar \cdot \frac{1}{2} \cdot \frac{\left[e^{+\frac{\gamma \cdot \hbar \cdot B_0}{2kT}} - e^{-\frac{\gamma \cdot \hbar \cdot B_0}{2kT}}\right]}{\left[e^{+\frac{\gamma \cdot \hbar \cdot B_0}{2kT}} + e^{-\frac{\gamma \cdot \hbar \cdot B_0}{2kT}}\right]}
\tag{6.5}
$$

where ρ is the number of protons per unit volume which is called the "proton density (PD)." (Make sure not to confuse it with the Archimedean density.) Using a Taylor series expansion, the exponential terms are approximated by the first term only, i.e.,

$$
\begin{cases}
e^x \approx 1 + x \\
e^{-x} \approx 1 - x
\end{cases}
\tag{6.6}
$$

Hence, Eq. 6.5 simplifies to

$$M_0 = \frac{\rho \cdot \gamma \cdot \hbar \cdot (1 + x - 1 + x)}{2(1 + x + 1 - x)} = \frac{\rho \cdot \gamma \cdot \hbar}{2} \cdot \frac{2x}{2} = \frac{\rho \cdot \gamma \cdot \hbar}{2} \cdot \frac{\gamma \cdot \hbar \cdot B_0}{2kT} \tag{6.7}$$

or

$$M_0 = \frac{\rho \cdot (\gamma \cdot \hbar)^2 \cdot B_0}{4kT} \tag{6.8}$$

As can be noted, the magnetization is linearly proportional to the scanner's main field B_0 and to the proton density ρ, but inversely proportional to the temperature T. These relations have important implications. First of all, since the MRI signal strength is proportional to the magnitude of M_0, it follows that higher magnetic fields for the scanner may lead to better SNR. It is therefore well understood why the MRI scanner fields have continuously been increased over the years. Secondly, it should be pointed out that the proton density is unrelated to the Archimedean density. The PD can be higher for less dense materials. For example, fatty tissue composition which has typically high number of hydrogen protons in a voxel will lead to stronger signals in PD-weighted images, as compared to muscle tissue which is denser. Finally, the reciprocal relation to temperature potentially offers a tool for noninvasive temperature mapping with MRI (other methods will be described in the following).

6.3.4 The Magnetic Moment and Its Response to an External Magnetic Field

The magnetic moment is another aspect of the magnetic field and can be conceived as the other face of the same coin. As we all know, a compass is actually a needle-shaped magnetic dipole which has the mechanical freedom for planar rotation about its center. When placed within an external magnetic field (such as that of Earth), it will rotate until its axis aligns with the direction of that field (see Fig. 6.8).

Fig. 6.8 A compass with a magnetic moment $\overline{M}$ subjected to an external magnetic field $\overline{B}$ which is not aligned with its axis will experience a torque $\overline{T}$ which will cause it to rotate

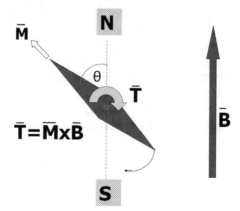

In order to obtain a rotational motion, a torque (force · distance) must be applied on the compass needle. The stronger the magnetic field of the needle or the external field, the stronger is the torque applied. The magnetic moment is therefore a vector which quantifies this relation. Hence, its units are [force · distance/magnetic field strength]. To describe the phenomenon in quantitative terms, consider a compass needle which has a magnetic moment $\overline{M}$ [N · m/T] = [J/T], subjected to an external magnetic field $\overline{B}$ [T] which is not aligned with its axis. The magnitude of the resulting torque $\overline{T}$ [N · m] is given by

$$\overline{T} = \overline{M} \times \overline{B} \qquad (6.9)$$

The cross-vector product indicates that the torque is proportional to $\sin(\theta)$ where θ is the inclination angle between the two magnetic fields. Thus, at $\theta = 90°$, the torque will be maximal, and its value will be zero when the two fields align, i.e., $\theta = 0$.

Following the above, we realize that when the tissue voxel has been placed within the magnetic field of the MRI scanner $\overline{B}_0$, it will have a magnetization (magnetic moment per unit volume) of $\overline{M}_0$. Thus, under equilibrium conditions, after the spins had enough time to adjust to the scanner field, the two fields will be parallel to each other, i.e., $\overline{M}_0 \| \overline{B}_0$. It should be noted that when using the commonly accepted MRI coordinate system, the direction of the scanner field $\overline{B}_0$ is always taken as the vertical coordinate, namely, axis $\hat{z}$.

If somehow (as will be explained in the following) we manage to apply a perturbation that will incline the magnetization by an angle θ relative to $\overline{B}_0$, a torque will be applied on the magnetization and will cause it to return to its equilibrium state, meaning $\overline{M}_0 \| \overline{B}_0$.

6.3.5 Precession

Proceeding with the simple classical physics model, it is recalled that in addition to its electrical charge, the proton has also a mass which spins about its axis. Therefore, it has a moment of inertia and an angular momentum $\overline{L}$ [J · s]. This implies that when no external torque is applied to it, it will preserve the spatial orientation of its rotation axis (in the same manner as gyro or rotating bicycle wheels do). It should be recalled that the angular momentum has magnitude and direction and therefore is treated as a vector.

When subjected to a torque $\overline{T}$, the angular momentum of a spinning object will change. The relation can be derived from Newton's second law and is given by

$$\int \overline{T} dt = \overline{L} + \text{Const.} \qquad (6.10)$$

where the constant represents the angular momentum of the object before applying the torque. In the context of MRI, it is preferable to use the derivative formulation of this relation. This relation which will be used in the next sections is given by

$$\frac{d\overline{L}}{dt} = \overline{T} \tag{6.11}$$

Considering a voxel with its multitude of spinning protons (still using the description of the classical physics model), we can assume that the summation of all its angular momentums can be presented by a single angular momentum $\overline{L}$ which direction coincides with that of its magnetization vector $\overline{M}_0$. Consequently, the torque generated by the magnetic field will alter the voxel's angular momentum as well. The resulting relation can be found by substituting Eq. 6.9 into Eq. 6.11 yielding

$$\frac{d\overline{L}}{dt} = \overline{T} = \overline{M} \times \overline{B} \tag{6.12}$$

As recalled, the gyromagnetic ratio, γ, of a system is the ratio of its magnetic moment to its angular momentum [3], i.e.:

$$\overline{M} = \gamma \cdot \overline{L} \tag{6.13}$$

Therefore, the motion of the magnetization vector can be described by the following differential equation:

$$\frac{d\overline{M}}{dt} = \gamma \cdot \overline{M} \times \overline{B} \tag{6.14}$$

This is a very important relation, as it can predict the response of the magnetization vector which is the source of the MRI signal, to changes in the magnetic field. However, it does not provide the full motion description yet! As will be introduced in the following sections, there are also two relaxation phenomena which affect the magnetization vector and should be accounted for.

As can be observed from Eq. 6.14, any change in the magnetic field and/or the magnetization orientation will result in a rotational motion of the latter. However, as the voxel tries to retain its angular momentum as well, it will behave like a spin top. Thus, in response to a perturbation from the equilibrium state, it will not realign with the magnetic field as a compass needle does. Instead, it will have a precession motion. The precession has an angular frequency ω_0 which is known as the "Larmor frequency" [4]. For a static magnetic field B_0, the Larmor frequency is actually the NMR resonance frequency and is given by

$$\omega_0 = \gamma \cdot B_0 \tag{6.15}$$

6.3.6 The Rotating Reference Frame

As will be explained in the following, there are two relaxation mechanisms that affect the magnetization motion. These mechanisms will cause the diminishing of the transverse components of the magnetization, M_x and M_y, and rebuilt its vertical component M_z. If we could imagine the magnetization vector as a stick rotating in three dimensions, then its axis would spiral to its original orientation (Fig. 6.9), similar to a "Dreidel" (a Jewish toy) but in reversed motion (rising up instead of falling).

This spiraling motion is rather complicated to describe in the stationary laboratory reference frame. Things look much simpler if the motion is described in a rotating reference frame. If the reference frame is taken to rotate at exactly the Larmor frequency ω_0, and if the magnetization vector is taken to be within the $y' - z'$ plane, then there will be no precession, and its motion would portray a simple rotation about the x'-axis as depicted schematically in Fig. 6.10.

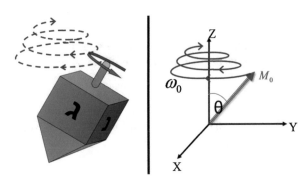

Fig. 6.9 (Left) Playing in reverse mode a video of a falling spinning toy resembles the precession motion of the reorienting magnetization vector (Right)

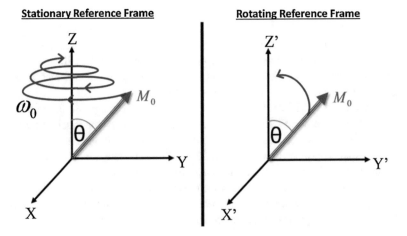

Fig. 6.10 (Left) The three-dimensional motion of the magnetization vector returning to its equilibrium state in a stationary reference frame. (Right) The same motion looks much simpler in a rotating reference frame

Fig. 6.11 The magnetic field B_1 is generated by transmitting from two orthogonal antennas along the X and Y directions of the scanner coordinate system. Naturally, it is also orthogonal to the main field B_0

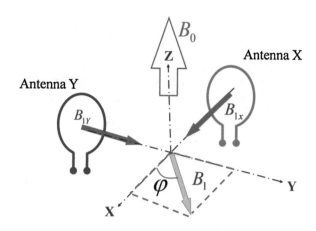

6.3.7 The Magnetic Field B_1

Thus far, only the static field of the MRI scanner B_0 has been described. As stated above, this field direction is defined as the Z-axis. At this point we add to the system a pair of RF antennas. The first is oriented along the X-axis and the second along the Y-axis as depicted in Fig. 6.11. By connecting the antennas to an electric power source, two orthogonal fields are generated. (This process is referred to as "RF transmission.") The first field is called B_{1x}, and the second one is called B_{1y}. These two fields are also orthogonal to the main field B_0.

As recalled, magnetic fields are subjected to vector summation. Therefore, the net outcome of this transmission will be a field called B_1 which orientation is defined by the ratio of the two fields and is orthogonal to B_0.

Next, consider a situation where the magnitudes of the two fields vary in time according to

$$\begin{cases} B_{1_x} = B_1 \cdot \cos\left(\omega \cdot t\right) \\ B_{1_y} = -B_1 \cdot \sin\left(\omega \cdot t\right) \end{cases} \tag{6.16}$$

This will result in a clockwise rotation of B_1 about the Z-axis with an angular frequency ω since

$$\overline{B}_1 = B_{1_x} \cdot \widehat{x} + B_{1_y} \cdot \widehat{y} \tag{6.17}$$

Accordingly, the instantaneous phase of B_1 will be given by

$$\varphi = -\omega \cdot t \tag{6.18}$$

And the field magnitude equals

$$|B_1| = B_1 \cdot \sqrt{\cos^2(\omega \cdot t) + \sin(\omega \cdot t)^2} = B_1 \cdot 1 = B_1 \qquad (6.19)$$

It should be pointed out however that the magnitude of B_1 can also be modulated in time, i.e., $B_1 = B_1(t)$. In any case, the combined magnetic field which the magnetization will be subjected to has now three components:

$$\overline{B} = B_1 \cos(\omega \cdot t) \cdot \widehat{x} - B_1 \sin(\omega \cdot t) \cdot \widehat{y} + B_0 \cdot \widehat{z} \qquad (6.20)$$

Substituting this term into Eq. 6.14 yields the following differential equation:

$$\frac{d\overline{M}}{dt} = \gamma \cdot \overline{M} \times \overline{B} = \gamma \begin{vmatrix} \widehat{x} & \widehat{y} & \widehat{z} \\ M_x & M_y & M_z \\ B_1 \cos(\omega \cdot t) & -B_1 \sin(\omega \cdot t) & B_0 \end{vmatrix} \qquad (6.21)$$

which can be presented as a set of three mutually dependent differential equations

$$\begin{cases} \dot{M}_x = \gamma \cdot M_y B_0 + \gamma \cdot M_z B_1 \sin(\omega \cdot t) \\ \dot{M}_y = -\gamma \cdot M_x B_0 + \gamma \cdot M_z B_1 \cos(\omega \cdot t) \\ \dot{M}_z = -\gamma \cdot \left[M_x B_1 \sin(\omega \cdot t) + M_y B_1 \cos(\omega \cdot t) \right] \end{cases} \qquad (6.22)$$

where $\dot{M}_x, \dot{M}_y, \dot{M}_z$ are the temporal derivatives of the magnetization along the x, y, and z directions, respectively. Without getting into an elaborated derivation, it can be shown that if the magnetic field B_1 is transmitted with an angular frequency ω which equals the Larmor frequency, i.e., $\omega = \gamma B_0$, the magnetization vector will be tilted by an inclination (flip) angle θ which is given by

$$\theta(t) = \gamma \int_0^t B_1(t) dt \qquad (6.23)$$

Moving to the rotating reference frame which rotates also at the Larmor frequency about the $z' \| z$-axis, the magnetic field B_1 direction is observed as fixed along the x' direction. The magnetization vector will thus rotate within the $y' - z'$ plane.

The implication of Eq. 6.23 is highly significant! It implies that we can manipulate the orientation of the magnetization vector $\overline{M}$ by turning ON and OFF the field $\overline{B}_1$ and we can control the flip angle θ by modulating the transmission amplitude and duration.

6.3.8   Off-Resonance and the Effective Field B_1 (For Advanced Reading)

The above methodology assumes ideal conditions whereby the B_1 field rotates at exactly the resonance frequency $\omega_0 = \gamma B_0$. However, in practice the transmission frequency of the B_1 field which is designated here on as ω_{RF} may be slightly at offset, and there would be an off-resonance frequency of $\Delta\omega = (\gamma B_0 - \omega_{RF})$. As a result, the rotating reference frame will rotate faster or slower than the Larmor frequency. Consequently, precession of the magnetization will be observed in the rotating reference frame as well. This precession can be conceived as stemming from a virtual magnetic field (see Fig. 6.12), which magnitude is given by [5]

$$\bar{B}_{Virtual} = \left(\bar{B}_0 - \frac{\bar{\omega}_{RF}}{\gamma}\right) = \frac{\bar{\omega}_0}{\gamma} - \frac{\bar{\omega}_{RF}}{\gamma} = \frac{1}{\gamma}(\bar{\omega}_0 - \bar{\omega}_{RF}) \tag{6.24}$$

or

$$\bar{B}_{Virtual} = \frac{1}{\gamma}(\Delta\omega) \cdot \hat{z} \tag{6.25}$$

Resulting from the virtual field, the magnetization will sense an effective field which direction is determined by the vector summation of $\bar{B}_1$ and $\bar{B}_{Virtual}$. And since $\bar{B}_1$ is commonly taken along the $\hat{x}'$ direction and $\bar{B}_{Virtual}$ is taken along the $\hat{z}'$, the effective field $\bar{B}_{Effective}$ will be located in the $x' - z'$ plane of the rotating reference frame (see Fig. 6.13), and its magnitude is given by

Fig. 6.12 An off-resonance RF transmission yields a virtual magnetic field in the rotating reference frame

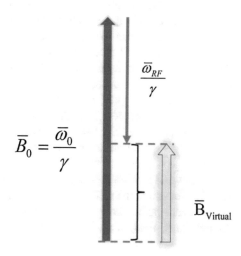

Fig. 6.13 Resulting from
the off-resonance and its
corresponding virtual field,
the magnetization will sense
an effective field $\bar{B}_{\text{Effective}}$
which is located in the
$x' - z'$ plane of the rotating
reference frame

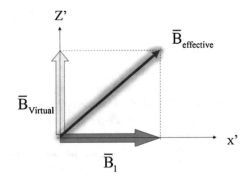

$$|\bar{B}_{\text{eff}}| = \sqrt{\left(\frac{\Delta\omega}{\gamma}\right)^2 + B_1{}^2} \tag{6.26}$$

As a result, the magnetization will rotate around the effective field $\bar{B}_{\text{Effective}}$ instead of around $\bar{B}_1$. This motion will portray a cone around the effective field. The angular frequency Ω in this case is given by

$$\Omega = \gamma \cdot B_{\text{eff}} = \sqrt{\Delta\omega^2 + (\gamma B_1)^2} \tag{6.27}$$

This phenomenon has two implications. On the negative side, it implies that if the transmission frequency is inaccurate, the magnetization will rotate in 3D in an unpredicted manner – even in the rotating reference frame. Consequently, the flip angle calculated by Eq. 6.23 will be erroneous.

On the positive side, it can be used to manipulate the orientation of the magnetization in a different manner by using long frequency varying RF excitation pulses which are called "adiabatic pulses." The idea behind this approach is as follows: (i) The pulse starts with a very large off-resonance $\Delta\omega$. As a result the effective field $\bar{B}_{\text{Effective}}$ is very close to the z'-axis. The magnetization $\bar{M}$ will have precession around it depicting a tightening conical motion. (ii) Given enough time [5] to allow the magnetization to follow $\bar{B}_{\text{Effective}}$, the frequency is swept continuously toward on-resonance, i.e., $\Delta\omega$ becomes smaller. As a result, the effective field $\bar{B}_{\text{Effective}}$ moves in the $x' - z'$ plane of the rotating reference frame until it aligns with the $\bar{B}_1$ field. During the process, the magnetization will keep its tight precession motion around the effective field $\bar{B}_{\text{Effective}}$. Consequently, it will be "dragged" in the $x' - z'$ plane, and its spatial orientation can be controlled.

This type of pulses is much less common than the regular pulses. The interested reader can find a more detailed description and design principles of such pulses in [6]. To summarize the difference between the two types, it is noted that the regular on-resonance pulses along the x' direction cause rapid nutation of the magnetization around the $\bar{B}_1$ field, i.e., within the $y' - z'$ plane, if we start from equilibrium state.

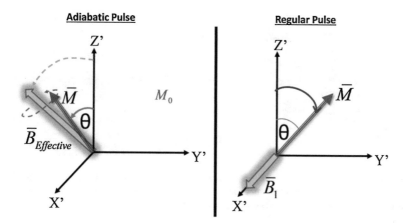

Fig. 6.14 (Left) Adiabatic pulses move the effective field $\overline{B}_{\text{Effective}}$ by changing the off-resonance frequency in the $x' - z'$ plane. The process is sufficiently slow to allow the magnetization to follow. (Right) In a regular pulse, the magnetization rotates rapidly around the field $\overline{B}_1$, i.e., within the $y' - z'$ plane in this case

Fig. 6.15 Flipping the magnetization $\overline{M}$ by an angle θ yields two components: the vertical magnetization, M_z, and the transverse magnetization, M_{xy}. The phase is marked as φ.

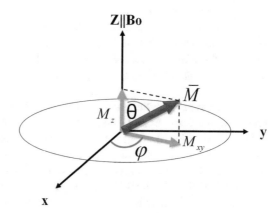

The adiabatic pulses on the other hand cause relatively slower motion but within the $x' - z'$ plane. This comparison is depicted graphically in Fig. 6.14.

6.3.9 The MRI Signal Source

Regardless of the method applied to flip the magnetization, the corresponding outcome is that the magnetization has now two distinct components (1) the vertical magnetization, M_z, and (2) the transverse magnetization, M_{xy}, as shown in Fig. 6.15,

As recalled, the physical principle which is the basis for electrical generators is the electromagnetic induction phenomenon, also defined as Faraday's law of

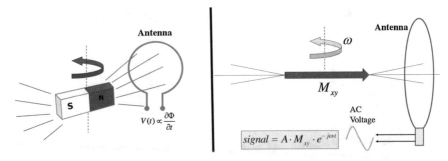

Fig. 6.16 (Left) A magnet rotating near a wire loop will induce voltage at its ends. (Right) Similarly, the rotating transverse magnetization, M_{xy}, will induce voltage in an antenna which axis is within the x-y plane. This is actually the MR signal

induction. It states that when the magnetic flux Φ through a wire loop changes, an electromotive force is generated in the wire. The magnitude of the electromotive force is proportional to the flux's temporal derivative (Fig. 6.16). Similarly, the rotating transverse magnetization M_{xy} will induce voltage in an antenna which axis is within the x-y plane, as shown in Fig. 6.16.

The induced voltage is actually the source of the MR signal. Stemming from the above, it can be deduced that the signal strength is proportional to the magnitude of $\overline{M}_{xy}$ and, therefore, the maximal signal will be obtained for a flip angle: $\theta = 90°$ or $\theta = 270°$. On the other hand, the signal will be nulled if the flip angle equals $\theta = 0°$ or $\theta = 180°$.

In order to retrieve the phase φ as well, two orthogonal antennas are positioned similar to the configuration used for transmitting the field B_1 (see Fig. 6.11). One antenna is positioned along the x-axis, and its detected signal is designated as the *real* part of the signal. The other antenna is positioned along the y-axis and is considered to detect the *imaginary* part of the signal. Thus, each data point collected from the MR signal is treated as a complex number.

Alternative explanation to the MR signal source using quantum physics description is as follows: The excitation of the tissue by the field B_1 changes the proton distribution between the two energy levels (see Eq. 6.4). Some protons absorb the RF photons and change their state from the lower energy level into the higher energy level (see Fig. 6.7). Consequently, the energy level partition is altered relative to the equilibrium state. When the field B_1 is turned off, the system (i.e., the tissue in the voxel) has excessive energy and is at an unstable state. The protons then start to return back to their equilibrium distribution by releasing RF photons at Larmor frequency. This relaxation process resembles the discharging of a loaded capacitor.

6.3.10 The Spin-Lattice and Spin-Spin Relaxation Processes

Following RF transmission, i.e., by turning ON the field B_1 and flipping the magnetization by an angle of $\theta = 90°$, for example, and then turning the field B_1 OFF, the MR signal is generated as described above. If the signal is acquired and plotted as a function of time, it is observed that it decays in time as shown schematically in Fig. 6.17. This is commonly referred to as the "free induction decay" or FID for short.

Careful examination of the decay process reveals that it stems from two different mechanisms. The first mechanism is called the "spin-lattice relaxation." Using classical physics it corresponds to the realignment of the magnetization $\overline{M}$ with the main field B_0, i.e., along the z-axis. Similar to the manner by which a compass needle realigns with the north-south direction of the Earth's magnetic field, the inclination angle is reduced with time, and the vertical magnetization component M_z is increased, as depicted schematically in Fig. 6.18. The process has a characteristic exponential behavior with a time constant called T_1.

The second process is called the "spin-spin relaxation." It stems from the growing incoherency of phase between spins in the same voxel. It can be understood as a micro-level process. As recalled, each voxel contains a huge number of protons located in water or in other molecules, where many of which are moving about in a random Brownian pattern, tumbling and bouncing and clashing into each other. As a result, the local magnetic field around each proton will vary randomly in time. Although the change may be very small and abrupt, it is significant enough to slightly change the local resonance frequency and insert a small phase change. Certain molecules will sense somewhat higher magnetic field and rotate faster, while others will sense somewhat lower field and rotate slower. As the pattern is random, these changes are unpredictable but cumulative. Consequently, the phase discrepancy between the spins will continuously grow. Actually, this may be considered as a manifestation of the second law of thermodynamics which states that entropy always increases and the process is irreversible. As phase incoherence between the spins will continuously grow, the transverse magnetizations of the spins

Fig. 6.17 Schematic depiction of an MR signal generated after a $\theta = 90°$ pulse. The amplitude of the signal which is referred to as the FID (free induction decay) declines rapidly with time

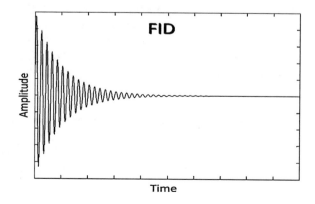

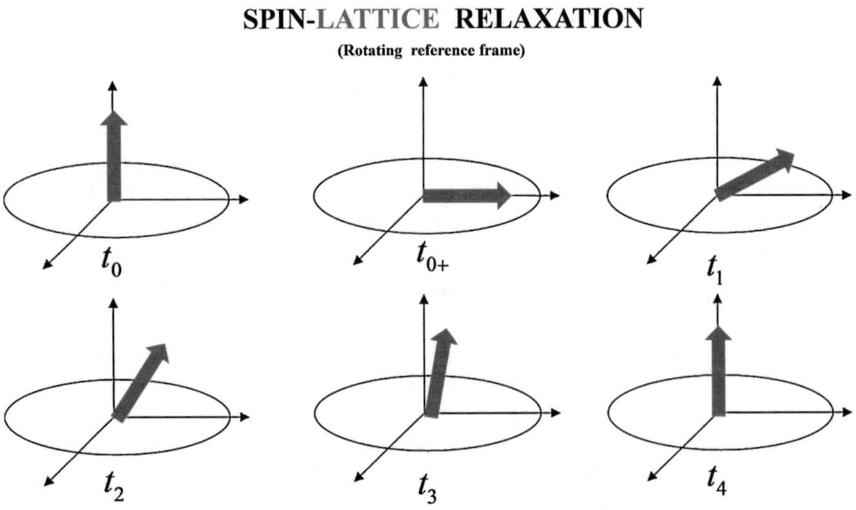

SPIN-LATTICE RELAXATION
(Rotating reference frame)

Fig. 6.18 The spin-lattice relaxation process describes the rebuilding of the longitudinal magnetization following a short RF excitation at time t_{0+} (in this case with a flip angle of 90°). It stems from the realignment of the magnetization vector (designated by the red arrow) with the main magnetic field $\overline{B}_0$ along the z-axis

will fan out. As a result, their vector summation which equals the magnitude of the transverse magnetization M_{xy} will continuously decrease. And when the spins will totally fan out into 360°, M_{xy} will be completely nulled. It is important to note that despite the fact that each fraction of the voxel may still have significant transverse magnetization, the total MR signal from the voxel will be zero! The process is explained graphically in Fig. 6.19. This process also has a characteristic exponential behavior but with a different time constant called T_2.

Importantly, it should be noted that the two relaxation processes occur simultaneously. That implies that if T_2 is not significantly shorter than T_1, the actual transverse magnetization will diminish faster. In the classical physics model, the transverse magnetization is comprised of projections of the magnetization vectors $\overline{M}$ onto the $x - y$ plane. Therefore, when the spin-lattice relaxation process is completed, there are no transverse magnetization components. Hence, it follows that spin-spin relaxation must be faster or equal to the spin-lattice relaxation and $T_2 \leq T_1$. Some exemplary relaxation values are outlined in Table 6.2. As can be noted in most cases, T_2 is significantly shorter than T_1. Also it should be noted that both time constants depend on the magnitude of the field.

Finally, another effect should be accounted for. Stemming from imperfect manufacturing of the scanner's main magnetic field B_0, there is usually a small inhomogeneity in the field of each voxel. This inhomogeneity varies from one location to another but is fixed in time. This inhomogeneity is designated as $\Delta B_0 = \Delta B_0(x, y, z)$. Its effect is to accelerate the de-phasing process. Resulting

SPIN-SPIN RELAXATION
(Rotating reference frame)

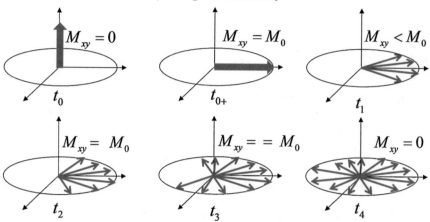

Fig. 6.19 Following a short RF pulse with a flip angle of 90^0 at time t_{0+}, the transverse magnetization is maximal, i.e., $M_{xy} = M_0$. At later times the sub-voxel magnetization de-phases. Consequently, their vector sum which constitutes the transverse magnetization is diminished until it is completely nulled

Table 6.2 Some representative T1 and T2 values for several tissues at 3 Tesla[a]

	T1[msec]	T2[msec]
Fat	378	123
Muscle	1248	33
Liver	777	32
Brain WM	974	70
Brain GM	1446	96
Heart	1187	48
CSF	4658	2000[b]

[a]These values were obtained by averaging some of the values reviewed in [7]
[b]From Ref. [8]

from that effect, the MR signal decays faster than the natural T_2. The effective exponential decay time constant is marked as $T_2{}^*$, and its value is given by [5]

$$\frac{1}{T_2{}^*} \simeq \frac{1}{T_2} + \frac{\gamma \Delta B_0(x, y, z)}{2} \tag{6.28}$$

Consequently, it will be shorter than the spin-spin relaxation time, i.e., $T_2{}^* \le T_2$. It is important to note that unlike T_2 which is a natural coefficient for each tissue, and its value remains the same regardless of the scanner used (for a given field strength), $T_2{}^*$ stems from engineering quality. That is to say that $T_2{}^*$ will change from one scanner to another! If the specific scanner inhomogeneity ΔB_0 is large, T_2^* will be much shorter than T_2, and the MR signal will decay much faster.

6.3.11   The Bloch Equations

In order to theoretically predict the effects of different magnetic field applications and also to account for the two relaxation mechanisms, Felix Bloch [9] has written a set of three differential equations describing the three magnetization components, M_x, M_y, M_z. The equations were set up by modifying Eq. 6.22 as follows:

$$
\begin{cases}
\dot{M}_x = \gamma \cdot M_y B_0 + \gamma \cdot M_z B_1 \sin\left(\omega \cdot t\right) - \dfrac{M_x}{T_2} \\[2mm]
\dot{M}_y = -\gamma \cdot M_x B_0 + \gamma \cdot M_z B_1 \cos\left(\omega \cdot t\right) - \dfrac{M_y}{T_2} \\[2mm]
\dot{M}_z = -\gamma \cdot \left[M_x B_1 \sin\left(\omega \cdot t\right) + M_y B_1 \cos\left(\omega \cdot t\right) \right] - \dfrac{(M_z - M_0)}{T_1}
\end{cases}
\tag{6.29}
$$

where $\dot{M}_x$, $\dot{M}_y$, $\dot{M}_z$ are the temporal derivatives of the magnetization along the x, y, and z directions, respectively.

This set of equations is applicable to a stationary reference frame, i.e., in the scanner coordinate system. However, as it is more convenient to use *a rotating reference frame*, the equations can be modified to

$$
\begin{cases}
\dot{M}_{x'} = -(\gamma \cdot B_0 - \omega) \cdot M_{y'} - \dfrac{M_{x'}}{T_2} \\[2mm]
\dot{M}_{y'} = (\gamma \cdot B_0 - \omega) \cdot M_{x'} - \gamma \cdot B_1 \cdot M_{z'} - \dfrac{M_{y'}}{T_2} \\[2mm]
\dot{M}_{z'} = -\gamma \cdot B_1 \cdot M_{y'} - \dfrac{(M_{z'} - M_0)}{T_1}
\end{cases}
\tag{6.30}
$$

where the term $(\gamma \cdot B_0 - \omega)$ is actually the off-resonance frequency, i.e., $\Delta\omega = (\gamma \cdot B_0 - \omega)$. If however the RF transmission is exactly on resonance, and the B_1 field is turned off, the equations simplify substantially to

$$
\begin{cases}
\dot{M}_{x'} = -\dfrac{M_{x'}}{T_2} \\[2mm]
\dot{M}_{y'} = -\dfrac{M_{y'}}{T_2} \\[2mm]
\dot{M}_{z'} = -\dfrac{(M_{z'} - M_0)}{T_1}
\end{cases}
\tag{6.31}
$$

The Bloch equations provide a very powerful analytical tool!
Using them we can predict the orientation of the three components of the magnetization M_x, M_y, M_z as a function of time and their response to changes in the magnetic fields and RF transmissions.

6.4 MRI Contrast Sources

As explained in Chap. 1, two types of information are needed for each voxel in order to build an image; these are (i) contrast, i.e., assigning different gray level values to different tissues, and (ii) spatial mapping, i.e., knowing the coordinates of each voxel, as shown schematically in Fig. 6.20.

There are many potential contrast sources in MRI. The main reason for that stems from our ability to manipulate the magnetization in various sequences of events which combine different types of RF pulses with different timings for transmissions and readouts as well as inserting temporal changes in the magnetic field. This carefully designed series of *accurately* timed events is called "pulse sequencing" (PSQ). The pulse sequence also incorporates gradient magnetic fields which will be introduced later. Resulting from the pulse sequence, certain type of information which manifests a certain property of the scanned tissue is collected at each time point. Herein, only the three most prominent contrast sources will be briefly introduced.

Relating to Eq. 6.8, it can be noted that the value of the equilibrium magnetization M_0 depends linearly on the proton density (*PD*) of the tissue within the voxel (assuming uniformity of course). Hence, proton density is the first natural source of contrast. The more protons there are in the voxel, the stronger is the magnetization M_0, and the stronger is the potential signal that can be obtained from that voxel (Fig. 6.21).

Fig. 6.20 In order to reconstruct an image, we need to assign an appropriate gray level (i.e. generate a visual contrast) and know the address of each voxel

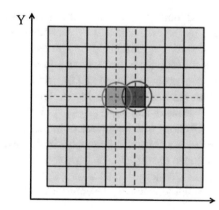

Fig. 6.21 The average proton density in each voxel can serve as a source of contrast in the MR image

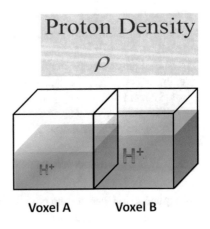

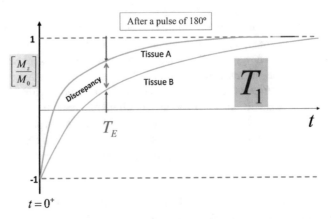

Fig. 6.22 Following a pulse of 180°, the longitudinal magnetization is inverted. Then, it starts to recover exponentially to its equilibrium magnitude M_0. The rate of recovery is determined by T_1. Therefore, there will be a discrepancy between two different materials which differ by their T_1. Importantly, the readout time TE will determine the size of this discrepancy in the image

The second natural source of contrast is spin-lattice relaxation which is characterized by the time constant T_1. Nonetheless, its relative contribution to the contrast depends on the specific PSQ implemented. For example, if we have two voxels with different tissue types that we wish to differentiate based on their spin-lattice relaxation, we need to design a PSQ that will be sensitive to the buildup of their longitudinal magnetization M_z. A simple PSQ would be to apply an RF pulse of 180°. Following that pulse (at time point which is marked as $t = 0^+$), the longitudinal magnetization will be inverted for all the voxels, i.e., their magnitude will be $M_z = -M_0$. It should be emphasized, however, that M_0 will be different for each voxel due to the differences in proton densities (as explained above). Then, resulting from the spin-lattice relaxation mechanism, their longitudinal magnetization will exponentially recover to its equilibrium magnitude M_0 as depicted in Fig. 6.22. The

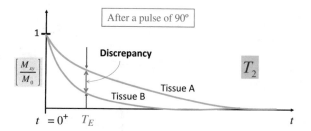

Fig. 6.23 Following a pulse of 90°, the transverse magnetization is maximal. Then, it starts to decay exponentially. The rate of decay is determined by T_2. Therefore, there will be a discrepancy between two different tissues which differ by their T_2. Importantly, the readout time TE will determine the size of this discrepancy in the image

rate of this recovery is determined by T_1 for that voxel. Consequently, there will be a discrepancy between two different materials which differ based on their T_1 values. Importantly, the readout time, which is commonly referred to as *echo time* (or time to echo) and designated as TE, will determine the observed magnitude of this discrepancy in the image. Thus, if TE is too short or too long, the discrepancy in terms of the longitudinal magnetization may be too minor.

The third natural source of contrast is the spin-spin relaxation which is characterized by the time constant T_2. Similar to the spin-lattice relaxation, its relative contribution to the contrast also depends on the specific PSQ implemented. If, for example, an RF pulse of 90° is applied along the x-axis of the rotating reference frame, the transverse magnetization will be $M_{xy} = M_y = M_0$, where again it should be noted that M_0 will be different for each voxel due to the differences in proton densities. Then, resulting from the spin-spin relaxation, their transverse magnetization will de-phase, and its magnitude will decay exponentially, as depicted schematically in Fig. 6.23. The decay is characterized by the corresponding T_2 value for each voxel. Consequently, there will be a discrepancy between the transverse magnetizations of two different tissue types which differ by their T_2 values. In reality however, stemming from the inhomogeneity in the scanner's main magnetic field, the actual decay may be determined (depending on the PSQ) by T_2^* as explained above (see Eq. 6.28). Consequently, the contrast quality may vary from scanner to scanner even when using the same PSQ.

Again, it should be noted that the readout time TE will determine the observed magnitude of the discrepancy in gray levels within the image. Thus, the selection of the appropriate PSQ and the optimal TE is crucial.

6.5 Spatial Mapping: The Field Gradients

In order to spatially map the voxels in the image, it was suggested to add to the main magnetic field $B_0 \cdot \hat{z}$ (which is designed to be uniform throughout the scanner's central zone) additional magnetic fields also directed along the same direction which

Fig. 6.24 Adding a gradient field to the main field will create a magnetic field which magnitude changes linearly with the location -in this case along the x-axis

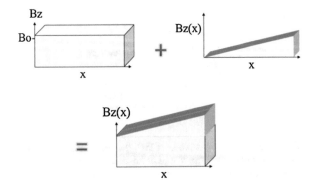

magnitude varies with the location. For clarity, let us consider first a field which linearly changes along the x-axis, in the stationary reference frame, as depicted in Fig. 6.24.

The gradient of this field G_x is defined as

$$G_x \equiv \frac{\partial B_z}{\partial x} \tag{6.32}$$

Note that the units of the gradient are [Tesla/m]. Thus, in order to determine the field strength at a certain location, the gradient has to be multiplied by its distance from the scanner's central point. (This point is considered the origin of the coordinate system.) It is important to note that the field gradient can be negative as well. Consequently, the overall field becomes location dependent, i.e., the strength of the magnetic field is a linear function of coordinate x. The combined field magnitude is thus given by

$$B_z(x) = (B_0 + G_x \cdot x) \tag{6.33}$$

Following the fact that the magnitude of the field becomes linearly dependent on the coordinate x, the Larmor frequency within the field will also become linearly dependent on the location along the x coordinate,

$$\begin{aligned} \omega(x) &= \gamma \cdot B_z(x) = \gamma \cdot (B_0 + G_x \cdot x) \\ &= \gamma \cdot B_0 + \gamma \cdot G_x \cdot x = \omega_0 + \gamma \cdot G_x \cdot x \end{aligned} \tag{6.34}$$

Or by presenting the frequency in Hertz units,

$$f(x) = f_0 + \frac{\gamma}{2\pi} \cdot G_x \cdot x \tag{6.35}$$

With the aid of this field gradient (sometimes referred to as *gradient* for short), we can now determine the x location of all transmitting voxels based on their frequency.

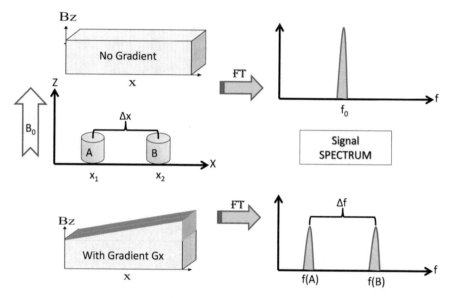

Fig. 6.25 Two glasses of water placed within an MRI scanner will emit signals after an RF pulse of flip angle θ is applied. Without a gradient in the field (Top), they will both transmit at the same frequency; hence, the spectrum will have only one peak. When a gradient G_x is applied (bottom), two smaller peaks will appear on the signal spectrum. The frequency distance Δf is proportional to the spatial distance Δx between the two glasses

To better understand the concept, consider a case where we have two small glasses filled with water, glass A and glass B, positioned within an MRI scanner at a distance Δx apart, as shown schematically in Fig. 6.25. Without any field gradients, both glasses will emit identical signals with the same frequency of $f_0 = \frac{\omega_0}{2\pi}$. Thus, when studying the spectrum of the emitted signal, a single peak will appear at f_0. On the other hand, when the field gradient G_x is applied, each glass will transmit at a different frequency. Therefore, two smaller peaks will appear on the signal's spectrum.

The frequency gap between the two spectral peaks Δf will be linearly proportional to the distance Δx between the two glasses, and its value is given by

$$\Delta f = \left(\frac{\gamma}{2\pi} \cdot G_x\right) \cdot \Delta x \tag{6.36}$$

Thus, by measuring Δf, we can calculate the distance Δx.

In a similar manner, two additional field gradients G_y and G_z can be applied to the field yielding the following set of gradients:

$$\begin{cases} G_x \equiv \dfrac{\partial B_z}{\partial x} \\[2mm] G_y \equiv \dfrac{\partial B_z}{\partial y} \\[2mm] G_z \equiv \dfrac{\partial B_z}{\partial z} \end{cases} \tag{6.37}$$

It is important to note that for all three gradients, the field changes are in the value of B_z, i.e., parallel to the main field B_0. In the general case where all three gradients are applied simultaneously, the field gradient can be presented in vector notation, i.e.:

$$\overline{G} = \frac{\partial B_z}{\partial x} \cdot \widehat{x} + \frac{\partial B_z}{\partial y} \cdot \widehat{y} + \frac{\partial B_z}{\partial z} \cdot \widehat{z} \tag{6.38}$$

Thus, the overall magnetic field will be 3D location dependent and given by

$$\begin{cases} B_z(x, y, z) \equiv Bo + \overline{G} \cdot \overline{R} \\ \text{where } \overline{R} = x \cdot \widehat{x} + y \cdot \widehat{y} + z \cdot \widehat{z} \end{cases} \tag{6.39}$$

6.6   Spatial Mapping: Slice Selection

The first implementation of the field gradients is to select a slice for imaging. Starting with a simple case where we wish to image an *axial* plane through the patient's head, i.e., parallel to the x-y plane of the stationary reference frame as depicted schematically in Fig. 6.26, a gradient G_z is applied. As a result, the magnetic field will change linearly along the z coordinate. Consequently, the corresponding Larmor frequency will also change linearly along the z coordinate. Thus, if at this state we transmit an RF pulse with a specific frequency $f(z_{slice})$, only the spins within the plane located at z_{slice} for which this is the Larmor frequency will be flipped according to Eq. 6.23.

This simple approach allows us to select planes which are perpendicular to one of the main axes, i.e., by using a gradient G_x, a *sagittal* plane can be selected, and by using a gradient G_y, a *coronal* plane can be selected. However, MRI can offer much more flexibility. As recalled from analytic geometry, the description of a general plane in space is given by the equation

$$Ax + By + Cz = D \tag{6.40}$$

where A, B, C and D are constants. If the constants are replaced by the field gradients, the following relation can be obtained:

Fig. 6.26 Applying a gradient field along the z-axis changes the magnetic field along that axis. Accordingly, the Larmor frequency will also change linearly along the z-axis. By transmitting an RF pulse at frequency $f(z_{slice})$, only the corresponding slice will be excited

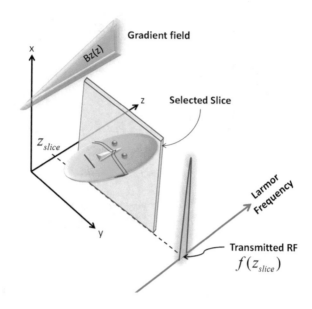

$$G_x \cdot x + G_y \cdot y + G_z \cdot z = \frac{\omega}{\gamma} \qquad (6.41)$$

As can be noted, this equation defines a plane in 3D space. Thus, by transmitting an RF pulse with a frequency ω, only the spins in the plane for which the Larmor frequency (in Hertz units) equals $\frac{\omega}{2\pi} = f(z_{slice})$ will respond. And since the gradients can be set arbitrarily, we can select any arbitrarily oriented slice using this relation as demonstrated schematically in Fig. 6.27.

The above described approach is applicable for exciting a very thin slice. Nonetheless, the slice thickness is also important, and we want to be able to control it too. Referring to Eq. 6.40, it is recalled that by varying the constant D, *parallel* planes are defined. Thus, if the upper face of the slice has a Larmor frequency of ω_2 and the bottom surface of the slice has a Larmor frequency of ω_1, then in order to excite the entire slice, a frequency band of

$$\Delta\omega = \omega_2 - \omega_1 = \gamma \cdot G_{slice} \cdot T \qquad (6.42)$$

has to be transmitted, where G_{slice} is a gradient orthogonal to the slice and T is the slice thickness. This is also demonstrated schematically in Fig. 6.27.

There are several approaches for slice selection using different RF transmissions. The simplest approach is to modulate the B_1 amplitude (see Eq. 6.19) using a SINC function in time domain. For example, if the input voltage to one of the transmitting antennas, $RF(t)$ is given by

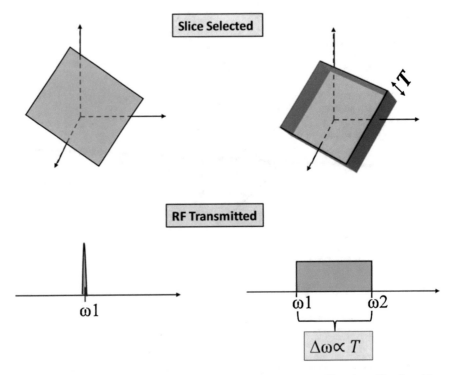

Fig. 6.27 An RF transmission (Bottom left) with a single frequency will excite a thin plane (Top left) according to Eq. 6.41. (Bottom right) A broadband transmission will select a thick slice (Top right). The slice thickness T is proportional to the band width

$$\text{RF}(t) = Vo \cdot \text{SINC}\left(\frac{\Delta\omega \cdot t}{2}\right) \cdot \text{SIN}(\omega_0 \cdot t) \tag{6.43}$$

Then, its manifestation in the frequency domain $\text{RF}(\omega)$ will be given by

$$\text{RF}(\omega) = A \cdot \mathbb{F}\left\{\text{SINC}\left(\frac{\Delta\omega \cdot t}{2}\right)\right\} \otimes \mathbb{F}\{\text{SIN}(\omega_0 \cdot t)\} \tag{6.44}$$

where A is the amplitude in the frequency domain, ω_0 is the Larmor frequency at the central plane of the slice, and the operator $\mathbb{F}$ designates the Fourier transform. As recalled, the Fourier transform of the SINC function is a RECT of width $\Delta\omega$, and the Fourier transform of the SIN function is a pair of delta functions. Thus, the convolution between the two (ignoring the imaginary delta) is a RECT of width $\Delta\omega$ centered around the frequency ω_0 as shown schematically in Fig. 6.28.

The above description is merely a generic presentation intended to explain the basic concept of slice selection. In reality things are slightly more complicated, and more sophisticated designs are needed in order to ensure optimal slice profiling and homogeneity. One issue that can be easily noted is that there is an inherent

Fig. 6.28 A SINC modulation of the RF transmission (B1 field) will yield a RECT in the frequency domain. The frequency band $\Delta\omega$ will determine the slice thickness as explained above

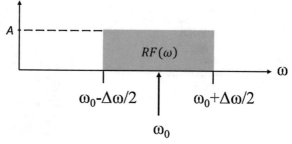

Fig. 6.29 During the RF transmission, the spins within the selected slice de-phase. In order to refocus them, a negative gradient is applied for half the transmission duration

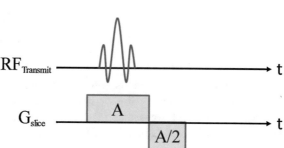

discrepancy in Larmor frequencies between different locations within the slice. The gap between each face of the slice and the center is of course $\Delta\omega/2$. That implies that spins within the same slice will be rotated relatively slower or faster than the center. Furthermore, as the pulse duration may be around 1 msec, the spins in the different planes within the slice will have time to de-phase. As a result, their vector sum will be smaller (similar to the effect of the T_2^* mechanism). Consequently, the magnitude of the magnetization will be reduced and so will be the emitted signal. Fortunately, there is a simple remedy for this problem, by applying a negative gradient with half the integrated duration of that applied during the RF pulse. For example, if the slice selection gradient is a RECT of magnitude G_{slice} and duration of Δt, then its temporal integral is given by $G_{slice} \cdot \Delta t$. The needed refocusing gradient is therefore given by $-G_{slice} \cdot \Delta t/2$. This is shown schematically in Fig. 6.29.

MRI also offers the option of multi-slice imaging; this can be achieved simply by applying several consecutive RF pulses with their proper slice selection gradients. Thus, the time delay between RF transmission and readout can be used to prepare and collect data from different spatial locations.

6.7 K-Space Formulation

Thus far, we have learned how to obtain an MRI signal and select a slice. In order to learn how images can be reconstructed, another building block is needed. For that aim, a very useful tool is to define the relation between the gradients and K-space (the Fourier domain).

Fig. 6.30 A single voxel generates a signal at its local Larmor frequency ω. The signal magnitude is proportional to the magnitude of its M_{xy} which has also an instantaneous phase φ

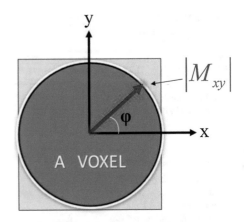

Consider first a single voxel (Fig. 6.30) located at position $\overline{R}$. Following RF transmission, i.e., applying the B_1 field, the magnetization is tilted according to Eq. 6.23. As a result, transverse magnetization M_{xy} is formed. The magnetization will rotate at the local Larmor frequency $\omega(\overline{R}, t)$. The signal $q_{\text{voxel}}(\overline{R}, t)$ obtained from that voxel is thus given by

$$q_{\text{voxel}}(\overline{R}, t) = A \cdot M_{xy}(\overline{R}, t) \cdot e^{-j\omega(\overline{R}, t) \cdot t} \tag{6.45}$$

where A is the amplitude of the detected signal (determined by the local proton density and the scanner's amplifications) and the transverse magnetization $M_{xy}(\overline{R}, t)$ changes as a function of time as described by the Bloch equations (Eq. 6.30). Naturally, the signal is complex as it has both magnitude and phase φ.

If during the signal acquisition process the magnitude of the magnetic field changes, the Larmor frequency changes as well. Hence, the local instantaneous phase is given by

$$\varphi = \int \omega(\overline{R}, t) \cdot dt \tag{6.46}$$

As commonly MR images display only the voxel magnitude, the signal magnitude of the voxel located at position $\{x, y\}$ for the 2D case could be designated as $f(x, y)$. Thus, Eq. 6.45 may be presented as

$$q_{\text{voxel}}(x, y, t) = f(x, y) \cdot e^{-j \int \omega(x, y, t) \cdot dt} \tag{6.47}$$

where $f(x, y)$ is the image to be reconstructed.

After selecting a slice as described above, the whole tissue within the slice will transmit signals. Each point will have its own magnitude and instantaneous phase. The total signal $Q(t)$ that will be detected by the antennas is therefore the sum of all the signals emitted from that slice, i.e.,

$$Q(t) = \int\limits_{-\infty}^{\infty} \int\limits_{-\infty}^{\infty} f(x,y) \cdot e^{-j \int \omega(x,y,t) \cdot dt} \cdot dxdy \qquad (6.48)$$

The integration over an infinite range is applied to match the Fourier transform as will be shown later. Next, we recall that the Larmor frequency is determined by the local magnetic field. Hence, the local magnetic field when gradient fields are present is given by Eq. 6.39. Therefore, the local Larmor frequency is given by

$$\begin{cases} \omega(x,y,t) = \gamma \left(Bo + \overline{G}(t) \bullet \overline{R} \right) = \gamma Bo + \gamma \overline{G}(t) \bullet \overline{R} = \omega_0 + \gamma \overline{G}(t) \bullet \overline{R} \\ \text{where } \overline{R} = x \cdot \hat{x} + y \cdot \hat{y} \end{cases} \qquad (6.49)$$

It is important to note that the derivation outlined herein is conducted in two dimensions for clarity, but it can be easily expanded to 3D. Next, moving to a presentation in the rotating reference frame, Eq. 6.48 can be written as

$$Q(t) = \int\limits_{-\infty}^{\infty} \int\limits_{-\infty}^{\infty} f(x,y) \cdot e^{-j \int \gamma \overline{G}(t) \cdot \overline{R} \cdot dt} \cdot dxdy \qquad (6.50)$$

Or if written more explicitly,

$$Q(t) = \int\limits_{-\infty}^{\infty} \int\limits_{-\infty}^{\infty} f(x,y) \cdot e^{-j \left[\int \gamma G_x(t) \cdot x \cdot dt + \int \gamma G_y(t) \cdot y \cdot dt \right]} \cdot dxdy \qquad (6.51)$$

Now, let us study one of the exponential terms and its physical units,

$$\int \gamma G_x(t) \cdot x \cdot dt = \left[\int \gamma G_x(t) \cdot dt \right] \cdot x = [\text{Rad}/\text{Tesla}/\sec][\text{Tesla}/\text{m}][\sec] \cdot x$$
$$= [\text{Rad}/\text{m}] \cdot x = K_x(t) \cdot x$$
$$(6.52)$$

As can be noted the term in the square brackets has the physical units of a spatial frequency, in this case along the x direction. Similarly, we can derive the same conclusion to the other exponential components and write the following set of equations:

$$
\begin{cases}
K_x(t) = \gamma \int\limits_0^t G_x(t)dt \\[2ex]
K_y(t) = \gamma \int\limits_0^t G_y(t)dt \\[2ex]
K_z(t) = \gamma \int\limits_0^t G_z(t)dt
\end{cases}
\tag{6.53}
$$

Or using vector notation,

$$
\vec{K} = \gamma \int\limits_0^t \vec{G} \cdot dt
\tag{6.54}
$$

(Kindly note that the spatial frequency is defined as explained in Chap. 1, as $K = \frac{2\pi}{\lambda}$ [Radians/unit distance].)

This set of equations is one of the most important relations in MRI. To understand its implications, let us compare it to the Fourier transform (for simplicity only for the 2D case)

$$
F[K_x, K_y] = \int\limits_{-\infty}^{\infty} \int\limits_{-\infty}^{\infty} f(x,y) e^{-jK_x \cdot x} e^{-jK_y \cdot y} dx dy
\tag{6.55}
$$

which is identical to the detected signal, i.e.:

$$
Q(t) = \int\limits_{-\infty}^{\infty} \int\limits_{-\infty}^{\infty} f(x,y) \cdot e^{-j\left[\int \gamma G_x(t) \cdot x \cdot dt + \int \gamma G_y(t) \cdot y \cdot dt\right]} \cdot dx dy =
$$
$$
= \int\limits_{-\infty}^{\infty} \int\limits_{-\infty}^{\infty} f(x,y) \cdot e^{-j\left[K_x(t) \cdot x + K_y(t) \cdot y\right]} \cdot dx dy = F[K_x(t), K_y(t)]
\tag{6.56}
$$

The conclusion is that:

> The MRI signal at time t is equal to the value of the Fourier transform of the image to be reconstructed for the coordinates $K_x(t)$, $K_y(t)$!

It is not difficult to show that this relation applies for the three-dimensional case as well.

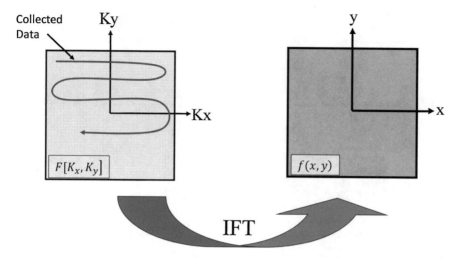

Fig. 6.31 As the MRI signal actually samples the Fourier transform of the image to be reconstructed, all we have to do in order to reconstruct the image $f(x, y)$ is to collect enough data for estimating $F[K_x, K_y]$ and then apply to it the inverse Fourier transform (IFT)

It follows then that in order to reconstruct the MR image $f(x, y)$, all we have to do is collect sufficient amount of data that will allow us to estimate the entire Fourier transform of the image $F[K_x, K_y]$ and then apply to it the inverse Fourier transform as shown schematically in Fig. 6.31.

The evident question is: how to cover the K-space properly with efficient data collection trajectories?

6.7.1 Traveling in K-Space

In order to design sampling trajectories in K-space for MRI data collection, we must first be aware of the implications of Eq. 6.53. The analysis at this point is done on 2D but it can be easily expanded to 3D.

Starting with the equilibrium state, and when no gradient fields are applied, our virtual pointer for collecting the data is positioned at the center of K-space, i.e., $F[K_x = 0, K_y = 0]$. When G_x is turned ON, the virtual pointer (our sampling point) will start moving (due to the temporal integration) along the K_x coordinate. If the gradient is positive, then it will move toward higher positive values of K_x. On the other hand, if it is negative, it will move toward more negative K_x positions. If G_x is turned OFF, the pointer will stop moving and will retain its location along the K_x coordinate. Similarly, when G_y is turned ON, the virtual pointer will start moving along the K_y coordinate, and when turned OFF, it will stop moving and will retain its last position along the K_y coordinate.

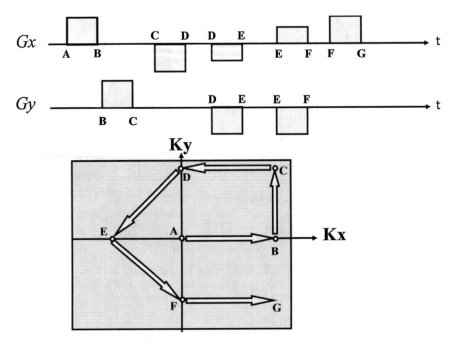

Fig. 6.32 (Top) Applying the gradients G_x and G_y allows us to move the virtual sampling point in K-space (Bottom). For example, in this series of applied gradients, the first positive gradient G_x moves the pointer horizontally from point A to point B. And the first positive gradient G_y moves the pointer vertically from point B to point C and so forth. It is important to note that data can be collected at any time during the journey

If both gradients are turned ON simultaneously, the pointer in K-space will move diagonally. In fact, by turning ON and OFF the gradients and by varying their values as a function of time, we can move the virtual pointer anywhere in K-space and along almost any desired trajectory. This is demonstrated schematically in Fig. 6.32. For simplicity RECT gradients were used.

As explained above, in order to reconstruct the MR image, all we have to do is collect sufficient amount of data in K-space and estimate the full $F[K_x, K_y]$. Then, we have to apply the inverse Fourier transform and reconstruct $f(x, y)$. By designing a sequence of gradient fields in time, we can move the sampling point along efficient trajectories in K-space to address the clinical need.

There are numerous methods for covering K-space in 2D and in 3D. These include variations with straight lines, spirals, circles, rosettes, and even random trajectories. Some exemplary trajectories are schematically depicted in Fig. 6.33. The simplest sampling scheme which is most commonly used today is a series of straight horizontal lines (Cartesian sampling). This pattern allows systematic and full sampling of the K-space and does not require any interpolation or other complementary computations before applying the inverse Fourier transform as will be discussed in the following for other trajectories.

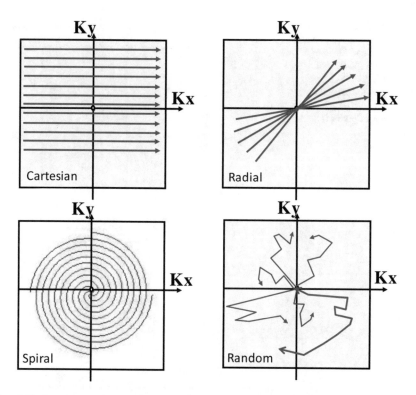

Fig. 6.33 Some exemplary K-space sampling trajectories used in MRI. There are numerous options. However, the Cartesian pattern (Top left) is currently the most frequently implemented scanning scheme

The above mathematical derivation was done for a continuous space presentation. However, in practice the signals are digitized and the images are depicted in a discrete format, i.e., using pixels. Furthermore, the image is limited to certain field of views (FOV) along the horizontal FOV_x and vertical FOV_y directions, and each pixel has a dimension of Δx and Δy along the horizontal and vertical directions, respectively, as shown schematically in Fig. 6.34.

The relation between the spatial resolution and the K-space dimensions is derived from the Nyquist-Shannon sampling theorem and is given by

$$
\begin{cases}
K_{xMAX} = \dfrac{K_{xSampling}}{2} = \dfrac{1}{2}\dfrac{2\pi}{\Delta x} = \dfrac{\pi}{\Delta x} \\[3mm]
K_{yMAX} = \dfrac{K_{ySampling}}{2} = \dfrac{1}{2}\dfrac{2\pi}{\Delta y} = \dfrac{\pi}{\Delta y}
\end{cases}
\tag{6.57}
$$

The number of pixels is simply derived by dividing the FOV to the pixel size, i.e.:

Fig. 6.34 (Top) In image
space, each pixel has a size
of $\{\Delta x; \Delta y\}$ and the image
size is set by the field of
view $\{FOV_x; FOV_y\}$.
(Bottom) In K-space, the
size is $\{2K_{xmax}; 2K_{ymax}\}$ and
the number of pixels is
the same

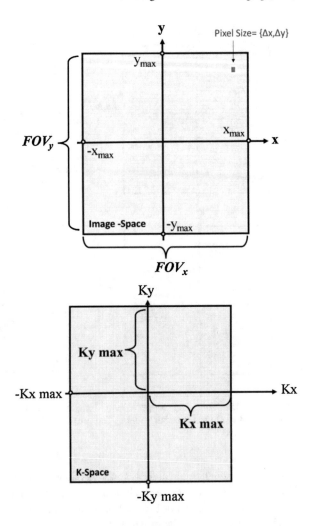

$$
\begin{cases}
Nx = \dfrac{\text{FOV}_x}{\Delta x} \\[2ex]
Ny = \dfrac{\text{FOV}_y}{\Delta y}
\end{cases}
\tag{6.58}
$$

where Nx is the number of columns and Ny is the number or rows. Accordingly the
image will have a matrix size of $Nx \cdot Ny$ pixels in *both spaces*. Therefore, the pixel
size in K-space is given by

$$
\begin{cases}
\Delta k_x = \dfrac{2 \cdot K_{xMAX}}{Nx} = \dfrac{2 \cdot \pi}{Nx \cdot \Delta x} = \dfrac{2\pi}{FOV_x} \\[2ex]
\Delta k_y = \dfrac{2 \cdot K_{yMAX}}{Ny} = \dfrac{2 \cdot \pi}{Ny \cdot \Delta x} = \dfrac{2\pi}{FOV_y}
\end{cases}
\tag{6.59}
$$

Again kindly recall that the spatial frequency is defined here as explained in Chap. 1, as $K = \frac{2\pi}{\lambda}$ [Radians/unit distance]. (Also note that other textbooks may use [1/unit distance] units instead, and hence certain equations may differ by a 2π factor.)

6.8   Imaging Protocols and Pulse Sequences

Following the above long introduction, we are finally ready to learn how imaging protocols are built in MRI. The blueprints of any imaging protocol in MRI are somewhat similar to the orchestra's conductor score (the book that combines all the simultaneous notes of all the music instruments in the orchestra). In MRI, there are basically five synchronized channels that compose the "pulse sequence," which is the imaging protocol scheme. A generic pulse sequence diagram which applies to a protocol called "gradient echo" is depicted in Fig. 6.35.

The first channel is the RF transmission channel. It describes schematically the timing for activating the B_1 field and its corresponding flip angle (in degrees or radian units). If not indicated otherwise, the B_1 field is applied along the x' coordinate of the rotating reference frame. For simplicity, the RF pulses are sometimes marked by a thin RECT with the corresponding flip angle written next to it. The time TE, known as *echo time* (or "time to echo"), designates the central time point of the data acquisition window. The time TR also known as *repetition time* (or "time to repeat") designates (if relevant) the time when the same set of RF and gradient events is repeated.

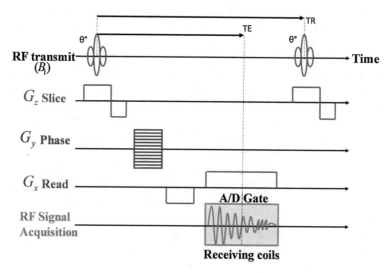

Fig. 6.35 A generic pulse sequence diagram used for "gradient-echo" imaging. See explanation in the text

The second channel, which is marked as G_{slice}, designates the slice selection gradients (as explained above). In the simple case where an axial plane is imaged, it equals G_z. In the more general case, it is comprised of three gradients as outlined in Eq. 6.41.

The third channel, which is marked as G_y or as the "phase encoding gradient," is used to set the K_y coordinate in K-space (see Eq. 6.53). It is important to note that in the case where an arbitrary slice orientation is selected, then coordinate y is the vertical coordinate within the imaged plane. As can be noted, this gradient is marked schematically as a ladder. This symbol indicates that its value is to be changed with every repetition time (TR). Using computer programming semantics, it designates a task loop. For example, something like

$$\left\{ \text{for } K\_\text{Line} = 0 : N_y; G_y = G_0 - K\_\text{Line} \cdot \Delta G; \text{end} \right\}.$$

That is to say, we start with a maximal gradient of magnitude $G_y = G_0$, and in every TR cycle, we reduce it by an increment ΔG.

The fourth channel, marked as G_x, is the "readout gradient" which is sometimes referred to as the "frequency encoding gradient." It is used to move the K-space sampling point (see above) along the K_x direction. Typically, it is turned ON during data collection (hence, the name "readout gradient").

The last channel describes the data acquisition time scheme. It indicates when to turn ON and OFF the data acquisition gate and use the analog to digital (A\D) circuits for storing the information. It should be noted that in most cases, special coils which are different from the transmission coils are used for signal detection.

In order to demonstrate how the pulse sequence diagram is implemented, we should also draw the corresponding trajectories in K-space. For example, relating to the "gradient-echo" imaging sequence shown above, the corresponding K-space trajectories for three exemplary TR periods are depicted in Fig. 6.36 along with their specific setups.

Starting from the first acquisition cycle (TR#1), it can be noted that following the slice selection step with a flip angle of θ°, the phase encoding gradient G_y is set to its maximal value (marked by the thick arrow in the figure) while simultaneously setting G_x to its minimal (negative) value. This results in motion of the K-space sampling point from the center (point A) to the upper left corner (point B). Then, G_x is applied for twice its previous duration while sampling the data. This yields the data from the uppermost line in K-space.

On the second step (TR#2), the magnitude of G_y is slightly reduced and is applied again simultaneously with the minimal magnitude for G_x. As a result, the sampling point in K-space moves from the center to a slightly lower coordinate along K_y. Then, G_x is applied again for twice its previous duration while sampling the data. This yields the data corresponding to the second uppermost line in K-space. Thus, at this point we have collected data from the two uppermost K-lines stored in the scanner's computer.

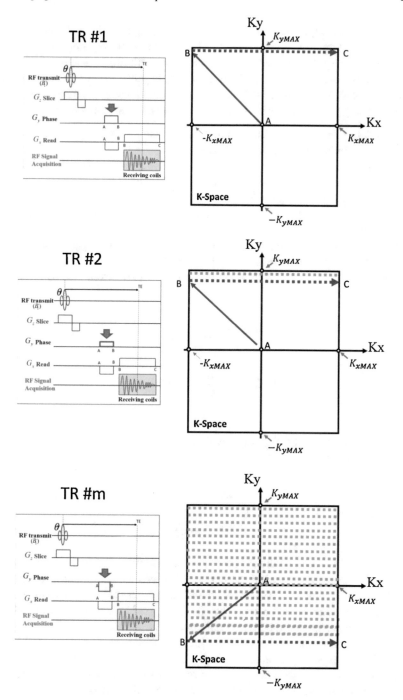

Fig. 6.36 Schematic gradient echo pulse sequence (left) and its corresponding K-space trajectories (right) protocol for three different TRs. (Top) On the first TR, the top raw in K-space is sampled using the maximal G_y gradient (thick arrow). Then by reducing G_y in each consecutive TR, additional K-lines are acquired until the K-space is fully sampled. The dotted lines designate the sampled data

The process is repeated over and over while varying in each TR the magnitude of G_y. After collecting N_y K-lines, the entire K-space matrix is sampled. In order to obtain the corresponding image, the inverse Fourier transform is applied.

The above explanation is qualitative in nature; however, for practical applications, we need to determine each parameter precisely using quantitative terms. In order to design the sequence in details, one has to track to following steps:

1. First determine the spatial orientation of the plane to be imaged (Eq. 6.40) and calculate the required G_{slice} (Eq. 6.41).
2. Given the slice thickness T, calculate the needed frequency band (Eq. 6.42).
3. For the corresponding Larmor frequency and frequency band, design the transmission pulse (e.g., Eq. 6.43).
4. Using Eq. 6.23 determine the $B_1(t)$ amplitude and duration to set the required flip angle θ.
5. Given the required spatial resolution Δx and Δy, calculate $K_{x\text{max}}$ and $K_{y\text{max}}$ using Eq. 6.57 and set the corresponding K-space dimensions.
6. From Eq. 6.53 the needed duration of the gradients is determined as follows:

$$
\begin{cases}
K_{x\,\text{max}} = \dfrac{\pi}{\Delta x} = \gamma \displaystyle\int_0^{\Delta t_x} G_x(t)\,dt \\[2em]
K_{y\,\text{max}} = \dfrac{\pi}{\Delta y} = \gamma \displaystyle\int_0^{\Delta t_y} G_y(t)\,dt
\end{cases}
\tag{6.60}
$$

If, for example, a RECT gradient with magnitude G_0 is applied for all directions, then it follows that the durations are given by

$$
\begin{cases}
\Delta t_x = \dfrac{\pi}{\gamma \cdot \Delta x \cdot G_0} \\[1.5em]
\Delta t_y = \dfrac{\pi}{\gamma \cdot \Delta y \cdot G_0}
\end{cases}
\tag{6.61}
$$

where Δt_x and Δt_y are actually the times needed to move the sampling point from the center of K-space to the end of the horizontal or vertical coordinates, respectively. Applying both gradients together will move the pointer to the upper right corner of K-space. Applying negative gradients for both gradients together will naturally move the pointer toward the bottom left corner, etc.

Importantly, it should be pointed out that RECT gradients are realistically impossible to produce. This stems from the infinite temporal derivative of their magnetic field and the resulting reaction of the coils. In practice, the gradient rise time is limited by the hardware "slew rate" which is defined as

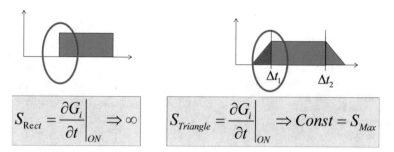

Fig. 6.37 (Left) A RECT-shaped gradient field has an infinite temporal derivative and therefore is impractical. (Right) Using the maximal slew rate for gradient field buildup leads to trapezoidal gradients

$$S_{\text{max}} = \left. \frac{\partial G_i}{\partial t} \right|_{\text{Max}} \tag{6.62}$$

Thus, trapezoidal gradients are used as schematically shown in Fig. 6.37.

In such case the rise time of the gradient field has to be accounted for, and the sampling point location during readout, i.e., between Δt_2 and Δt_2, should be calculated as follows:

$$Kx(t) = \gamma \left[S_{\text{max}} \cdot \frac{\Delta t_1^2}{2} + Gx_{\text{max}} \cdot t \right] \quad \text{for } t = \Delta t_1 \div \Delta t_2 \tag{6.63}$$

Thus, the sampling point should be moved beyond the K-space matrix.

7. Given the horizontal FOV_x and vertical FOV_y fields of view the number of pixels $Nx \cdot Ny$ can be determined from Eq. 6.58.
8. Then, the phase encoding gradient increment can be set from $\Delta G_y = 2G_0/(N_y - 1)$.
9. Solving the Bloch equations for this "gradient-echo" protocols yields the following estimated signal [10]:

$$\text{Sig} = A \cdot \rho \cdot \frac{1 - \exp\left(-\text{TR}/T_1\right)}{1 - \cos\left(\theta\right) \cdot \exp\left(-\text{TR}/T_1\right)} \sin\left(\theta\right) \cdot \exp\left(-\text{TE}/T_2*\right) \tag{6.64}$$

where A is a constant and ρ is the proton density. If we have some a priori knowledge about the properties of the tissue to be imaged (which is usually the case), the appropriate values of TR and TE can be estimated. For example, if we have a tissue of interest with $T_1 = 500$[msec] and $T_2^* = 50$[msec] and the flip angle was set to be $\theta = 30°$, then by using charts as the one shown in Fig. 6.38, we can note that the contribution to the signal is minor for TR longer than about 350 [msec].

Fig. 6.38 The expected signal from a tissue with $T_1 = 500$ [msec] and $T_2^* = 50$ [msec] following a flip angle of $\theta = 30°$, as a function of TR and TE. The thick dashed line designates the TR value beyond which the signal does not increase significantly

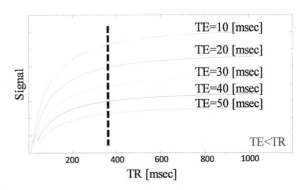

6.9   Some Important Pulse Sequences (Advanced Reading)

As stated above, there are numerous options for designing imaging protocols. Some of the more prevalent pulse sequences are introduced herein.

6.9.1   Inversion Recovery

This sequence is used for measuring T_1 or obtaining T_1-weighted images and also as a preparatory pre-sequence. For example, it is used in a pulse sequence called "fluid-attenuated inversion recovery (FLAIR)," to distinguish between tissues and fluids as will be explained in the following.

Basically, the generic code can be written as $180° \text{-} \tau \text{-} 90°$-Read. As can be noted it consists of three stages: The first is an inversion pulse, i.e., $\theta = 180°$, which flips the magnetization toward the negative z-axis. The inversion is followed by a delay of duration τ which allows the magnetization to respond according to its T_1. The third step is a $\theta = 90°$ (or other angle), which enables readout. Alternatively, the last pulse can be replaced by an additional imaging protocol. The basic idea is better explained using the graphical description outlined in Fig. 6.39.

It starts with an inversion pulse (1) which flips the magnetization toward the negative side of the z-axis. The spins then undergo two simultaneous processes: They de-phase due to the spin-spin relaxation mechanism, and in addition they return to their equilibrium state due to the spin-lattice relaxation mechanism. The result is a conical bundle of spins which widens like an opening umbrella. The longitudinal magnetization M_z therefore continuously shrinks. At a certain time point (4) most of the spins are located within the horizontal plane and the longitudinal magnetization is nulled, i.e., $M_z = 0$. As the net transverse magnetization is zero throughout the recovery process (due to the symmetrical spread of the spins), the magnetization sort of "disappear" at that moment. Afterward, the spins continue to

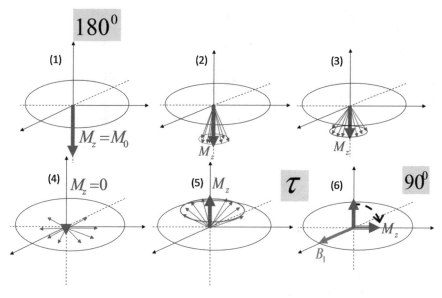

Fig. 6.39 Inversion recovery sequence. At step (1) the magnetization is inverted. Then the spins start to de-phase (due to the spin-spin relaxation mechanism) while returning upward to their upright equilibrium state (due to the spin-lattice relaxation mechanism). This forms a conical bundle which vector sum is the longitudinal magnetization M_z and which magnitude depends on T_1. At a certain time point (4) the magnetization is located within the horizontal plane and its sum is nulled along all directions. Then, it rebuilds again along the positive side of the z coordinate. If flipped by a $90°$ pulse (or implementing an imaging protocol), the obtained *transverse* magnetization will be proportional to the *longitudinal* magnetization at time τ

rise toward the positive side of the z-axis, and the conical bundle narrows like a closing umbrella. The longitudinal magnetization rebuilds and its magnitude is determined by T_1 (see Eq. 6.65). If at time τ a pulse of $\theta = 90°$ is applied, the *transverse magnetization* will equal the magnitude of the *longitudinal magnetization* just before the pulse, i.e., $M_{xy} = M_z(\tau)$. Hence, the signal obtained immediately after this pulse will map the T_1 effect on that voxel. Alternatively, an imaging sequence can be applied instead of the $90°$ pulse, and the image will be T_1 weighted. That is to say that the variations in T_1 values will be emphasized in the picture.

Using the Bloch equations, it can be shown that the longitudinal magnetization after the inversion is given by

$$M_z(t) = M_0\left(1 - 2e^{-\frac{t}{T_1}}\right) \tag{6.65}$$

This magnitude will determine the magnitude of the detected signal following the $\theta = 90°$ pulse. If we can find the exact time τ_0 for which $M_z(\tau_0) = 0$, then the value of the longitudinal relaxation T_1 can be determined from

Fig. 6.40 A "FLAIR" pulse
sequence starts with an
inversion pulse of 180°;
after a delay time *TI*, a
regular imaging protocol
can be applied. (In this case
a "gradient echo" sequence)

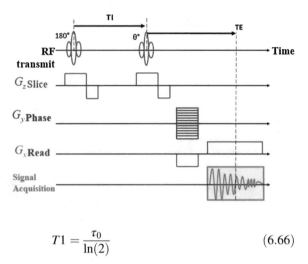

$$T1 = \frac{\tau_0}{\ln(2)} \tag{6.66}$$

One important application of inversion recovery is in a "fluid-attenuated inversion recovery (FLAIR)" protocol. An exemplary pulse sequence scheme is shown in
Fig. 6.40. It should be pointed out that there are other imaging options that can be
combined with inversion pulses.

The FLAIR utilizes the fact that the spin-lattice relaxation of body fluids is
typically much longer than that of tissues. For example (see Table 6.2), for the
cerebrospinal fluid (CSF) at 3 T, $T_1 = 4, 658$[msec]. This is very long compared to
the brain white matter $T_1 = 974$[msec] or gray matter $T_1 = 1, 446$[msec]. Thus, after
applying the inversion pulse of $180°$, the longitudinal magnetization of the tissue will
recover much faster than that of the fluid, as shown schematically in Fig. 6.41.
Therefore, by setting the inversion time to be IT $= T_1 \cdot \ln(2)$ for the fluid, signals
will be obtained only from the tissue, and the fluid will be blackened in the image.
A demonstrative FLAIR image is depicted in Fig. 6.42.

Another useful implementation of inversion pulses is in a method that depicts
"black-blood." Commonly the blood appears white in standard MR images. However, in cardiac applications, for example, it is important to delineate the inner wall
of the heart chambers (the endocardium). By blackening the blood, the myocardial
walls (the heart's muscle) are clearly depicted. Simple inversion recovery may be
useless, since the blood flows continuously into and out of the heart. The trick is to
implement two consecutive inversion pulses. The first one is non-selective, i.e., it is
applied to the entire scanned volume. This will invert all the spins of blood and
tissue. The second one is applied selectively to the slice of interest; this will restore
the magnetization of the spins within the chosen slice. After waiting for a proper
delay, the blood within the selected slice will be washed out and replaced by new
blood, which is inverted. Consequently, when applying an imaging protocol, the
blood will appear black in the MR image.

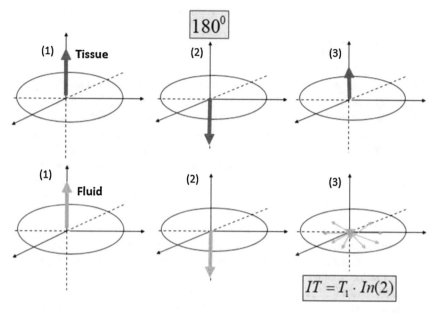

Fig. 6.41 The inversion recovery pre-pulse effect. (Top row) a tissue of interest subjected to the inversion will recover faster than the fluid (Bottom row) due to the longer T_1 of the fluid. If imaging is commenced at $IT = T_1 \cdot lan(2)$ for the fluid, its magnetization will be nulled. Hence, no signal will be obtained from the fluid

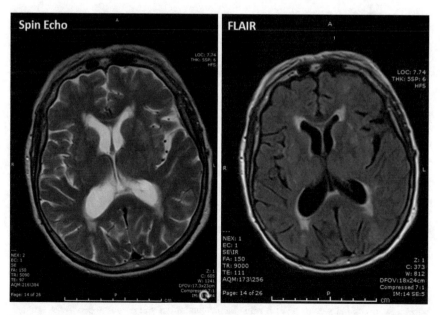

Fig. 6.42 A comparison between a regular Spin Echo image of a brain (Left), and a FLAIR image of the same cross section (Right). Note the dark appearance of the ventricles and all other fluid containing structures in the FLAIR image. (Images were provided courtesy of Prof. Moshe Gomori, The Hebrew University of Jerusalem)

6.9.2 Spin Echo

This sequence is probably the most popular one. It allows one to obtain high-quality images by overcoming the field inhomogeneity effects which results in the T_2^* decay. As recalled from Eq. 6.28, the main magnetic field inhomogeneity $\Delta B_0(x, y, z)$ accelerates the spin-spin de-phasing process. The problem is that field inhomogeneity varies from one scanner to another and also from one location to another even in the same scanner. Consequently, image quality is reduced and quantitative assessment of transverse relaxation is not possible. The spin-echo pulse sequence suggested by Han [11] offers a remedy for the problem using an elegant trick.

Basically, the generic code for the spin-echo sequence is 90°- τ- 180°- τ- Read. As the mathematical derivation for this sequence is slightly cumbersome, the idea behind this sequence is explained herein graphically. The steps are depicted in Fig. 6.43 in the rotating reference frame.

Referring to Fig. 6.43, we start at the equilibrium state at time $t = 0$. All the individual spins within the voxel, marked by the thin arrows, will be (on the average) aligned with the z'-axis of the rotating reference frame. The longitudinal magnetization will be M_0, and there will be no transverse magnetization. Applying a 90°

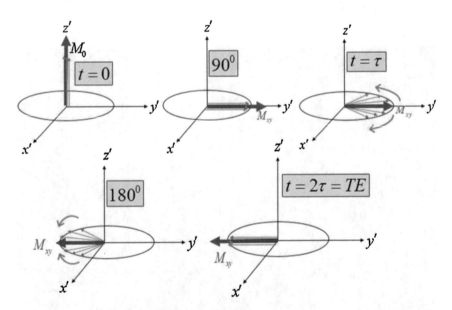

Fig. 6.43 The "spin-echo" spin propagation in a rotating reference frame. At equilibrium state the entire magnetization is aligned with the z'-axis. Following a 90° pulse, the magnetization is now aligned with the y'-axis. After a while the magnetization will de-phase due to the spin-spin relaxation mechanism. Some spins will rotate faster and some slower than the average Larmor frequency. Consequently, the signal will become weaker. Applying a 180° at time τ will revert their orientation toward the negative side of y', but not their rotational direction! After another time τ, the signals will realign along the negative y-axis. The signal will be strong again

pulse will flip the magnetization so that it will be aligned with the y'-axis. The transverse magnetization will be initially $M_{xy} = M_0$. Then, due to the spin-spin mechanism and due to the inhomogeneity in the main field, the spins will de-phase. Referring only to the field inhomogeneity, some of the individual spins will sense a stronger magnetic field and will rotate faster than the others (higher Larmor frequency). On the other hand, some of the individual spins will sense a weaker magnetic field and will rotate slower than the others (lower Larmor frequency). As a result, the spins will fan out. Their spread angle will be increased, and the transverse magnetization which is given by their vector sum M_{xy} will be reduced.

Applying a $180°$ at time τ will revert their orientation toward the negative side of y'. (Note that the blue arrows and the green arrows in Fig. 6.43 retain their position relative to the y' coordinate.) However, since each individual spin will continue to sense the same magnetic field as before, the faster spins will continue to rotate faster, and slower ones will continue to rotate slower. This implies that they will retain their rotational directions! As a result, after another time period of τ, all the individual spins will realign again along the negative y'-axis. The transverse magnetization will peak and the signal will be strong again. This signal is referred to as the "echo."

To complete the description, consider the pulse sequence diagram along with its corresponding K-space trajectory as shown in Fig. 6.44.

It starts with a $90°$ pulse, combined with the slice selecting gradient. Then, similar to the gradient-echo protocol, the sampling point in K-space is moved to the beginning of a new K-line, using the phase encoding and readout gradients as explained before. However, there is one important difference. The sampling point

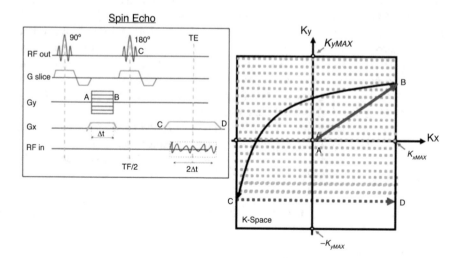

Fig. 6.44 (Left) The "spin-echo" diagram. It starts with a $90°$ pulse while applying a slice selection gradient. The sampling point is then moved to its new position using the gradient fields. The following $180°$ pulse starts the refocusing mechanism. (Right) Note that in K-space (shown for an arbitrary TR), this flips the sampling point to its symmetrical location in the plane. The last stage is identical to the gradient echo protocol where data is collected along the chosen K-line

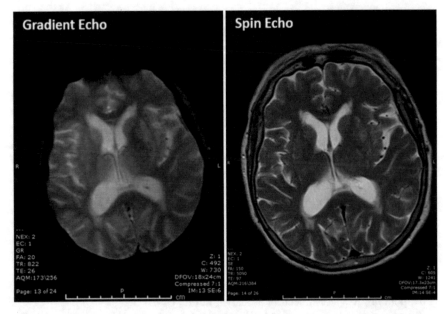

Fig. 6.45 A comparison between a gradient-echo image of a brain (Left) and a spin-echo image of the same cross section (Right). Note the much improved contrast and better image quality depicting the anatomical details obtained with the spin-echo sequence. (Images were provided courtesy of Prof. Moshe Gomori, The Hebrew University of Jerusalem)

is moved not to the beginning of the desired K-line, but to its symmetrical location in the plane. That means that instead of moving the sampling point to $\{-K_{xmax}, -K_{y0}\}$, for example, it is moved to $\{K_{xmax}, K_{y0}\}$. The following 180° pulse will take it back to the desired location as shown schematically in Fig. 6.44 (shown here for an arbitrary TR). Of course the 180° pulse has to be applied at $t = \text{TE}/2$ in order to obtain an echo at TE. The last step is identical to the gradient-echo protocol. Note that the signal sampling has to start slightly before and continue slightly after the echo. The sampling time is calculated using Eq. 6.61.

The spin-echo images are superior to the gradient-echo images, since their signal decay is affected only by T_2 and not by T_2^*. This is demonstrated in Fig. 6.45.

Finally, by solving the Bloch equations for the spin-echo sequence, the corresponding predicted signal is given by

$$\text{Sig} = A \cdot \rho \cdot e^{-\frac{\text{TE}}{T_2}} \left(1 - e^{-\frac{\text{TR}}{T_1}} \right) \tag{6.67}$$

where A is a constant and ρ is the proton density and TE and TR are the echo time and repeat time, respectively. Note that the signal strength depends on T_1 and on T_2 and not on T_2^* (which was the case for the gradient echo).

Studying the spin-echo signal, it is obvious that we cannot isolate any one of the natural properties of the tissue, i.e., proton density T_1 and T_2. Nonetheless, by

Table 6.3 How to obtain different weights in spin echo

	TE	TR
Proton density weighted	*Short*	*Long*
T_1 weighted	*Short*	*Short*
T_2 weighted	*Long*	*Long*

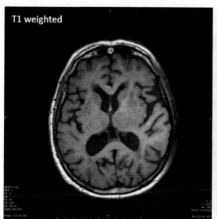
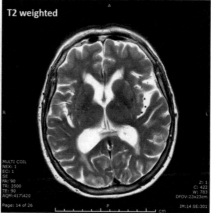

Fig. 6.46 (Left) A T1-weighted axial cross-sectional image of a brain. Note especially the dark appearance of the ventricles. (Right) A T2-weighed image of the same brain. Note the white appearance of the ventricles and other structures. (Images were provided courtesy of Prof. Moshe Gomori, The Hebrew University of Jerusalem)

properly setting TE and TR, we can enhance significantly the contribution made by one of these parameters. This is called *weighting* the image. Accordingly, there are three weighting options as outlined in Table 6.3.

Naturally, weighting can be obtained also using other MR imaging protocols. A demonstrative comparison between T_1- and T_2-weighted images is depicted in Fig. 6.46.

6.9.3   Fast or Turbo Spin Echo

Although the spin-echo sequence yields high-quality images, it is time-consuming. The total acquisition time is determined by the number of K-lines (phase encoding lines) Ny needed to reconstruct the entire image and the duration of the TR period, i.e.,

$$\text{Image Scan Time} = \text{TR} \cdot \text{Ny} \tag{6.68}$$

Thus, if we could acquire several K-lines (say m) within a single TR period, then the total acquisition time will be shortened accordingly, i.e., Scan time $= \text{TR} \cdot \text{Ny}/m$.

A simple approach to obtain this goal is to implement the method suggested by Carr and Purcell (CP) [12]. It utilizes the spin-echo refocusing pulse as a building block. The idea is to extend the data acquisition period by allowing the spins to de-phase and then refocus them again several times within the same TR. The generic code for the CP sequence is

$$90^0_x - \tau - 180^0_x - \tau - \boxed{\text{read}}$$

$$\tau - 180^0_x - \tau - \boxed{\text{read}}$$

$$\tau - 180^0_x - \tau - \boxed{\text{read}}$$

$$\tau - 180^0_x - \tau - \boxed{\text{read}}$$

$$\cdots\cdots$$

As can be noted, the flip angles are marked with a subscript x. This was done in order to indicate that the field B_1 is applied along the x' direction of the rotating reference frame. This sequence was later improved by Meiboom and Gill [13]. They showed that by applying the refocusing pulses along the y' direction of the rotating reference frame, i.e., using 180°_y instead of 180°_x, better signal is obtained (by overcoming inaccuracies in the flip angle application). This sequence is known as the CPMG method. However, as the CP sequence is easier to explain, we shall refer only to the CP here.

The CP sequence starts with a regular spin-echo segment. Consider the last part of the spin diagram shown graphically in Fig. 6.43. We shall continue after the first echo signal is obtained at time 2τ (Fig. 6.47).

After obtaining the first echo at time $t = 2\tau$ (designated here as TE#1), the spins are allowed to de-phase for a period of time which equals τ. Note that they still retain their angular direction. Then, at $t = 3\tau$, another 180° pulse is applied, and the spins are inverted around the x'-axis (the flipping time is considered negligible). As they retain their angular velocity, they will realign again (as happened for the first echo) at time $t = 4\tau$. The second echo is designated here as TE#2. The spins are then allowed to de-phase again, until another refocusing pulse of 180° is applied at $t = 5\tau$. The third echo signal will be obtained at $t = 6\tau$ and so forth.

The advantage of this sequence is obvious; instead of reading only one K-line per TR, we can have several echo signals which allow us to measure several good-quality K-lines, i.e., with signals which are not affected by the field inhomogeneity and not subjected to T_2^*. This will naturally shorten the total acquisition time as explained above. Thus, the "fast spin-echo" sequence is superior in quality relative to gradient echo and faster than spin echo. The drawback is that the signals decay as

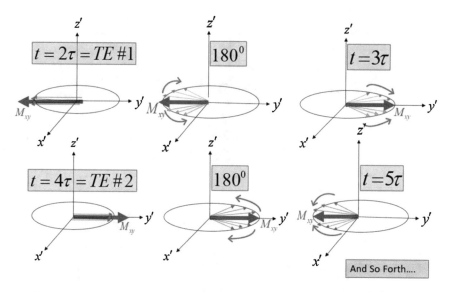

Fig. 6.47 Fast (turbo) spin-echo spin propagation. After acquiring the first echo at time 2τ, the spins are allowed to de-phase. At time 3τ another 180° pulse is applied. As a result the spins flip horizontally and refocus at time 4τ yielding the second echo. The spins are allowed to de-phase again, and another 180° pulse is applied at time 5τ. This will yield a third echo at time 6τ, and so forth...

the acquisition is prolonged, i.e., the number of echoes is increased. The decay is mainly attributed to the T_2 relaxation which is not cancelled by the refocusing procedure. The pulse sequence for three echoes in an arbitrary TR and their corresponding K-space trajectories are depicted in Fig. 6.48.

6.9.4 Echo Planar Imaging (EPI)

Following the rational of the fast spin-echo sequence and Eq. 6.68, the obvious question is: why not acquire the entire K-space in a single TR? And indeed the idea was implemented in a pulse sequence called: "echo planar imaging" (EPI). The simplest implementation is to extend the gradient-echo sequence and "blip" the phase encoding gradient G_y at the end of every scanned K-line so that the sampling point is moved to the next K-line. The scheme which is referred to as "gradient-echo echo plannar imaging" or GE-EPI is depicted graphically in Fig. 6.49.

As can be observed we start by moving the sampling point to the bottom left corner of the K-space. After waiting for a short while to obtain the desired contrast, the first K-line is read by applying G_x; upon completion of data acquisition, a small "blip" of G_y is applied. This moves the sampling point to the next K-line. The sign of the readout gradient G_x is inverted, and the second K-line is read. Then, another

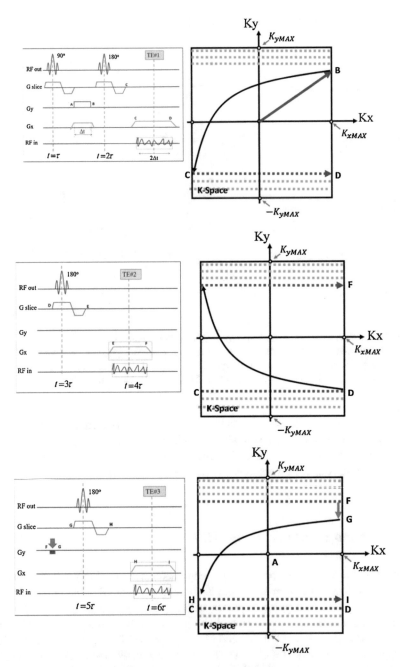

Fig. 6.48 "Fast (turbo) spin-echo" sequence. (Top) At a certain TR, we send the sampling point to the end of an unsampled K-line (point B). The 180° focusing pulse sends it to its symmetrical K point (point c). The first K-line is sampled during the formation of echo #1. (Middle) After sampling that K-line (point D). The second 180° refocusing pulse sends it to its symmetrical K point. The second K-line is sampled during the formation of echo #2. (Bottom) Upon reaching the end of the line (point F), a small gradient (marked by an arrow) sends the pointer to a new K-line. The 180° refocusing pulse sends it to its symmetrical K point (point H). The third K-line is sampled during the formation of echo #3. And the process can continue to acquire additional K-lines

small "blip" of G_y is applied. The sign of the readout gradient G_x is changed again and the third K-line is read and so forth.

The advantages of the gradient-echo EPI (GE-EPI) sequence are as follows: (i) it is very fast and (ii) it covers the entire K-space in a Cartesian manner, which allows a simple reconstruction. The disadvantages are as follows: (i) The signal decays significantly as the readout is prolonged. Hence, image quality is not high especially considering the fact that process is governed by T_2^*. (ii) Furthermore, in order to complete the readout within an acceptable time, very strong gradients and high slew rates are needed. This imposes a challenge in hardware terms.

An improvement is suggested by combining the EPI with the spin-echo sequence. The spin-echo planar imaging (SE-EPI) is schematically depicted in Fig. 6.50.

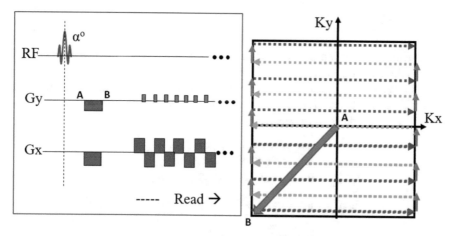

Fig. 6.49 (Left) A gradient-echo EPI pulse sequence scheme. (Right) The corresponding K-space trajectories. For simplicity only the RF pulse and the two planar gradients G_x and G_y are depicted

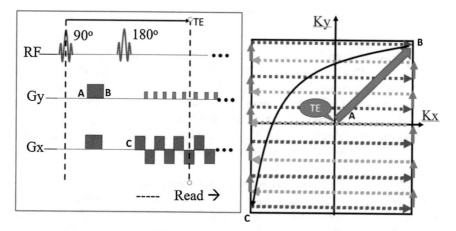

Fig. 6.50 Schematic depiction of the "spin-echo echo planar imaging" (SE-EPI). (Left) The sequence begins like a regular spin echo, but instead of reading only one K-line, all the K-lines are read within the same TR. (Right) The corresponding K-space trajectory

Following the 90° pulse, the sampling point is moved to the upper right corner of K-space. After waiting for TE/2, the refocusing 180° pulse is applied. As a result, the sampling point is moved to the lower left corner. The K-space is then rapidly scanned as done in the GE-EPI. However, as the trajectory is designed to reach the K-space center at exactly TE, the spins continuously refocus, and the signal quality is improved. After passing the K-space center, it will deteriorate again due to the spin de-phasing. It should be also recalled that simultaneously the signal decays as in a regular spin echo according to T_2 and also T_1 if acquisition time is relatively long. This PSQ provides better images than the GE-EPI.

6.9.5 Steady State

The methods suggested above for shortening the image acquisition time is focused on maximizing the number of K-lines acquired within each TR. However, by examining Eq. 6.68, it can be noted that image acquisition time can also be reduced if the TR is shortened. This approach is implemented in steady-state imaging protocols. The generic code for steady state is

$$\boxed{\text{Preparatory Stage}}$$

$$\theta - T_R - \theta - T_R - \theta - T_R - \theta - T_R - \theta - \dots$$

Until Reaching Steady State – Then continue:

$$\boxed{\text{Acquisition Stage}}$$

$$\dots - \theta - \tau - Read - \theta - \tau - Read \dots$$

$$\overset{\longleftarrow T_R \longrightarrow}{}$$

As can be observed, it comprises two stages. During the first stage, an RF pulse of angle θ is repeatedly applied every TR. It can be shown [10] that this leads to a steady state of the spins. That means that the longitudinal magnetization is exactly the same at the end of every TR prior to the next RF pulse. This happens even for short TR, shorter even than T_2^*. The implication is that we can rapidly collect data in a consistent manner during every TR period. Once the steady-state condition has been reached, data acquisition can be applied at time τ after every RF pulse (to obtain proper contrast).

There are several options and variations for implementing the steady-state concept in imaging protocols (see, e.g., [3, 6]). The main challenge stems from residual transverse magnetization and retaining phase coherence between pulses. In general, at the steady-state condition, a regular FID appears following each RF. This yields a signal which we can use for reading the needed data. However, stemming from the residual transverse magnetization prior to the next pulse, an echo can appear (a train of RF pulses may yield stimulated echoes). The properties of the echo signal differ from that of the FID. But in order to obtain usable images, consistency of the signal properties is desired.

One suggested protocol applies a fully balanced steady-state sequence. That is to say that every gradient is applied twice: once before and once after readout. Moreover, the gradients are applied so that the sum of all the positive gradient parts along one direction is cancelled by the sum of the corresponding counter negative gradients. Balancing is applied to all three gradients, i.e., slice selection, phase encoding, and readout. Hence, the net sum of all the gradients within one TR period is zero. Furthermore, the flip angle is applied with an alternating sign. Consequently, de-phasing is minimized, and the magnetization components are nearly identical before each RF application. The corresponding pulse sequence scheme for the fully refocused steady-state protocol is shown in Fig. 6.51. Kindly note that the diagram corresponds to the part which is applied after reaching the steady-state condition.

6.9.6 Advanced Rapid Imaging Methods

The MR imaging methods described above are merely a brief representation of the full spectrum of possibilities suggested by many investigators. Other techniques

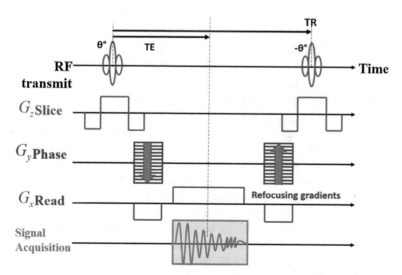

Fig. 6.51 Fully refocused steady-state sequence. Data is acquired after reaching the steady-state condition. Balancing is applied to all three gradients, i.e., slice selection, phase encoding, and readout. Hence, the net sum of all the gradients within one TR period is zero. Consequently, the magnetization components are nearly identical before each RF pulse

suggest new methods for data acquisition, new contrast sources, and functional imaging. However, since MRI is considered a relatively slow imaging modality, the quest for faster imaging has yielded a bundle of options.

The first option which is termed "parallel imaging" made use of the multi-coil arrays used for image acquisition. Although all the coils in an array acquire the data simultaneously, each coil has its own sensitivity function. Thus, it "sees" only a part of the image. This coil sensitivity is location dependent and hence can be used as an additional source of information. Consequently, if only a part of the K-space data is sampled, the missing information can be retrieved using various mathematical derivations. Thus, if, for example, only one quarter of the image K-lines is needed to reconstruct the full image, acquisition time can be shortened by a factor of 4. Acceleration factors of 3 and more are now commonly used in the clinic. There are many methods for implementing parallel imaging. Some exemplary methods are SENSE [14] which is routinely applied in many scanners, GRAPPA [15], SMASH [16], and SPID [17].

The second option is to implement data retrieval methods using "compressed sensing." The idea is to transform the image into another domain, e.g., wavelet, where the information representation is sparse. Thus, only a small amount of coefficients is needed in order to fully reconstruct the image. The basic procedure is iterative, where the estimated image is transformed into the sparsifying domain, modified (commonly by using a threshold), and transformed back while retaining the originally acquired data. The concept was first suggested by Lustig et al. [18], who have shown that high acceleration factors can be obtained. Many works that followed have suggested modifications to the basic concept, for example, [19–21].

6.10   Three-Dimensional (3D) Imaging

Thus far, we have discussed only planar (2D) imaging protocols. However, MRI can be easily implemented to scan objects in three dimensions. It should be emphasized that real 3D acquisition does not refer to the data obtained by scanning multiple planar images (like a stack of cards), but rather to acquisition in actual 3D K-space. Referring to Eq. 6.53, we note that in addition to K_x and K_y, we can control the sampling location along K_z as well. Thus, instead of using G_z for slice selection, we can apply RF pulses to the entire volume and use G_z for adding the third dimension to our sampling trajectory in K-space. An exemplary 3D sequence is depicted in Fig. 6.52 along with its corresponding 3D K-space trajectories.

The applied trajectories in 3D K-space do not necessarily have to be Cartesian. They can be radial or spiral as demonstrated schematically in Fig. 6.53. The drawback in that case is the need to interpolate into a Cartesian K-space or use non-uniform Fourier transform [22].

The advantages of 3D scanning are as follows: (i) The signal is strong as the spins in the entire volume contribute to it; hence, the SNR is high. (ii) The reconstruction is truly 3D in nature. Hence, different views and arbitrary cross sections can be

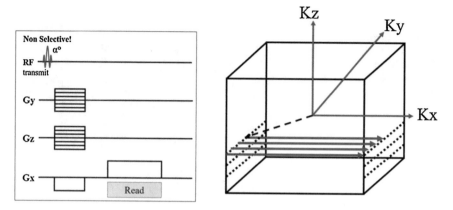

Fig. 6.52 3D gradient-echo imaging. A nonselective RF pulse is applied to the entire volume. The sampling trajectories are set in 3D K-space by applying different combinations of the three gradients

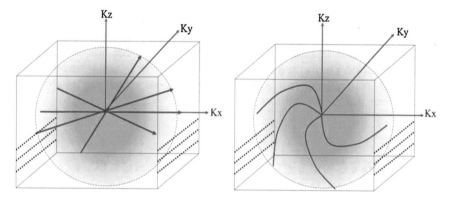

Fig. 6.53 Non-Cartesian sampling trajectories in 3D K-space. (Left) 3D radial lines may be used to scan a spherical volume. (Right) Alternatively, a set of spiral trajectories may be used

obtained. The disadvantages are as follows: (i) Stemming from the fact that a large amount of data is needed, the scanning time is relatively long. (ii) Acquisition is more prone to artifacts from patient motion. A small motion artifact can corrupt the entire 3D image.

6.11 Magnetic Resonance Angiography (MRA)

One of the many advantages offered by MRI is the ability to image blood vessels and quantify blood flow. This field is termed magnetic resonance angiography (MRA). There are several optional techniques that can be implemented for MRA. In this

Fig. 6.54 Bright blood MRA can be obtained by injecting a contrast-enhancing material to the blood or by suppressing the signal from the surrounding tissue

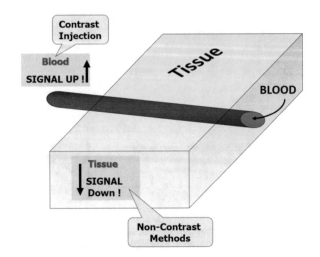

section we shall focus only on "bright blood" techniques, i.e., techniques that make the blood look white on the MR image. As shown schematically in Fig. 6.54, there are two options to see only the blood vessels embedded within an imaged slice: (i) Make the signal from the blood much stronger than that obtained from the surrounding tissue. For that aim contrast-enhancing materials (CEM) are commonly used. (ii) Alternatively, make the signal from the tissue much weaker than that obtained from the blood.

6.11.1   MRA Using Contrast-Enhancing Material (CEM)

There are several contrast-enhancing materials (CEM) offered today for clinical MRA. The first and most prominent is a family of gadolinium-based compounds. These CEM are paramagnetic in nature, and they substantially shorten the longitudinal relaxation. Hence, after injection into the body, gadolinium-based CEM will cause the blood to appear bright on T_1-weighted images. The signal from the background tissue can be suppressed by applying a subtraction technique. In order to do that, one set of images is acquired prior to CEM injection and another set post injection. By subtracting the paired images, only the blood for which the signal has changed will appear on the final image.

Another option to display only the contrast-enhanced blood vessels is to apply a technique called "maximal intensity projection" (MIP). The idea is to make use of the fact that CEM containing voxels will have relatively stronger signal than their surroundings. Hence, a 3D volume is initially scanned, either by acquiring a set of parallel planar slices or by using 3D imaging protocols as described above. Then, the volumetric information is projected onto a 2D image, but by using only the brightest voxel in each projected ray. The concept is depicted schematically in Fig. 6.55.

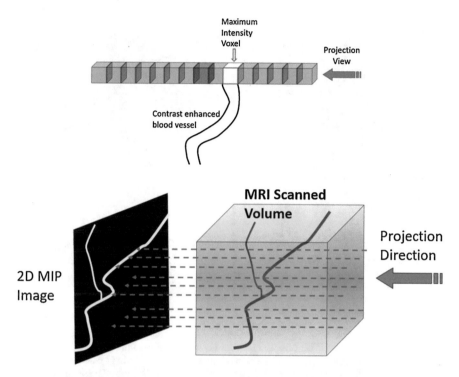

Fig. 6.55 Maximal Intensity Projection (MIP) is obtained by displaying only the brightest voxel along each ray (Top). A 3D volume will thus be projected onto a 2D image (Bottom)

In addition to gadolinium, the use of other materials for MRI CEM, such as manganese, has been examined. More recently, the use of nanoparticles (NPs) has been suggested. These very tiny molecules have unique physical properties that may induce changes in the MR image. Among those, superparamagnetic iron oxide (SPIO) NPs have been cleared by the FDA for clinical use as an MRI CEM. However, SPIO presence reduces the local T_2-weighted signal making its near by surrounding look dark in the image. Its use for MRA is currently very limited. One exemplary use was to darken the liver tissue in order to improve portal vein visibility [23].

6.11.2   Time-of-Flight (TOF) MRA

If the use of CEM is to be avoided, then we can use special imaging protocols in order to reduce the signal from the surrounding tissue without affecting the blood. The most prevalent non-CEM MRA protocol is called time-of-flight (TOF). The basic idea is to distinguish between tissue and blood based on the motion of the

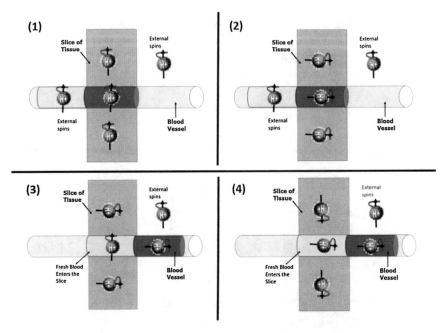

Fig. 6.56 Time-of-flight (TOF) MRA. (1) At the beginning, all the spins are in the equilibrium state. (2) The first RF pulse flips the spins within the slice volume. (3) During a waiting period, fresh blood enters the slice replacing the excited blood. (4) The second RF pulse inverts the stationary tissue spins while maximizing the transverse magnetization of the fresh blood. Thus, if image is acquired at this point, the blood vessel will look bright relative to the tissue

latter. There are several options to implement TOF MRA. The generic concept is explained schematically in Fig. 6.56.

Starting at the equilibrium state, the spins within the slice to be imaged and its surroundings are all at the upright position with no transverse magnetization. Applying the first slice selective RF pulse, only the spins within the slice, including blood and tissues, are flipped (say, e.g., by 90°). At this point we wait for a while. Due to the inflow of fresh blood, unexcited spins enter the slice volume, while the excited blood with its flipped spins leave the slice. Now, the blood vessels within the slice contain unexcited spins which differ from the excited stationary tissue spins. Applying the second RF pulse inverts the *stationary* spins, but maximizes the transverse magnetization of the blood. Acquiring an image at this point will yield pictures with bright blood vessels. An example of a TOF-MRA image is depicted in Fig. 6.57 along with a contrast enhanced MIP image.

The advantage of the TOF method is that it does not require any injection of contrast-enhancing materials. However, TOF has two major disadvantages: (i) It cannot depict in-plane flow. Hence, it is applicable mostly to blood vessels which are perpendicular to the imaged plane. (ii) It is not quantitative. Moreover, it cannot distinguish between arterial and venal flow.

6.11.3 Phase Contrast MRA

In order to allow quantitative flow imaging, a method called "phase contrast" was developed. This method utilizes the fact that MRI yields complex images and measures the motion-induced changes in the voxel's phase. As recalled, in the presence of a gradient field $\overline{G}(t)$, the phase (in a rotating reference frame) for a specific particle is given by

$$\varphi(t) = \int\limits_{0}^{t} \gamma \cdot \overline{G}(t) \cdot \overline{R}(t) \cdot dt \tag{6.69}$$

where $\overline{R}(t)$ is the time-related position.

If a bipolar gradient is applied prior to imaging along one axis, say the x-axis, as schematically shown in Fig. 6.58, then there will be a phase difference between stationary and non-stationary particles.

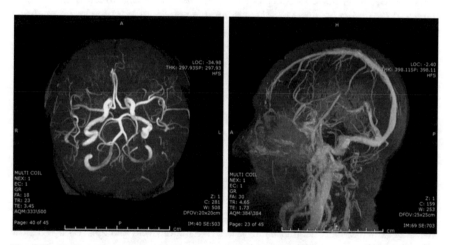

Fig. 6.57 (Left) A 2D magnetic resonance angiography (MRA) of a brain obtained using the time-of-flight (TOF) method. (Right) A maximal intensity projection (MIP) obtained from a 3D magnetic resonance venography (imaging of the veins) following injection of a contrast-enhancing material. (Images were provided courtesy of Prof. Moshe Gomori, The Hebrew University of Jerusalem)

Fig. 6.58 The bipolar gradient applied in a phase contrast flow measurements

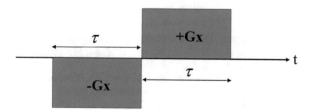

Clearly the integral in Eq. 6.69 will be zero for a stationary particle, and so will be the accumulated phase,

$$\Delta\varphi(t) = 0 \qquad \forall X(t) = \text{Const} \tag{6.70}$$

On the other hand, it can be easily shown that for a particle moving with a velocity V_x, the accumulated phase (neglecting the acceleration) is given by

$$\begin{cases} \text{for } X(t) = X_0 + V_x \cdot t \text{ and for } t > 2\tau \\ \Delta\varphi(t) = [\gamma G_x \tau^2] \cdot V_x \end{cases} \tag{6.71}$$

As can be noted the change in phase due to the bipolar gradient is linearly proportional to the velocity along the x direction. Naturally, the direction of the gradient can be set along any spatial direction, and the velocity will be measured along that direction as well. A full phase contrast velocity mapping can be conducted by scanning the object four times. The first acquisition is done without the bipolar gradient and serves as a reference. During the other three following acquisitions, the bipolar gradient is applied along each of the axes. It is important to note that if trapezoidal or other gradient profiles are implemented, Eq. 6.71 has to be modified accordingly. Either way, the velocities are linearly related to the phase change resulting from the bipolar gradient. Thus, by subtracting the reference phase from the obtained encoded phase, the velocity along all three directions is given by

$$\begin{cases} V_x = \Delta\varphi_1(t)/A \\ V_y = \Delta\varphi_2(t)/A \\ V_z = \Delta\varphi_3(t)/A \end{cases} \tag{6.72}$$

where A is the corresponding constant.

The phase contrast method has several advantages: (i) It does not require the use of CEM. (ii) It provides flow direction information. In fact, the velocity can be measured along any arbitrary direction. (iii) It is quantitative, thus enabling accurate flow assessments. On the other hand, it has two major disadvantages: (i) It is a slow process. (ii) Phase wrapping can induce errors.

References

1. Calculation of the NMR frequencies were done using the site of Brigham Young University. http://bio.groups.et.byu.net/LarmourFreqCal.phtml.
2. Engler N, Ostermann A, Niimura N, Parak FG. Hydrogen atoms in proteins: positions and dynamics. PNAS. 2003;100(18):10243–8.
3. Haacke EM, Brown RW, Thompson MR, Venkatesan R. Magnetic resonance imaging physical principles and sequence design. New York: John Wiley & Sons Inc.; 1999.
4. Mansfield P, Morris PG. NMR imaging in biomedicine. New York: Academic Press Inc; 1982.

5. Farrar TC, Becker ED. Pulse and fourier transform NMR introduction and theory. New York: Academic Press Inc; 1971.

6. Bernstein MA, King KF, Zhou XJ. Handbook of MRI pulse sequences. San Diego: Elsevier Academic Press; 2004.

7. Bojorquez JZ, Bricq S, Acquitter C, Brunotte F, Walker PM, Lalande A. What are normal relaxation times of tissues at 3 T? Magn Reson Imaging. 2017;35:69–80.

8. Spijkerman JM, Petersen ET, Hendrikse J, Luijten P, Zwanenburg J. T2 mapping of cerebrospinal fluid: 3 T versus 7 T. Magma (New York, NY). 2017;31(3):415–24.

9. Bloch F. Nuclear induction. Phys Rev. 1946;70:460–74.

10. Wehrli FW. Fast-scan magnetic resonance principles and applications. New York: Raven Press; 1991.

11. Hahn EL. Spin echoes. Phys Rev. 1950;80:580–94.

12. Carr HY, Purcell EM. Effects of diffusion on free precession in nuclear magnetic resonance experiments. Phys Rev. 1954;94:630.

13. Meiboom S, Gill D. Effects of diffusion on free precession in nuclear magnetic resonance experiments. Rev Sci Instrum. 1958;29:688.

14. Pruessmann KP, Weiger M, Scheidegger MB, Boesiger P. SENSE: sensitivity encoding for fast MRI. Magn Reson Med. 1999;42(5):952–62.

15. Griswold MA, Jakob PM, Heidemann RM, et al. Generalized autocalibrating partially parallel acquisitions (GRAPPA). Magn Reson Med. 2002;47(6):1202–10.

16. Sodickson DK, Manning WJ. Simultaneous acquisition of spatial harmonics (SMASH): fast imaging with radiofrequency coil arrays. Magn Reson Med. 1997;38(4):591–603.

17. Azhari H, Sodickson DK, Edelman RR. Rapid MR imaging by sensitivity profile indexing and deconvolution reconstruction (SPID). Magn Reson Imaging. 2003;21(6):575–84.

18. Lustig M, Donoho D, Pauly JM. Sparse MRI: the application of compressed sensing for rapid MR imaging. Magn Reson Med. 2007;58(6):1182–95.

19. Jung H, Sung K, Nayak KS, Kim EY, Ye JC. k-t FOCUSS: a general compressed sensing framework for high resolution dynamic MRI. Magn Reson Med. 2009;61(1):103–16.

20. Feng L, Grimm R, Block KT, et al. Golden-angle radial sparse parallel MRI: combination of compressed sensing, parallel imaging, and golden-angle radial sampling for fast and flexible dynamic volumetric MRI. Magn Reson Med. 2014;72(3):707–17.

21. Lustig M, Pauly JM. SPIRiT: iterative self-consistent parallel imaging reconstruction from arbitrary k-space. Magn Reson Med. 2010;64:457–71.

22. Dutt A, Rokhlin V. Fast approximate Fourier transforms for nonequispaced data. SIAM J Sci Comput. 1993;14(6):1368–93.

23. Reimer P, Marx C, Rummeny EJ, Müller M, Lentschig M, Balzer T, Dietl KH, Sulkowski U, Berns T, Shamsi K, Peters PE. SPIO-enhanced 2D-TOF MR angiography of the portal venous system: results of an intraindividual comparison. J Magn Reson Imaging. 1997;7(6):945–9.

Chapter 7
Ultrasound Imaging

Synopsis: In this chapter the reader is introduced to the basic physics of acoustic waves. The reader will learn about the interaction between acoustic waves and matter and will learn about attenuation, reflection, and speed of propagation. The reader will be introduced to ultrasonic transducers and acoustic fields, will learn about beam forming and focusing, and will learn about the different acquisition modes used in medical ultrasound and the methodologies implemented for image formation. The reader will also learn how the Doppler effect is utilized for flow imaging and color flow mapping.

The learning outcomes are: The reader will comprehend the mechanisms of ultrasonic wave propagation and the factors that affect their attenuation and reflection, will know how to analyze and utilize the Doppler shift effect, and will be able to generate ultrasound images from reflected echoes and through-transmission waves.

7.1 Introduction

Ultrasound is probably the most cost-effective medical imaging modality available today. Although ultrasonic scanners in major clinical centers may not be too cheap, good-quality systems for private clinics are affordable for a few thousand dollars. Moreover, ultrasound is by far the most portable imaging modality. Palm-size scanners and even add-on to cell phones are available on the market. Ultrasound is the second most popular modality in use (x-ray imaging is currently the most popular one). Furthermore, stemming from its high availability and from the fact that it is

© Springer Nature Switzerland AG 2020
H. Azhari et al., *From Signals to Image*,
https://doi.org/10.1007/978-3-030-35326-1_7

considered hazardless, with almost no adverse effects and no radiation exposure, it is highly popular in pregnancy monitoring.

The clinical information that can be obtained with ultrasound is versatile. It enables quick visualization of the anatomy and allows geometrical measurements in 1D, 2D, 3D, and also 4D (space and time). It enables the detection of many tissue abnormalities and cysts, as well as gal and kidney stones. Additionally, it offers real-time visualization of the beating heart with the ability to provide quantitative functional evaluation of its function. Using the Doppler effect, it enables the visualization and quantification of blood flow in vessels and within the heart chambers. In addition, ultrasound imaging is also popular in guiding minimal intervention procedures, such as insertion of biopsy devices and fine needle aspiration (FNA) and thermal ablation applicators.

Despite all the above-listed advantages, ultrasound has three major drawbacks. The first and most important one is that imaging quality is highly operator dependent. The skills of the technician or the physician operating the scanner are a major factor affecting the clinical value of the obtained images. The second major drawback is its inadequate ability to penetrate bones and image tissues located behind them, e.g., the skull bone and the brain, and also its limited ability to pass through air gaps, e.g., as in the lungs. The third drawback is its relatively low SNR and its characteristic speckles contaminated images.

7.2 Physical Phenomena

Ultrasound is based on the physical phenomena of mechanical waves. When applying pressure or stress (in solids) on materials, they move and/or deform. This deformation and motion are actually manifestations of energy stored mechanically within the substance. Motion is a manifestation of the kinetic energy, and deformation is a manifestation of the contained potential energy. If the application of pressure is localized and rapid, the inserted excessive energy will dissipate in the form of propagating mechanical waves. This phenomenon is very familiar to us and can be easily demonstrated by throwing a pebble into a pond. The local perturbation induced by the impact of the pebble is the energy source, and the emitted waves (the ripples) are the expression of the dissipating energy. Similar phenomenon occurs when we clap our hands or hit an object. The perturbation in the local pressure field is the energy source, and the sound waves carry the excessive energy away.

However, there is one important difference between the hand clap and the example given above of ripples in the pond. While in the latter case the motion of the water molecules is (generally) up and down and orthogonal to the wave propagation direction, in the case of sound waves, the motion of the air molecules (although we cannot see it) is along the wave propagation direction. (Both types can be demonstrated by moving a "Slinky" spring toy up and down or back and forth.) If the molecules move orthogonally to the wave propagation direction, the wave is termed "shear" or "transverse" wave, and if they move parallel to the wave

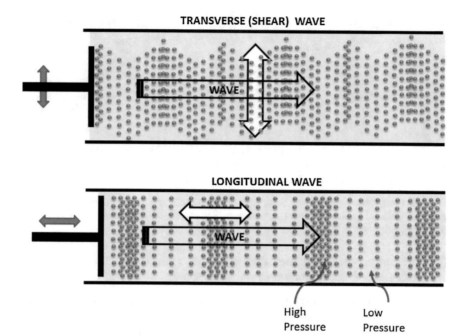

Fig. 7.1 (Top) During the passage of a transverse (shear) wave, the molecules (designated by the small spheres) move up and down (white arrow) perpendicular to the direction of the wave propagation. (Bottom) During the passage of a longitudinal wave, the molecules move back and forth parallel to the direction of the wave propagation, thus forming regions of high and low pressures

propagation direction, the wave is termed a "longitudinal" wave as shown schematically in Fig. 7.1.

Shear waves can exist in soft tissues, in viscous fluids, and in bones. However, their velocity in soft tissues is very slow, it is in the order of several meters per second, compared to the velocity of longitudinal waves which is in the order of kilometers per second. These shear waves are difficult to transmit into the body and are hardly used in medical imaging. Their main application is currently in a subfield termed ultrasonic "elastography". In elastography, tissue stiffness is mapped in order to detect pathologies, such as tumors. (This methology will be presented toward the end of the chapter).

In addition to the above distinction between longitudinal and shear waves, the geometry of the wave propagation front is also important. The basic three geometries are (see Fig.7.2):

(i) Planar – where the wave front is an infinite (theoretically of course) plane which propagates away from a planar source.
(ii) Spherical – where the wave front is a sphere with a continuously increasing radius which propagates away from a point source. It should be noted that the direction can also be reversed yielding imploding waves.

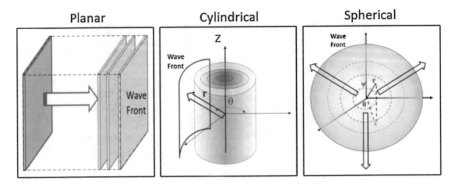

Fig. 7.2 (Left) Planar waves have a flat front which is perpendicular to the wave propagation direction. (Middle) Cylindrical waves have an expanding cylindrical front and an axis of symmetry. (Right) Spherical waves expand symmetrically from a point source

(iii) Cylindrical – where the wave front is a cylinder with a continuously increasing radius which propagates away from a line source.

For practical reasons, it is more convenient to assume in medical imaging that the transmitted waves are approximately planar. However, when calculating the acoustic fields from ultrasonic transducers, we have to relate to Huygens's principle which states that every point on a wave front may be considered as a source of spherical waves. The interference between the numerous waves emitted from all these tiny sources produces the observed wave front geometry.

7.2.1   Speed of Sound

Assuming that soft tissues can be represented by a homogenous solid material, the speed of sound for shear waves C_{shear} is given by [1]

$$C_{\text{shear}} = \sqrt{\frac{\mu}{\rho}} \qquad (7.1)$$

where μ is the shear modulus for the medium and ρ is the Archimedean density [Kg/m^3]. This equation can be further approximated by

$$C_{\text{shear}} \approx \sqrt{\frac{E}{3\rho}} \qquad (7.2)$$

where E is called "Young's modulus." This elastic coefficient relates the amount of strain, which is the relative change in the dimension of an object, to the stress applied on that material along the same direction. In other words, it characterizes the stiffness of the medium. Hence, stems its applicability to "elastography" (see Sect. 7.6.2).

Longitudinal waves on the other hand are much more relevant to medical imaging. They are easy to generate and to transmit into the body, and their velocity is high (about 1500 [m/s] in soft tissues). This high velocity, considering the size of the body, implies that many data collecting transmissions can be conducted within a short fraction of time and real-time imaging can be obtained.

In the context of medical imaging, we can approximate soft tissues by a fluidic medium. Accordingly, its response to a change in the external pressure is characterized by an index called "compressibility." This compressibility index, β, is defined as

$$\beta = -\frac{1}{V}\left(\frac{\Delta V}{\Delta P}\right)_T \tag{7.3}$$

where ΔV is the change in the volume of an object with an initial volume, V, subjected to an external pressure elevation of ΔP. The negative sign indicates that the pressure increases when the volume is reduced. The index T indicates that the process takes place under isothermal conditions, i.e., constant temperature.

In certain texts the index of compressibility is replaced by the "bulk" modulus, κ, which is simply the reciprocal value of the compressibility index, i.e., $\kappa = 1/\beta$. Either way, the speed of sound C for longitudinal waves in a fluidic medium is given by [1]

$$C_{\text{Longitudinal}} = \frac{1}{\sqrt{\beta\rho}} = \sqrt{\frac{\kappa}{\rho}} \tag{7.4}$$

7.2.2   Attenuation

The amplitude of the acoustic wave is commonly measured in units of pressure. When an acoustic wave with an initial *pressure amplitude* P_0 propagates through a homogeneous medium, it is observed that its amplitude decays exponentially with the distance (similar to x-rays –as described in Chap. 3). Part of the energy carried by the wave is absorbed within the tissue, thus causing temperature elevation. The relation is quantitatively given by

$$p(x) = p_0 \cdot e^{-\alpha \cdot x} \tag{7.5}$$

where x is the distance and α [1/cm] is the attenuation coefficient. The wave intensity $I(x)$ [watts/cm^2] is proportional to the square of its amplitude; therefore, it follows that

$$I(x) = I_0 \cdot e^{-2\alpha \cdot x} \tag{7.6}$$

where I_0 is the initial intensity.

As the change in amplitude and intensity is typically substantial even for a short travelling distance, it is convenient to describe it in Decibels units, which are defined as

$$dB = 20 \log_{10}\left(\frac{p}{p_0}\right) \tag{7.7}$$

where positive dB values indicate gain and negative values designate attenuation. Practically, it is worth noting that -3 dB indicates 50% reduction in *intensity* and -6 dB indicates reduction to 25% of the initial *intensity* and so forth.

In the range of medical ultrasonic frequencies, i.e., the low megahertz range, most attenuation coefficients are approximately related linearly to the frequency. Thus, a simple approximation for a certain frequency range can be

$$\alpha(f) = \alpha_1 \cdot f \tag{7.8}$$

Accordingly, the attenuation may also be expressed in units of α [dB/cm/MHz]. Thus, the coefficient has to be multiplied by the corresponding frequency and the distance to obtain the attenuation in dB units. The frequency effect of course is substantial. If, for example, the attenuation coefficient is $\alpha = 0.3$ [dB/cm/MHz], then, at a 10 cm distance from the source, the attenuation will be -3 dB for waves with a frequency of 1 MHz. This corresponds to 50% of the initial value, while for 3 MHz the attenuation will be -9 dB which corresponds to only 12.5% of the initial value. This is demonstrated graphically in Fig.7.3.

7.2.3   The Acoustic Impedance

As stated above the acoustic wave is actually a mechanical manifestation of propagating energy. The source of forces which cause the matter to locally vibrate and deform stems from an instantaneous pressure gradient along the wave propagation direction. When observing an infinitesimal matter element, this dynamic state can be modelled as an electric circuit, as shown schematically in Fig.7.4. Using this electric analogy, the local excessive pressure P_e which is associated with the wave passage through the element is analogous to voltage. The particle velocity u (which differs from the speed of the propagating wave) is analogous to electric current. And the ratio between the two (commonly marked by Z) is termed the *acoustic impedance*.

Without showing the full derivation (for more details see, e.g., [2]), it is stated here that for a planar wave, the acoustic impedance is given by

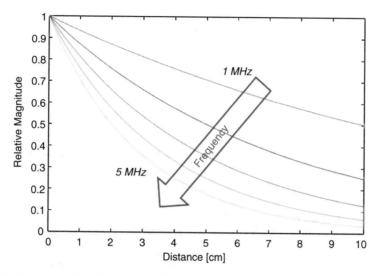

Fig. 7.3 Exemplary attenuation as a function of distance for frequencies ranging from 1 MHz to 5 MHz. Note the much more rapid decay for the higher frequencies

Fig. 7.4 In an analogy to Ohms law, the acoustic impedance Z of a tissue element is the ratio between the local excessive pressure P_e and the particles velocity U

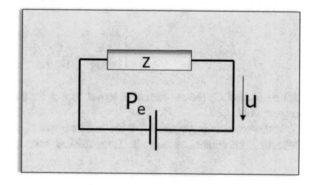

$$Z \triangleq \frac{P_e}{u} = \rho_0 \cdot c \tag{7.9}$$

where ρ_0 is the density at rest (when no waves exist) and c is the speed of sound in the medium.

The acoustic impedance plays an important role in medical imaging as will be explained in the next section. Another important feature is its relation to the ultrasonic intensity which is a crucial factor in ultrasonic safety. For a sinusoidal wave with amplitude P_0, the average intensity is given by

$$I = \frac{P_0^2}{2Z}$$

$\qquad\qquad\qquad\qquad\qquad\qquad\qquad\qquad\qquad\qquad\qquad$ (7.10)

7.2.4   Reflection Refraction and Transmission

When a propagating wave encounters a region with a different acoustic impedance, part of its energy is reflected, and the rest passes through the mutual boundary into the second region. The reflected waves which are actually *echoes* therefore contain information which can be utilized for imaging as will be explained in the following. The through-transmitted waves also undergo changes in terms of amplitude and propagation direction. Though much less popular, these informative changes may also be used to generate images.

Assuming a naive model where the different tissue regions have relatively flat boundaries, the angles of reflection and deflection can be estimated from Snell's law.

Referring to Fig. 7.5, each wave front is represented by a thick arrow. Let us designate the angle of incidence as θ_i and the reflection angle as θ_R and the angle of the through-transmitted wave as θ_T. It follows from Snell's law that

$$\begin{cases} \theta_i = \theta_R \\ \dfrac{C_1}{\sin \theta_i} = \dfrac{C_2}{\sin \theta_T} \end{cases}$$

$\qquad\qquad\qquad\qquad\qquad\qquad\qquad\qquad\qquad\qquad\qquad$ (7.11)

where C_1 and C_2 are the speed of sound values for the first and second mediums, respectively.

If the acoustic impedances for the mediums are Z_1 and Z_2, respectively, then the reflection coefficient R and the transmission coefficient T, which refer to the

Fig. 7.5 A planar wave (i) impinging upon the boundary between two materials for which the speed of sound is different will be partially reflected (R) and partially through transmitted (T). The angles can be calculated using Snell's law

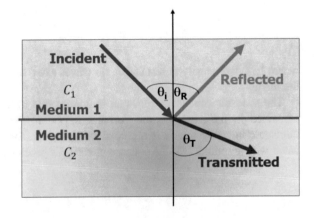

amplitude (pressure) of the reflected wave, i.e., the echo, and the *amplitude* of the through-transmitted wave normalized to the amplitude of the incident wave, are given by

$$R = \frac{P_R}{P_i} = \frac{Z_2 \cos \theta_i - Z_1 \cos \theta_T}{Z_2 \cos \theta_i + Z_1 \cos \theta_T}$$
$$T = \frac{P_T}{P_i} = \frac{2Z_2 \cos \theta_i}{Z_2 \cos \theta_i + Z_1 \cos \theta_T}$$
(7.12)

where P_i, P_R, and P_T are the *pressure amplitudes* of the incident, reflected, and through-transmitted wave, respectively.

In practice, for simplicity, one can assume that the incident wave is perpendicular to the boundary surface; hence, the terms become

$$R = \frac{P_R}{P_i} = \frac{Z_2 - Z_1}{Z_2 + Z_1}$$
$$T = \frac{P_T}{P_i} = \frac{2Z_2}{Z_2 + Z_1}$$
(7.13)

Studying these relations, it can be deduced that if $Z_2 = 0$, such as the case for air and gases relative to soft tissues, total reflection occurs, i.e., $R = -1$. (The negative sign indicates phase inversion.) This fact has two important implications:

(i) The first (negative) implication is that it would be difficult to image tissues located behind the lungs. Indeed, for cardiac (heart) imaging the available acoustic windows are consequently limited. Also, air trapped between the ultrasonic transducer and the skin, for example, in hairy locations, will block the waves. Therefore, a coupling gel is used to ensure good acoustic contact between the transducer and the skin. Also, shaving the skin may be needed in certain cases.

(ii) The second (positive) implication is that gas bubbles can be easily detected. This has led to the development of a family of contrast-enhancing materials which are based on gas-filled microbubbles. The small size allows them to systemically circulate in the bloodstream and pass the lungs. Their high reflectivity, on the other hand, allows good visualization of their distribution in the body post injection. This can allow the detection of perfusion problems.

Another aspect of Eq. 7.13 relates to calcified tissues. If $Z_2 \gg Z_1$, as is the case for bones which acoustic impedance is much higher than that of soft tissues, then the reflection coefficient is high, i.e., $R \rightarrow 1$. As a result, most of the wave energy will be reflected, and imaging behind them will be difficult. This effect may be manifested as an acoustic shadow which may appear behind them. This is also truth for gal and kidney stones. This, however, has also a positive aspect as it allows better detection of these stones.

Recalling the ratio given by Eq. 7.10 for the wave intensity, the following relations estimating the energy or power reflection and transmission can be derived:

$$R_{\text{Intensity}} = \frac{I_R}{I_i} = \frac{P_R{}^2}{2Z_1} \cdot \frac{2Z_1}{P_i{}^2} = \frac{P_R{}^2}{P_i{}^2} = R^2$$

$$T_{\text{Intensity}} = \frac{I_T}{I_i} = \frac{P_T{}^2}{2Z_2} \cdot \frac{2Z_1}{P_i{}^2} = \frac{Z_1}{Z_2} \cdot \frac{P_T{}^2}{P_i{}^2} = \frac{Z_1}{Z_2} \cdot T^2 \qquad (7.14)$$

$$= \frac{4Z_1Z_2}{(Z_1 + Z_2)^2}$$

where $R_{\text{Intensity}}$ is the reflection coefficient for the *intensity* and $T_{\text{Intensity}}$ is the transmission coefficient for the *intensity*.

7.2.5   The Doppler Shift Effect

When transmitting an ultrasonic wave with frequency f_0 toward a moving object, the reflected waves will have a different frequency stemming from the Doppler shift effect. The change in frequency Δf is proportional to the velocity of the target relative to the transmitting transducer. As described schematically in Fig. 7.6, this Doppler shift effect also occurs when transmitting an ultrasonic wave toward a blood vessel. The clouds of the moving blood cells reflect some of the wave energy. This reflection is very small but detectable. If the blood velocity is V and the angle of the ultrasonic beam relative to the blood vessel axis is θ, and the speed of sound is C, it can be shown [3] that the frequency shift is given by

$$\Delta f = \frac{f_0 \cdot 2V \cdot \cos\theta}{C - V} \qquad (7.15)$$

Considering the fact that the speed of sound in soft tissues and blood is about 1500 [m/s] while the bloodstream velocity under normal conditions is less than 1 [m/s], this equation can be modified to

Fig. 7.6 Stemming from the Doppler shift effect, the frequency of the echoes reflected from the blood cells will be different from the transmitted wave frequency

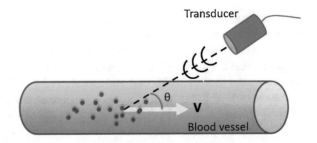

$$\Delta f = \frac{f_0 \cdot 2V \cdot \cos\theta}{C} \tag{7.16}$$

Thus, by measuring the Doppler frequency shift, Δf, the velocity of the blood can be calculated by

$$V = \frac{\Delta f \cdot C}{2f_0 \cos\theta} \tag{7.17}$$

7.3 Ultrasonic Transducers and Acoustic Fields

7.3.1 Piezoelectric Transducers

Ultrasound waves are most commonly produced by piezoelectric materials. These materials are characterized by a unique property. When pressure is applied onto such a material along a certain direction, it will deform and produce an electric voltage. On the other hand, when subjected to an electric voltage along that certain direction, it will deform and apply pressure upon its surroundings.

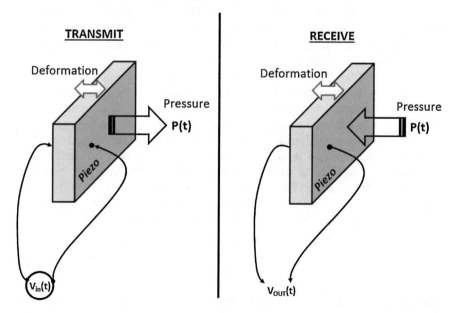

Fig. 7.7 (Left) A piezoelectric element can generate ultrasonic waves, by applying voltage onto its opposing surfaces. The voltage causes it to deform and vibrate and generate the pressure waves. (Right) The same element can also be used for detection of acoustic waves. The waves apply pressure on the element; the element deforms and yields electric voltage

Fig. 7.8 The generic
"sandwich" structure of an
ultrasonic transducer
consists of a matching layer
at the front of the
piezoelectric element and a
backing layer behind it

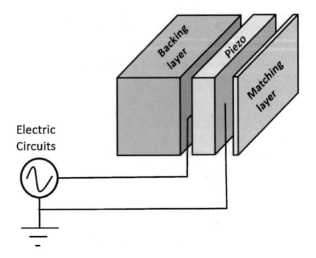

This dual property is used in building ultrasonic transducers. Considering a piezoelectric element as shown schematically in Fig. 7.7, two electric wires are soldered to it opposing faces. In order to transmit waves, voltage is applied between the two sides. (The transmitting voltage in medical imaging is commonly in the order of tens of volts.) As a result, the piezo element will vibrate and generate an acoustic wave. When operating in the receiver mode, the electric power source is replaced by a receiver box. The impinging acoustic waves will apply pressure on the element. It will consequently deform and yield voltage between its two faces, which can be detected, amplified, filtered, and registered. Amplification is usually needed since the received voltage is commonly less than 1 volt.

As recalled from Eq. 7.13, the reflection coefficient between two materials is determined by the discrepancy between their acoustic impedances. In the case of piezoelectric elements, the acoustic impedance of the element, Z_{peizo}, is much higher than that of soft tissues. As a result, if attached directly onto the body (assuming proper electrical isolation is used of course), most of the acoustic energy transmitted from the element will be reflected. Consequently, the element will heat up and eventually malfunction. In order to overcome this problem, a thin "matching layer" is attached to the surface which is in contact with the skin. It can be shown [4] that optimal transmission efficiency can be obtained if the thickness of the layer is set to be an odd multiple of $\lambda/4$ (where λ is the wavelength) and the acoustic impedance of the matching layer, $Z_{matching}$, is set to be

$$Z_{matching} = \sqrt{Z_{peizo} \cdot Z_{water}} \tag{7.18}$$

where Z_{water} is the acoustic impedance of water, which is a good approximation to the acoustic impedances of most soft tissues. In addition, in order to reduce reverberations, a backing layer is sometimes placed behind the piezoelectric element, yielding a sandwich configuration as shown schematically in Fig. 7.8.

7.3.2   Single Element Transducers

In medical applications, two generic configurations of transducers are commonly used. The first is a single element transducer and the second is an array transducer. The transducers are either attached to the body or moved about in or out of the body in order to collect the data needed for image reconstruction.

The most common single element transducer is a disc-shaped device. The acoustic pressure field for such a transducer is given by [5]

$$P(R, \theta) = \left[\pi a^2 \cdot A_0 \right] \cdot \left[\frac{e^{j(2\pi f \cdot t - k \cdot R)}}{R} \right] \cdot \left[\frac{2 \cdot J_1(ka \sin \theta)}{(ka \sin \theta)} \right] \tag{7.19}$$

where R is the distance from the transducer's center (the range) and θ is the azimuth to the target point, $P(R, \theta)$ is the pressure as a function of location, a is the radius of the transducer, A_0 is the transmission amplitude, $k = 2\pi/\lambda$ is the wave number, f is the frequency, and J_1 is a Bessel function of order 1.

The rightmost brackets are referred to as the "directivity function" and indicate the azimuthal variation of the field, the "lobes." This factor is very important for imaging as will be discussed in the following. In principle, the more "directed" is the beam, i.e., it has a thin main lobe, the better it is for imaging.

An exemplary simulated acoustic field is depicted in Fig. 7.9. Note that the field is actually three-dimensional. This map depicts a cross section of the field through the symmetry axis of the disc transducer. As can be noted the acoustic field has one strong main lobe (indicated by the brighter color) and many small side lobes (The pressure scale is logarithmic).

As can be noted from Fig. 7.9, the pressure field very close to the transducer varies substantially within a short distance. This part of the field is called the "near field." Beyond this zone the field varies in a more monotonic manner. This zone is

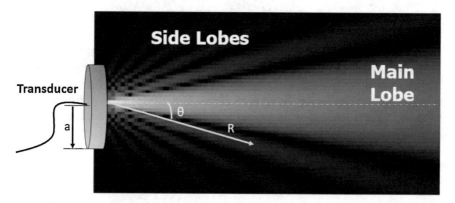

Fig. 7.9 The acoustic pressure field of a disc transducer (logarithmic scale). R is the distance coordinate from the transducer's center, and θ is the azimuth. Note the strong main lobe and the many small side lobes. (Higher pressure is indicated by whiter color)

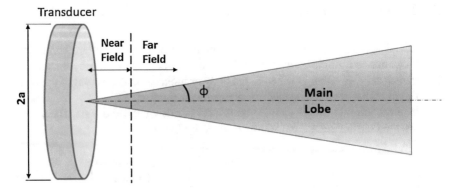

Fig. 7.10 In the far field the main lobe can be approximated by a cone which head angle ϕ is given by Eq. 7.21. The larger the transducer radius and the higher the frequency, the narrower the main lobe

called the "far field." For imaging purposes, it is preferable to work in the far field. The distance NF to the end of the near field and the beginning of the far field is given by

$$
\begin{cases}
\text{NF} = \dfrac{4a^2 - \lambda^2}{4\lambda} \\[2ex]
\text{if } a^2 >> \lambda^2 \;\Rightarrow\; \text{NF} = \dfrac{a^2}{\lambda}
\end{cases}
\tag{7.20}
$$

When working in the far field, the main lobe can be approximated by a cone which axis aligns with the axis of the transducer, as shown schematically in Fig. 7.10. The head angle for that cone, ϕ, is given by

$$
\phi = \sin^{-1}\left(\frac{3.83}{ka}\right) = \sin^{-1}\left(\frac{0.61\lambda}{a}\right)
\tag{7.21}
$$

The spread angle of the main lobe ϕ is important for imaging, as will be discussed in the following. In principle, the narrower the main lobe, the better the lateral resolution. In order to obtain a much narrower beam, focusing is needed. This can be achieved, either by shaping the transducer in a concaved manner or by adding an acoustic lens, as described schematically in Fig. 7.11. Either way, the diameter of the beam at the focal zone is approximately given by

$$
D_x \approx \lambda \cdot \left(\frac{F}{2a}\right)
\tag{7.22}
$$

where λ is the wavelength in the medium and the term in the brackets (focal distance/ aperture) is called the "F-number."

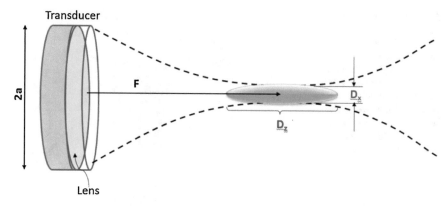

Fig. 7.11 Using a lens or a concaved transducer yields a focused beam. Although the beam is narrowed at the focal zone, the zones in front and behind it are substantially more diverged

The focal zone has a "cigar" shape, meaning its length D_z is larger than the diameter. The ratio between the two depends on the specific design, but D_z is commonly several times the diameter D_x. This implies that a substantial part of the beam can be within the focused zone. The advantage in the context of imaging is that better lateral resolution can be obtained and scanning objects within the focal zone will yield sharper images. However, the penalty for that is poorer resolution in the regions located before and after the focal zone. This limitation can be overcame by using phased array transducers and dynamic focusing.

7.3.3 Linear Phased Array Transducers

In most of the medical imaging applications, it is preferable to use array transducers. The array transducers offer flexibility in beam steering and allow dynamic focusing. The most common type in use is a linear array. This array is comprised of many small cubical piezo elements placed along a straight line adjacent to each other as shown schematically in Fig. 7.12.

Each element is wired separately and has its own electric circuit which controls its transmission time, its phase, and its amplitude. In order to ensure the absence of grating lobes, the elements should be small and close to each other. The distance Δx between the element centers should be smaller than the wave length λ. The wave transmitted from each element is roughly circular within the imaged plane. The interference of all the waves transmitted from all the elements together yields a cumulative wave front which can be manipulated electronically.

If all the elements transmit simultaneously with the same phase, the cumulative wave front is approximately planar as shown schematically in Fig. 7.12. However, if the elements are turned ON with a proper time delay (for a pulse transmission) or proper phase (for a continuous wave transmission), which is proportional to the element's location, the beam can be steered to point at different azimuths within the

Fig. 7.12 A linear phased array transducer is comprised of a row of many small piezoelectric transducers placed close to each other. Each element phase and amplitude is controlled separately. The interference between all the individual transmitted waves set the cumulative wave front

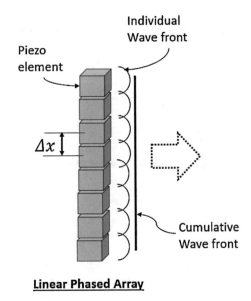

Linear Phased Array

imaged plane. The maximal steering angle β_{max} relative to a line perpendicular to the array center is given by

$$\beta_{max} = \sin^{-1}\left(\frac{\lambda}{\Delta x} - 1\right) \tag{7.23}$$

Moreover, if the time delays or phases are designed to emulate a concaved transducer, the beam can be focused as well, as shown in the acoustic field simulations shown in Fig. 7.13.

The phased array transducer thus allows us to control electronically the steering angle and the focal distance as well. This process is termed *beam forming*. A demonstration for three different beams with different steering angles and different focal distances are presented in Fig. 7.13. The ability to electronically control the transmission of each element enables us also to implement "dynamic focusing," which means that we can arbitrarily set many different focal distances along a certain direction without changing the hardware. Therefore, an object can be scanned by several virtual beams which focal zones are chained one after the other, to yield sharper images.

7.3.4 Implications on Image Resolution

Ultrasonic imaging is commonly characterized by an elliptic point spread function (PSF), with a wider *lateral* dimension (perpendicular to the beam axis). The quality of the acoustic field of the transducer primarily determines the lateral resolution of the obtained image. As the imaging process is fundamentally based on moving the

Fig. 7.13 Acoustic field simulations for a 128 element linear phased array in water. As can be noted the beam can be simultaneously steered and focused with varying angles and different focal depths by controlling the phase of each element. (The transmission is from the left side)

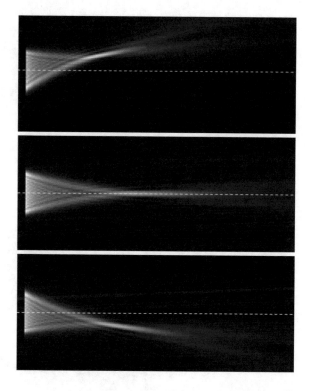

acoustic beam through the target and collecting the scattered waves, it follows that a focused narrow beam provides more localized information. On the other hand, an unfocused beam will collect echoes from the same target even when the beam has been moved to another location. This will result in a laterally smeared PSF. Moreover, as depicted schematically in Fig. 7.14, a wide beam may simultaneously collect echoes from multiple targets which may result in overlapping PSFs and poor lateral resolution. *Thus, sharper "directivity function," focused beam, and narrow main lobe are better for imaging.*

As for the axial resolution, the process is governed mainly by the temporal profile of the transmitted pulse. This stems from the fact that the images are typically reconstructed from collected reflections (echoes). Thus, our ability to distinguish between two targets located one behind the other is determined by our ability to differentiate between their echoes. Short bursts or pulses will yield better time-resolved echoes. As recalled, a narrow object in time domain has a wide band in the frequency domain. Also, fast changes in time are associated with high frequencies. This implies that in order to improve the *axial resolution*, the transmission pulses must be *broadband and of high frequencies*. (See graphic illustration in Fig. 7.15.)

Fig. 7.14 (Top) The width of the point spread function (PSF) of an image generated by moving the acoustic beam and collecting echoes is determined by the width of the beam. (Bottom) Wide PSF will result in poor lateral resolution

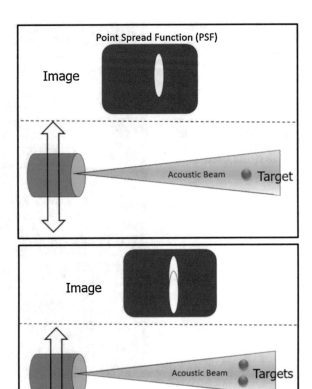

7.4   Imaging Modes and Image Formation

7.4.1   The Signal Source

Ultrasound imaging is fundamentally based on transmitting acoustic waves into the scanned object and collecting the scattered echoes or through-transmitted waves. Current clinical scanners are almost exclusively based on the pulse-echo technique. This means that a temporally short burst of acoustic waves is transmitted into the body. The propagating waves encounter different tissues along their path. These different tissues have different acoustic impedances. Consequently, scatter occurs and echoes are reflected back toward the transducer. The receiving transducer, which is usually the same transducer which was used for transmitting the waves, picks these echoes, and using a backend electronic system, they are registered as a function of their arrival time.

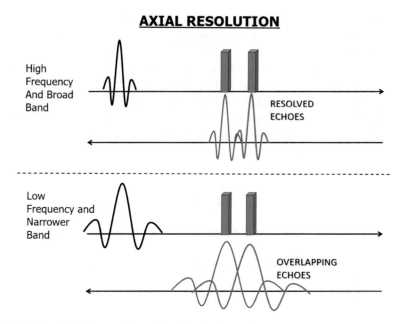

Fig. 7.15 (Top) High frequencies and broadband pulses yield narrow echoes in time which can be better resolved. (Bottom) Low frequencies and narrow band pulses yield temporally broad echoes which overlap and reduce the axial resolution

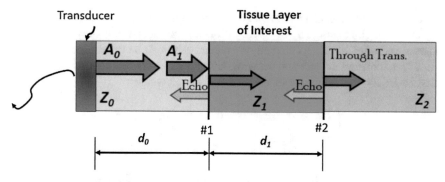

Fig. 7.16 A tissue layer of interest positioned between two different layers will reflect two echoes as explained in details in the text

In order to describe the process which yields the signals in quantitative terms, several alternative models have been suggested. Here, we shall implement a very simple model: the "multilayer plate model." Using this model, each tissue along the acoustic beam is represented by a plate which faces are perpendicular to the acoustic beam. For clarity, we shall initially discuss a "sandwich" comprising of only three layers of tissues, where the tissue of interest is located in the middle, as depicted schematically in Fig. 7.16.

If the amplitude of the transmitted wave is A_0, the amplitude of the wave impinging upon the first boundary will be $A_1 = A_0 e^{-\alpha_0 \cdot d_0}$, where α_0 and d_0 are the attenuation coefficient and thickness of the first layer, respectively. The amplitude of the first echo can be calculated according to Eq. 7.13, and accounting for the attenuation, the amplitude detected by the transducer is given by

$$\begin{cases} \text{Echo}_1 = A_0 e^{-2\alpha_0 \cdot d_0} \cdot R_{01} \\ R_{01} = \dfrac{(Z_1 - Z_0)}{(Z_0 + Z_1)} \end{cases} \tag{7.24}$$

Importantly, note the **2** in the exponential term, which stems from the fact that the waves have to travel back and forth.

As explained before, part of the wave will continue travelling through the boundary. Thus, according to Eq. 7.13, and accounting for the attenuation along the path, the amplitude reaching the second boundary is given by

$$\begin{cases} A_2 = A_0 e^{-\alpha_0 \cdot d_0} \cdot e^{-\alpha_1 \cdot d_1} \cdot T_{01} \\ T_{01} = \dfrac{2Z_1}{(Z_0 + Z_1)} \end{cases} \tag{7.25}$$

where T_{01} is the transmission coefficient from the first layer to the second.

The through-transmitted wave will encounter the second boundary between the second and third layers. As a result, the second echo will be generated. Accounting for attenuation and the fact that the wave has also to pass back and forth through the first boundary, the amplitude of the second echo is given by

$$\begin{cases} \text{Echo}_2 = A_0 e^{-2\alpha_0 \cdot d_0} \cdot e^{-2\alpha_1 \cdot d_1} \cdot T_{01} \cdot T_{10} \cdot R_{12} \\ R_{12} = \dfrac{(Z_2 - Z_1)}{(Z_1 + Z_2)}; \quad T_{10} = \dfrac{2Z_0}{(Z_0 + Z_1)} \end{cases} \tag{7.26}$$

where T_{10} is the transmission coefficient from the second layer to the first and R_{12} is the reflection coefficient between the second and third layers.

Following the same steps, the model can be expanded to any number of layers. Accordingly, the amplitude of the Nth echo is given by [5]

$$
\begin{cases}
\text{Echo}_N = A_0 \cdot \left[\prod_{n=0}^{N-1} e^{-2\alpha_n d_n} \cdot T_{(n-1,n)} \cdot T_{(n,n-1)} \right] \cdot R_{(N-1,N)} \\[2mm]
\text{where}: T_{(-1,0)} = T_{(0,-1)} = 1 \\[1mm]
\text{and for } n > 0; \\[2mm]
T_{(n,n-1)} = \dfrac{(2Z_{n-1})}{(Z_{n-1} + Z_n)}; T_{(n-1,n)} = \dfrac{(2Z_n)}{(Z_{n-1} + Z_n)} \\[3mm]
\text{and } R_{(N-1,N)} = \dfrac{(Z_N - Z_{N-1})}{(Z_N + Z_{N-1})}
\end{cases}
\tag{7.27}
$$

As for the arrival time of each echo, it depends on the cumulative speed of sound and thickness of all the layers in the acoustic path all the way to the target, i.e.,

$$
t_N = \delta \left(t - \sum_{n=0}^{N-1} \frac{2d_n}{C_n} \right)
\tag{7.28}
$$

where t_N is the arrival time of echo number N and C_n is the speed of sound in the nth layer.

The above model assumes that the transmitted wave is temporally very short as an impulse, i.e., a delta function. Thus, the echo train is represented by a series of reflected pulses; for clarity, we shall mark it as $E_{\text{impulses}}(t)$. In practice however, each transducer has its own impulse response and the transmitted pulses have a certain frequency band. A common model used for simulating a more realistic pulse is the Gaussian pulse, defined analytically by

$$
p(t) = A_0 \cdot e^{\left(-\beta \cdot t^2 \right)} \cdot \cos \left(2\pi \cdot f_0 \cdot t \right)
\tag{7.29}
$$

where $p(t)$ is the pressure profile of the pulse, β is a parameter characterizing the spectral band, and f_0 is the central frequency of the pulse.

According to the above, the echo train can be more realistically simulated by convolving the series of reflected impulses, $E_{\text{impulses}}(t)$, with $p(t)$, i.e., $E(t) = p(t) \otimes E_{\text{impulses}}(t)$. A demonstrative simulated echo train is depicted in Fig. 7.17.

7.4.2   The A-Line (A-Mode)

The fundamental building block for generating an image from echoes is called the *A-line*. The *A-line* is simply a one-dimensional time-related train of echoes obtained by transmitting a pulse into the scanned object. Usually (but not necessarily), the implicit assumption is that the acoustic beam is more or less propagating along a straight line. Thus, the obtained information can be related to changes occurring along its direction. Acquiring such a one-dimensional vector is also referred to as scanning in the *A-mode*. It allows one to measure noninvasively distances between

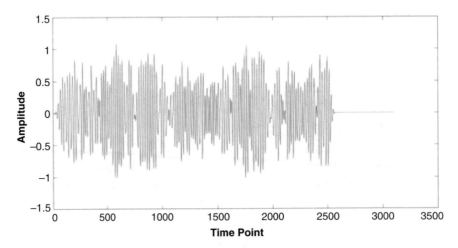

Fig. 7.17 A simulated (noiseless) more realistic echo train obtained using a Gaussian pulse

two points, for example, the inner diameter of the eye or the head diameters of an embryo in the womb.

Actual *A-line* signals resembling the one shown in Fig. 7.17 are referred to as the "RF signal" and are the raw input for generating an image. As can be observed, the signal seems visually chaotic. Furthermore, in reality the RF signals also contain noise from various sources, which further complicate its pattern. Indeed, at first glance it seems that generating an image from such signals is unfeasible. Thus, the first step is to transform the RF signal into something more comprehendible in visual terms.

In order to achieve the above goal, each A-line is converted into a displayable vector of gray levels. There are basically three fundamental steps in this conversion (see Fig. 7.18).

The first step is to "clean" the signal and prepare it to the second stage. This includes the application of a band-pass filter (or other filters) to remove undesired signals which probably did not stem from the acoustic process and to reduce the noise level. In addition, a time gain compensation (TGC), i.e., time-dependent signal amplification, is typically applied as will be explained later.

The second step is to extract the signal's envelope. This procedure substantially smooths the signal and allows better visual representation of the changes present in the signal's amplitude (see Fig. 7.18). The envelope extraction is achieved by taking the absolute value of the "Hilbert transform" of the signal. Mathematically, the envelope extraction procedure is given by

Fig. 7.18 The three
fundamental steps in turning
the raw RF signal into an
image. (Top) Filter and
"clean" the RF signal.
(Middle) Apply the Hilbert
transform to extract the
signal envelope. (Bottom)
Display the amplitude in
terms of gray levels

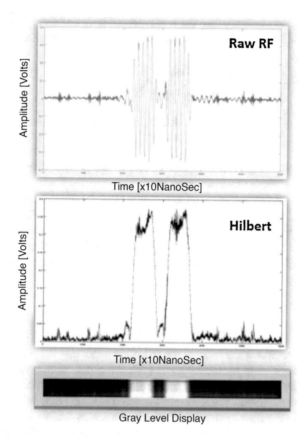

$$H(t) = \text{Envelope}\{S(t)\} = \left| S(t) + j\frac{1}{\pi} \int\limits_{-\infty}^{\infty} \frac{S(\tau)}{(t-\tau)} d\tau \right| \qquad (7.30)$$

where $S(t)$ is the raw signal and $H(t)$ is the obtained envelope signal.

The third step is to convert the envelope signal $H(t)$ into a row of pixels with different gray levels. Here, we need to set two parameters: The first is to set the spatial coordinate, and the second is to set the dynamic range. The latter procedure was already explained in details in Chap. 1 (see Sect. 1.3.1). As recalled, the lowest relevant physical value, which is in this case the echo amplitude, is displayed as a black-colored pixel. Naturally, every value smaller than this level will also be displayed as a black pixel. On the other side of the scale, the highest relevant value will be assigned a white color. Naturally, every value above this level will also be colored white. The scale between the highest and lowest physical values is then divided into N gray levels (typically N is a power of 2, i.e., 128, 512, 1024, etc.).

Every intermediate value is assigned a shade of gray according to its relative value within this scale, according to the following formula:

$$G(t) = \begin{cases} H(t) \leq L_{\text{low}} \Rightarrow \text{Black} \\ H(t) \geq L_{\text{high}} \Rightarrow \text{White} \\ \\ \qquad\text{else} \qquad\qquad H(t) = \text{round}\left\{ N \cdot \dfrac{(H(t) - L_{\text{low}})}{(L_{\text{high}} - L_{\text{low}})} \right\} \end{cases} \qquad (7.31)$$

where L_{low} and L_{high} are the lowest and highest displayed values, respectively, and $G(t)$ is the assigned gray level.

As for estimating the distance from the transducer to each reflecting target, there is a problem. In theory we should apply Eq. 7.28 and sum up all the layers' thicknesses. However, in reality we almost always do not know the speed of sound within each tissue layer. The applied solution is to assume an average speed of sound for all soft tissues, $\overline{C}$ (commonly about 1540 m/s), and implement the following relation:

$$x = \overline{C} \cdot \frac{\Delta t}{2} \qquad (7.32)$$

where x is the range to the reflecting target and the division by two is needed in order to account for the back and forth travel path of the wave. It is important to note that by implementing this solution, we inherently induce an error into our geometrical calculations. For example, the speed of sound in fatty tissue is about 1450 m/s, while in muscles it can reach 1590 m/s. Hence, assuming, for example, an error in speed of sound of ± 50 m/s along a certain A-line, the distance estimation produced by Eq. 7.32 will err by about 3%.

In a discrete format where the pixel size is set to Δ, the pixel address m along that direction is given by

$$m = 1 + \text{floor}(x/\Delta) \qquad (7.33)$$

Thus, at this point, we have established a procedure for producing a row or column of pixels each of which has a size Δ and a specific gray level as defined by Eq. 7.31.

7.4.3   The Time Gain Correction (TGC)

In most of the pulse-echo imaging applications, we are interested in displaying mainly the reflectivity map, i.e., $R(x, y)$. The textural variations in the image stemming from the numerous reflections which are visible are commonly associated with different tissues, and changes in these texture may indicate pathology. The geometry

Fig. 7.19 The amplitudes of echoes from identical reflectors positioned at an increasing distances from the transducer will become smaller and smaller due to attenuation and passage between layers

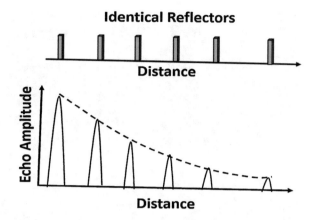

of the anatomy is revealed from the borders between such different textural zones. Furthermore, important clinical information is encompassed within the spatial variations of the echoes amplitudes. For example, a highly attenuating region may appear as relatively dark in in the image and reveal a suspicious finding. Naturally, strong echoes are assumed to be associated with strong reflectors and vice versa. However, studying Eq. 7.27, we note that the magnitude of the term

$$\left[\prod_{n=0}^{N-1} e^{-2\alpha_n d_n} \cdot T_{(n-1,n)} \cdot T_{(n,n-1)} \right] \leq 1 \qquad (7.34)$$

is always smaller than *1!* (The reader is challenged to prove it.)

Furthermore, its value will get smaller as the wave propagates deeper into the tissue and when passing from one layer to the other. Consequently, if we scan a row of identical reflectors, the echoes will diminish with the distance as shown schematically in Fig. 7.19.

In order to overcome this problem, almost all ultrasonic scanners have a built-in feature called *TGC* (time gain correction or compensation). The TGC allows the operator to manually amplify signals detected from deeper layers, which is equivalent to later arrival times. In most modern scanners, automatic TGC option is available. The TGC thus compensates for the loss of signal from deeper tissue layers and provides some equalization effect of the near and far echo amplitudes. Alternatively, it allows the operator to selectively enhance regions which are clinically more important. A demonstrative image displaying the TGC effect is depicted in Fig. 7.20.

7.4.4 The M-Mode

M-mode imaging is an extension of the *A-mode* into the time domain. Unlike the A-mode which provides a static echo train, the M-mode produces dynamic 2D

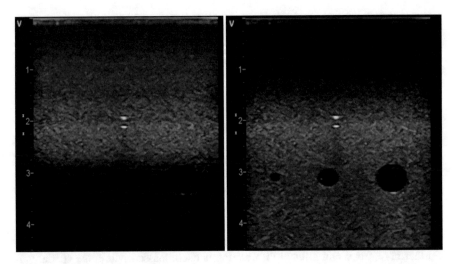

Fig. 7.20 A demonstration of the TGC effect on an image. (Left) Without gain compensation for the deeper regions, the bottom of the scanned phantom is dark. (Right) Proper TGC setting reveals three circular targets at the lower part of the image

images. The M in the name stems from the fact that motion is being displayed. The transmission and reception setup are similar to the A-mode. An acoustic beam is transmitted along a straight line, but in this case, the target is dynamically changing. For example, when acquiring a series of A-lines across the short axis of the heart's left ventricle, typical time-related changes in distances between echoes will be visible from one transmission to the other. These changes correspond to the temporal changes in the diameter of the contracting heart, which is an important parameter in cardiology. Using M-mode scanning, these changes can be displayed as a function of time.

The image formation in M-mode comprises the following steps:

(i)   At time point #1, an A-line is acquired along the target (as shown schematically in Fig. 7.21). The A-line is transformed into a vector of pixels with varying gray levels as explained in the previous section.

(ii) The vector is then rotated into vertical orientation, and the column of pixels is displayed at the leftmost part of the screen. The side proximal to the transducer is displayed at the top.

(iii) At time point #2, another A-line is acquired along the target. A second column of pixels is then generated and displayed alongside the first one.

(iv) The procedure is repeated N times until the entire image matrix is filled.

(v)   Following transmission number $N + 1$, a new column of pixels is generated. However, as the image is comprised of only N columns, one column is redundant. Thus, in order to include the fresh information in the display, the FIFO (*First In First Out*) procedure is applied, i.e., the first (oldest) column is deleted. Then, all the columns are shifted to the left by one step. Then, the new column is inserted into the rightmost column of the displayed matrix.

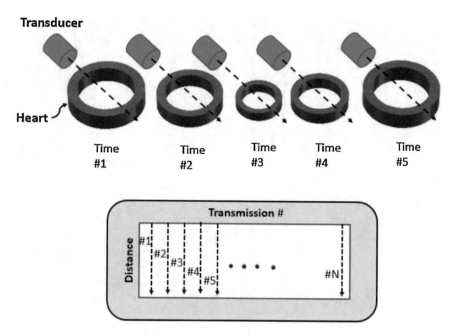

Fig. 7.21 The procedure of generating an M-mode image consists of acquiring consecutive A-lines. Each A-line is depicted as a column in the image matrix. After acquiring N columns, the first column is omitted, and the displayed columns are shifted one step to the left. The "fresh" column is inserted at the rightmost column of the image matrix. The display is dynamic

The procedure is continuously repeated. The pulse transmission rate is set so as to allow real-time tracking of the physiological changes. The display is thus dynamic and the impression is of a continuous flow of data. A clinical example of an M-mode image of the heart's mitral valve is depicted in Fig. 7.22.

7.4.5 The B-Scan (2D Imaging)

B-scan imaging is an extension of the *A-mode* into two dimensions (2D). The idea is quite simple. Instead of keeping the transducer or the beam stationary, the acoustic beam is steered mechanically or electronically through the scanned object to cover a planar cross section. At each transmission orientation, an A-line is acquired along the beam direction. The process explained above is applied, and a vector of pixels with gray levels corresponding to the post-processed and TGC-corrected echo amplitude is generated. The vector is then inserted into the image matrix according to its spatial location and orientation. The concept is depicted schematically for a Cartesian scan in Fig. 7.23.

Beam steering can be conducted along any scanning trajectory. The simplest approach is to implement a Cartesian scan. In such case the transducer or only the

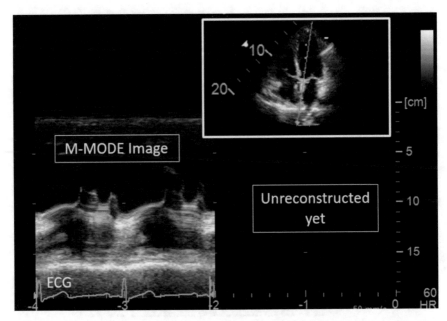

Fig. 7.22 (Top frame) A 2D image (see B-scan section) of the heart is used for selecting the orientation of the A-line along the mitral valve to be imaged (green line). (Bottom left) The corresponding M-mode image. (Bottom right) This part will be filled later column by column on the following acquisitions. The corresponding ECG signal is also displayed at the bottom. (Courtesy of Prof. Dan Adam- Technion)

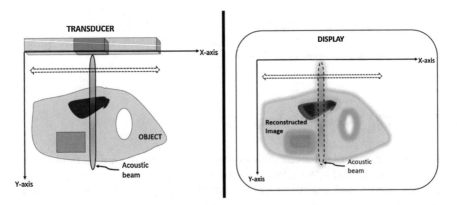

Fig. 7.23 Schematic depiction of a Cartesian B-scan. The beam is steered mechanically or electronically along the horizontal direction. A-lines are acquired along the selected range and at preset positions in order to cover the entire FOV. Each A-line is added as a column in the displayed image matrix

beam (when using a phased array transducer) is moved in a trajectory which is perpendicular to the acoustic beam (as in Fig. 7.23). The common standard in commercial scanners is that the transducer location is always taken at the top part of the displayed image (regardless of its actual physical configuration relative to the body). Accordingly, each A-line is displayed as a vertical line (along the y-axis) in the reconstructed image, and the spatial location of the point of transmission is moved along the horizontal direction (x-axis).

Once the entire planar field of view (FOV) has been scanned, the procedure is repeated, and each newly acquired A-line replaces its preceding counterpart which corresponds to the same spatial orientation. As recalled, the speed of sound in the body is sufficiently high to travel back and forth within a few tens of microseconds (depending of course on the distance). For example, if the imaged range is 75 mm, then the latest echo will be detected at about 100 µsec post transmission. Assuming that 100 A-lines are required to produce a single 2D image, a frame rate of 100 images per second can be easily achieved.

By using more sophisticated transmission protocols (e.g., using parallel transmissions and interleaving), even higher frame rates can be achieved. The implication is that B-scan can be used to produce "real-time" imaging. Thus, the operator can cover a large volume in the body within a short time by simply moving the transducer's location or orientation. In addition, B-scan provides an excellent means for guiding, in real-time, interventional procedures such as the insertion of biopsy needles.

Although Cartesian scans offer a simple reconstruction scheme, it is not always possible or practical to implement it. For example, if the transducer is several centimeters in length, and we wish to scan a large abdominal FOV, e.g., for pregnancy monitoring, the transducer will have to be moved many times back and forth in order to cover the entire FOV. Thus, instead of obtaining a single image which depicts the entire region of interest, several partial images will be obtained, and they will not be as informative as visualizing the entire FOV. Moreover, the coupling with the skin (using smeared gel) may be lost occasionally, and image quality may vary from one location to the other. Another example is cardiac scanning, which is also known as cardiac "echo." The heart is located within the ribs cage and between the two lungs. As explained above, both bones and air reflect strongly the waves and obstruct penetration of the waves into targets located behind them. This leaves a rather small acoustic window for wave transmission into the heart. As a result we can use only small transducers that can fit into the intercostal space, i.e., between two ribs. In such cases a curved transducer is commonly used which scans a fan-shaped FOV (somewhat like a pizza slice). The two configurations are shown schematically in Fig. 7.24, and two corresponding clinical scans are depicted in Fig. 7.25.

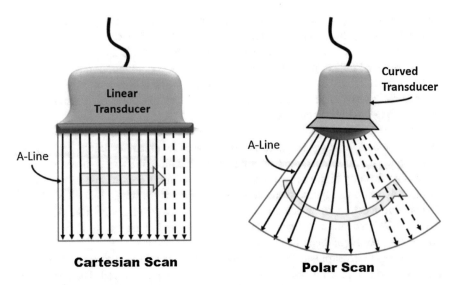

Fig. 7.24 (Left) A Cartesian scan configuration using a linear transducer. (Right) A polar scan configuration using a curved transducer. The newly scanned A-lines are depicted as dashed lines. The set of A-lines is continuously refreshed allowing dynamic imaging

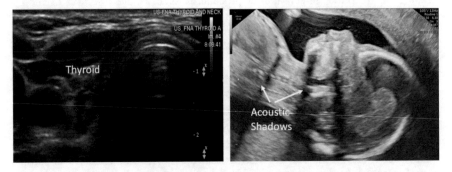

Fig. 7.25 (Left) A Cartesian B-scan of a neck depicting the thyroid gland. The transmission is from top to bottom. Soft tissues appear as white clouds. Blood vessels appear as black regions. Strong reflectors such as bones or cartilage may cast an acoustic shadow (as appears at the bottom right side). (Right) A polar B-scan of a baby (depicted using a golden color map). Note the diagonal acoustic shadows indicating the polar transmission direction

7.4.6   Three-Dimensional (3D) and Four-Dimensional (4D) Imaging

Imaging in 3D and 4D (space and time) is simply another extension of the same idea as the B-scan. Instead of acquiring A-lines within a planar cross section, the set of A-lines is set to cover a volumetric field of view. This can be achieved by using special 2D phased array transducers or by manipulating the beam's 3D orientation

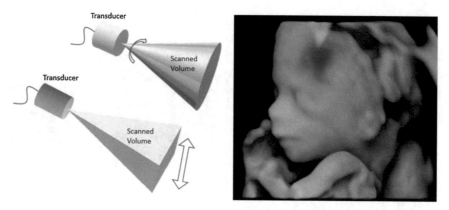

Fig. 7.26 (Left) Schematic depiction of two possible approaches for 3D imaging. (Upper figure) If fan-shaped B-scans are acquired at rotated planes, a conical volume is imaged. (Lower figure) If fan-shaped B-scans are acquired at a set of tilted planes, a pyramidal volume is imaged. (Right) A 3D scan of a 15-week-old baby in the womb. The image was post-processed to depict the external surfaces only

mechanically and/or electronically. If a reasonable frame rate is achievable, then temporal changes can be visualized as well, i.e., yielding 4D imaging. Two exemplary possible configurations for 3D imaging are schematically depicted in Fig. 7.26. Using one approach, a conical volume is scanned by rotating the plane of acquisition around the transducer's axis. In each scanned plane, a fan-shaped B-scan is applied. Consequently, the conical volume is actually comprised of many B-scan fan beam planes *rotated* relative to each other. Another option is to *tilt* the B-scanned plane. This yields a set of planes which fill a pyramidal volume as shown schematically in Fig. 7.26. It should be pointed out that other 3D scanning strategies are also available.

7.4.7 Through-Transmission Imaging

Although pulse-echo (B-scan) imaging is the dominant method which is implemented almost exclusively in all clinical scans, alternative approaches utilizing through-transmission (TT) waves have been suggested (e.g., [6]). TT waves offer two significant advantages. The first stems from the fact that their SNR is typically better than that of echoes. The second is that they allow, to some extent, quantitative estimation of acoustic parameters. Their main disadvantage is that they require a clear acoustic path between the transmitting and the receiving transducers. This means that the waves should not encounter any gas-containing regions and preferably no bones which will reflect them or attenuate them substantially. This practically leaves the breast as the preferable organ for TT scans.

Breast imaging with ultrasonic TT waves has two main modes. The first and simpler mode is projection imaging (e.g., [7]). The second mode is ultrasonic

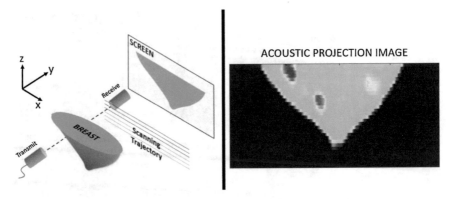

Fig. 7.27 (Left) Schematic depiction of an ultrasonic TT scan of the breast. The waves are transmitted from one side of the breast and detected on the other side. (Right) An exemplary clinical acoustic projection obtained by processing the TT waves. In this case a breast of a woman with three suspicious regions is displayed

computed tomography (UCT), which was suggested by Greenleaf et al. [8] already in 1974. With projection imaging, the transmitting transducer is placed on one side of the organ, i.e., the breast, and the receiving transducer is placed on the other side, as shown schematically in Fig. 7.27. Both the transducers and the organ are immersed in a coupling medium, commonly water. A projection is obtained by scanning the organ in a raster mode and measuring the properties of the TT waves, relative to a clear water path. The two most commonly mapped properties are the acoustic attenuation coefficient which is calculated based on the TT wave amplitude and the speed of sound which is calculated based on the wave's time-of-flight (TOF) from one side to the other.

Neglecting wave diffraction, the equations used for obtaining the amplitude projection image are given by

$$
\begin{aligned}
A(x, z) &= A_0 \cdot e^{-\int \alpha(x, y, z) dy} \\
\Rightarrow p(x, z) &= \ln\left(\frac{A(x, z)}{A_0}\right) = -\int \alpha(x, y, z) dy
\end{aligned}
\tag{7.35}
$$

where A_0 is the reference amplitude measured when passing through a clear water path, $A(x, z)$ is the amplitude measured at a certain horizontal (x) and vertical (z) location but after passing through the breast, $\alpha(x, y, z)$ is the attenuation coefficient as a function of location between the transmitter and the receiver, and $p(x, z)$ is the obtained projection as a function of the spatial position, which is actually the reconstructed image.

When measuring the time-of-flight between the transmitter and the receiver, $TOF(x, z)$, another type of acoustic projection depicting the refractive index can be obtained:

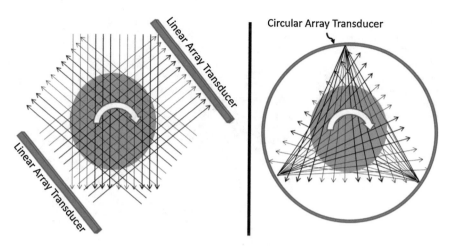

Fig. 7.28 (Left) Cartesian acquisition UCT. Data is collected along parallel rays using two linear array transducers facing each other. (Right) Fan beam acquisition UCT. Data is transmitted from one element located on a circular array and collected by a designated subgroup of transducers on the other side

$$p(x,z) = \text{TOF}(x,z) = \int \frac{1}{C(x,y,z)}dy \qquad (7.36)$$

where $C(x,y,z)$ is the speed of sound in the medium as a function of location.

The advantage of TT scan is that it can produce several alternative images, each depicting a different physical property of the tissue. For example, one can apply spectral analysis and study the frequency dependence of the attenuation coefficient or its non-linear properties. The disadvantage of projection imaging is that it yields an integrative value and hence one dimension (depth) is lost. As a partial remedy for that (as commonly done in x-ray mammography for example), at least two orthogonal views need to be acquired.

In order to obtain exact values of the acoustic properties of the tissue, ultrasonic computed tomography (UCT) setup can be applied. This requires the acquisition of TT projections from many different angles around the object, similar to the methods applied in x-rays (see Chap. 3). Two optional data collection configurations can be implemented (see Fig. 7.28). The first is collecting data from parallel rays. This can be achieved, for example, by placing two linear array transducers from both sides of the scanned object. The beam has to be as narrow as possible in order to reduce the effects stemming from diffraction. The second option is to transmit a wide beam from one point and detect the TT waves using many receivers located on the opposing side. This yields a fan-shaped coverage. A possible implementation for such scanner is by using a ring array (see Fig. 7.28). Each transmission is done from a small subgroup of piezoelectric elements (possibly only one) and detection by a designated subgroup on the other side. Rotation of the acquisition views is obtained

Fig. 7.29 An ultrasonic
computed tomography
(UCT) image of a breast.
The variations in colors
designate variations in the
acoustic properties on the
medium. A large papilloma
(benign tumor) was detected

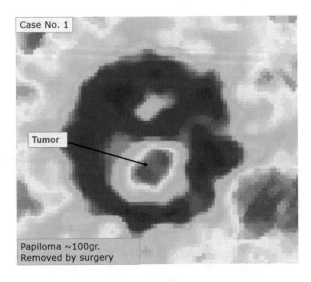

by switching electronically ON and OFF the relevant elements. No physical motion
is required. Hence, the acquisition time is very fast. In addition, spiral tomography
acquisition can also be implemented in order to obtain 3D imaging as demonstrated
in [9].

A clinical example of UCT image of a breast obtained using parallel beam
configuration and a small number of projections is depicted in Fig. 7.29. The colors
depict changes in the acoustic properties within the scanned FOV. Although the
spatial resolution in this case is very poor, the image clearly revealed a papilloma
(benign tumor) which was later removed by surgery. (As explained in Chap. 1, and
also demonstrated in Chap. 4, the quality of a medical image should be judged by its
clinical merit and not by its appearance!).

7.5   Doppler Imaging

7.5.1   Single Zone Velocity Imaging

One of the prominent benefits of ultrasound is its ability to measure and display flow
and motion. For that aim, as explained above, the Doppler effect is utilized. There
are several available modes for implementing Doppler imaging; the simplest one is
applied for measuring the blood velocity in a small specific region within a blood
vessel. The procedure is commonly preceded by acquiring a B-scan image of the
zone of interest. Once the specific blood vessel has been identified, using a computer
cursor, the region of interest (ROI) is marked. The distance from the transducer to the
center of that ROI is calculated using Eq. 7.32. Then, a first pulse is transmitted. The
reflected echoes are gated, and only the echo received from the ROI is registered, as

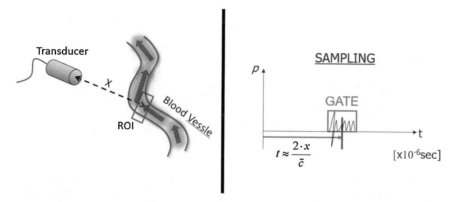

Fig. 7.30 (Left) In order to measure the blood flow within a specific location within a blood vessel, a ROI is marked on the corresponding B-scan of the area. (Right) The reflected echoes are registered only from a temporal gate which corresponds to the spatial location of the ROI

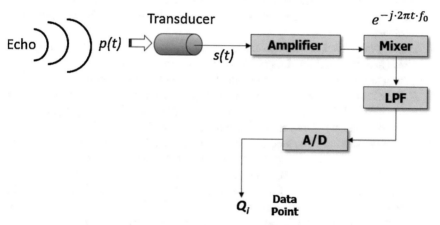

Fig. 7.31 The process applied for generating a complex data point Q_i for the Doppler analysis. The reflected echo applies pressure on the transducer, which converts it into an electric signal. The signal is then amplified and mixed as explained. Then, it is low-pass filtered and sampled (see text)

shown schematically in Fig. 7.30. Since the distances are commonly in the order of few centimeters, the time scale is in the order of tens of microseconds.

Following detection of the echo from the selected temporal gate, the signal undergoes several processing stages as depicted schematically in Fig. 7.31:

1. The signal is first amplified.
2. Then, it is mixed with an harmonic signal of the central transmitted frequency. The mixing is conducted along two parallel channels. Mathematically, it is equivalent to multiplication of the signal by $e^{-j \cdot 2\pi t \cdot f_0}$. This yields a complex signal, i.e., it has real and imaginary components.

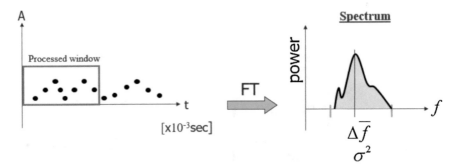

Fig. 7.32 (Left) After collecting a sufficient number of Q_i points, spectral analysis is applied. The obtained spectrum (Right) is used for calculating the velocities within the ROI via Eq. 7.17. $\Delta \bar{f}$ and σ^2 designate the mean frequency shift and its variance, respectively

3. The obtained complex signal is also comprised of high-frequency and low-frequency components. It can be shown that the low frequency is actually the sought Doppler shift. Therefore, a low-pass filter (LPF) is applied to retrieve it.
4. Finally, the signal is sampled using an analog to digital converter (A/D). This yields a single complex data point which is designated as Q_i.
5. Consecutive pulses are then transmitted at a set pulse repetition frequency (PRF). If the maximal expected frequency shift Δf_{max} is known, then the PRF is set to be at least PRF $\geq 2\Delta f_{max}$ to comply with the Nyquist-Shannon sampling theorem. The next pulse will have to be transmitted after at $T_{PRF} = 1/PRF$.
6. The above steps are repeated over and over until a sufficient number of Q_i points are collected. The term *sufficient* depends on the requirements of the spectral analysis procedure applied. Finally, spectral analysis is applied, for example, by implementing the Fourier transform (or faster estimators, as demonstrated in the next section), and the detected frequency shifts Δf within the processed window are estimated (Fig. 7.32). From these values, the corresponding velocities are calculated via Eq. 7.17.

This type of analysis yields a profile of the velocities distribution within the ROI corresponding to the mean time of the processed window. The velocities distribution is then converted into a column of pixels which brightness is proportional to the spectral power of the corresponding frequency. The column of pixels is then displayed on the screen. Upon acquisition of new Q_i points using the *FIFO* approach (see above in the M-mode section), an updated profile of the velocities distribution is obtained corresponding to the mean time of the new processed window. A new column of pixels is generated and displayed. The image generation and update is similar to the one used for M-mode imaging as explained above. A typical clinical image depicting the velocity profile near the heart's left ventricular valve (the mitral valve) as a function of time obtained by this procedure is shown in Fig. 7.33.

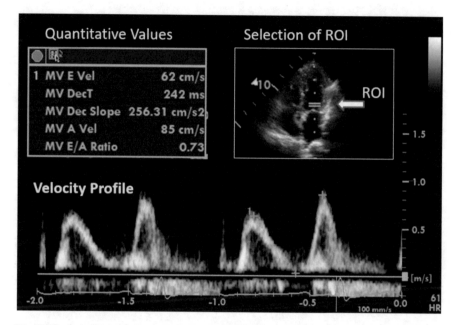

Fig. 7.33 A demonstrative Doppler image of a single zone velocity imaging. (Top right) A preliminary B-scan is used to select the region of interest (ROI) which is set between the two horizontal lines (see arrow). (Bottom) The velocity profile as a function of time. The velocity scale is displayed at the right side. (Top left) The user can obtain quantitative values by selecting specific points on the profile (see marks). (Courtesy of Prof. Dan Adam -Technion)

7.5.2 Two-Dimensional (2D) Doppler and Color Flow Mapping (CFM)

The main clinical merit of medical imaging is that it allows the physician to detect pathologies at one glance (more or less). In this context Doppler ultrasound offers wonderful tools. Using color coding, blood flow maps at 2D (and even at 3D) can be obtained and depicted atop the corresponding anatomical image acquired by simultaneous B-scan imaging (as depicted schematically in Fig. 7.34). Both the anatomical and flow maps are refreshed at real time allowing rapid inspection of problematic regions. One prominent application of this technique is cardiology. Color flow mapping (CFM) where red corresponds to flow toward the transducer and blue away from the transducer (the color can be switched by the operator) is obtained near the valves. Leaky valves can be immediately spotted by noting abnormal flow directions. With more sophisticated color coding, the variance of the spectrum is also estimated. High variance indicates flow turbulence and is indicative of a problem. The variance magnitude is added as a green color component to the image, yielding a full RGB composition for each pixel.

The technical challenge in implementing this method is to calculate rapidly (in real time) the Doppler shift and the corresponding velocity for each of the pixels

Fig. 7.34 In order to obtain Doppler color flow mapping (CFM), a region of interest (ROI) is selected within an acquired B-scan image. Flow velocities are calculated for every pixel within the ROI

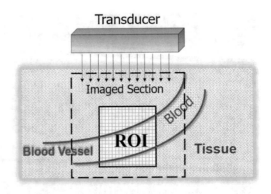

in the ROI. As each ROI may consist of few hundred pixels, the spectral analysis has to be very simple and fast. One of the approaches commonly applied for fast spectral estimation is to utilize the relation between the autocorrelation function of a signal and its spectrum. Consider the procedure described above applied for each pixel several times, so that a vector $\overline{Q}$ containing N complex Q_i values is collected. The autocorrelation function at zero shift $\mathbb{A}(0)$ can be approximated by

$$
\begin{aligned}
\mathbb{A}(0) \approx \frac{1}{2}[Q_1, Q_2. \ldots, Q_{N-1}] \cdot [Q_1*, Q_2 * \ldots, Q_{N-1}*]^T + \\
+ \frac{1}{2}[Q_2, Q_3. \ldots, Q_N] \cdot [Q_2*, Q_3 * \ldots, Q_N*]^T
\end{aligned}
\tag{7.37}
$$

and at one step shift $\mathbb{A}(1)$ is approximated by

$$
\mathbb{A}(1) \approx [Q_1*, Q_2 * \ldots, Q_{N-1}*] \cdot [Q_2, Q_3. \ldots, Q_N]^T
\tag{7.38}
$$

where the sign "$*$" indicates complex conjugation.

Without presenting the entire mathematical derivation (see, e.g., [3]), it can be shown that the velocity V can be approximated by

$$
V \approx C \cdot \frac{1}{4\pi \cdot f_0 \cdot T_{\mathrm{PRF}}} \cdot \arctan\left[\frac{\mathrm{Imag}\{\mathbb{A}(1)\}}{\mathrm{Real}\{\mathbb{A}(1)\}}\right]
\tag{7.39}
$$

where C is the average speed of sound and f_0 is the central frequency of the transmitted pulse and T_{PRF} is the time interval between two consecutives transmitted pulse (as explained above).

The corresponding variance can be estimated from

$$
\sigma^2 = \frac{2}{T_{\mathrm{PRF}}^2} \cdot \left[1 - \frac{|\mathbb{A}(1)|}{\mathbb{A}(0)}\right]
\tag{7.40}
$$

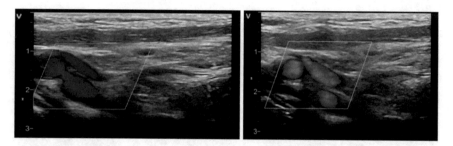

Fig. 7.35 (Left) An exemplary Doppler colored flow mapping image (CFM) depicting three blood vessels near the carotid bifurcation. Note that the flow at the upper vessel is opposite to the direction at the lower ones. (Right) Power Doppler image of the same region. As can be noted although there is no information on flow direction the three flow regions are clearly visible

The advantage of these approximations is that they can produce relatively accurate estimates even when the size of vector $\overline{Q}$, i.e., N, is very small containing only a few points. An exemplary color flow mapping image is depicted in Fig. 7.35. It is important to note that the flow velocity is measured *relative to the transducer* (see Eq. 7.17). That means that its value will be smaller (less red or blue) if the angle between the beam and the flow direction is increased. It will be maximal at $0°$ and zero at $90°$.

Another mode in the family of ultrasonic flow images is termed *Power Doppler* imaging. Using of the same procedure described above, the total power of the Doppler spectrum is calculated for each pixel of the ROI (instead of calculating the mean frequency). Consequently, the velocity magnitude and its directional information are lost. However, the information obtained by the power Doppler mode has several advantages. First, it provides a map of regions where flow exists; thus, it can detect even slow flow regions. Secondly, it is basically angle independent (unlike CFM). Thirdly, it is less susceptible to noise. The power Doppler image is displayed using a single pseudo color (typically golden shaded), where higher values are brighter (see Fig. 7.35).

7.6   Advanced Imaging Techniques

7.6.1   Contrast-Enhanced Ultrasound (CEUS) Imaging

In certain clinical applications, it is important to visualize blood vasculature and perfusion. However, blood being homogeneous hardly reflects any echoes. Thus, it appears mostly black on the displayed B-scan images. In order to see it, its acoustical properties must be altered substantially without affecting its other properties. For that aim and for other applications, a series of contrast-enhancing materials (CEM) has been developed [10].

Before Injection   **After Injection**

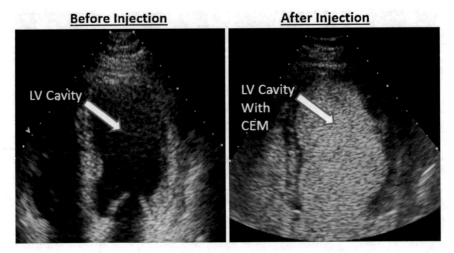

Fig. 7.36 (Left) A B-scan image of a heart depicting the left ventricular (LV) cavity which is filled with blood (black color). (Right) A B-scan image acquired following injection of contrast-enhancing material (CEM). The cavity which is now filled with the highly reflecting microbubbles appears as a bright region. (Courtesy of Prof. Jonathan Lessick – Rambam MC)

Typically, ultrasound CEM is comprised of micro- or even nano-sized bubbles. The small bubbles have a lipid, albumin, or polymeric shell and contain an inert gas, such as perfluorobutane (C_4F_{10}), for example. The size of these microbubbles is commonly about few microns in diameter. This allows them to circulate freely in the bloodstream. They sustain their structure for several minutes and then dissolve in the body. Although they are much smaller than the typical ultrasonic wavelength which is in the order of tenths of a millimeter, they are clearly visible in the B-scan image (see example in Fig. 7.36). This stems from two reasons: (i) Their acoustic impedance is much smaller than that of the blood; therefore, their reflection coefficient is very high. (ii) They are injected in a bolus that contains a huge number of bubbles. Thus, they are washed in the bloodstream as clusters of clouds which are larger than the wavelength.

The microbubbles can be visualized by several techniques. The simplest one is by setting the dynamic range and watching the augmented signal following their injection. They can increase the backscatter signal by up to 30 dB [11]. Alternatively, there are methods which utilize the fact that the microbubbles resonate and produce non-linear effects which yield subharmonic signals. Quantification of the changes that the CEM induce in the image may help improve the diagnosis. One common analysis tool is to study the temporal changes in the ROI following injection and quantify their "wash-in" and "wash-out" curves. The common pattern has a characteristic rise to a peak followed by signal decay which may be altered under pathological conditions .

7.6.2 Elastography

One of the oldest diagnostic methods utilized in medicine is palpation. It is well-known that pathological changes in tissues are also associated in many cases in changes in tissue elasticity. Breast tumors, for example, can be manually detected by feeling a stiff lump in the breast. However, palpation is non-quantitative; its diagnostic value depends on the skills of the physician and cannot be applied to tissues located deep in the body. Ultrasound elastography is a method which addresses this issue.

Presumably, the first to suggest this methodology were Ophir et al. [12] in 1991. Their concept was very simple. If pressure is applied onto the tissue of interest while backing it by a solid surface, then the tissue will deform in response (see Fig. 7.37). The deformation will be relatively larger for soft tissues and smaller for stiffer tissues (such as tumors). Stemming from the deformations, the distance between two proximal reflectors within the tissue will change. Thus, by measuring the displacement of their corresponding echoes, one can assess the deformation of the tissue and detect stiffer regions.

Several alternative approaches were suggested afterward (see, e.g., review in [13]), and some are commercially available as an add-on to standard ultrasonic scanners. Some of the suggested methods implement dynamic cyclic deformations, some implement transient vibrations, and some use the Doppler effect. But mostly, they rely on deformation assessment and analysis.

One prominent descendant method is called acoustic radiation force impulse (ARFI) imaging [14]. This approach utilizes a non-linear acoustic effect whereby body force is generated within the tissue. This force pushes the tissue along the acoustic beam direction. If the beam is focused, then the acoustic radiation force at the focal zone is strong enough to push and deform the tissue at this location. This

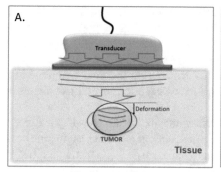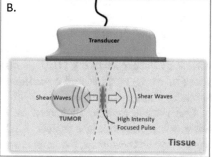

Fig. 7.37 (Left) Ultrasound elastography based on deformation analysis stemming from pressure applied mechanically or acoustically on the tissue. (Right) Elastography which is based on measurement of shear waves speed (see text). The shear waves are generated by transmitting a high-intensity focused pulse. Rapid imaging using longitudinal waves is then used to track the shear waves and measure their velocity

allows the selective induction of deformation in a noninvasive manner (a "virtual finger").

Following the idea of using ARFI, another elastographic approach was developed. This approach utilizes the relation of the elastic coefficient which is called Young's modulus E and the speed of sound for shear waves C_{Shear} given by Eq. 7.2. The basic concept is depicted schematically in Fig. 7.37. If a high-intensity pulse is transmitted into the body with a focal zone near the tissue of interest, then acoustic radiation force will rapidly push the tissue relative to its surroundings. As a result, shear waves will be generated. These shear waves will propagate from both sides of the focal zone along direction perpendicular to the beam. By using fast B-scan imaging, the displacements associated with the shear waves can be detected, and the local shear wave velocity C_{Shear} can be assessed. Rearranging Eq. 7.2, the elastic coefficient E (Young's modulus) is given by

$$E \approx 3\rho \cdot C_{shear}^2 \tag{7.41}$$

Using pseudo colors, the stiffness at each location at the ROI is assigned colors. Commonly, red designates high stiffness which is associated with pathology.

7.6.3   Ultrafast Imaging

One of the advantages offered by ultrasound is its ability to provide "real-time" imaging. That term implies that ultrasound can produce several dozens or even hundreds of frames per second. This allows imaging of the contracting heart, flow in blood vessels, and monitoring of invasive procedures such as needle insertion, laparoscopic surgery, and more. The quest for faster ultrasonic imaging continued persistently along with the development of the field. Already in 1977, Bruneel et al. [15] suggested a methodology which combines acoustic and optical devices to produce 1000 frames per second. However, the technology at that time was not developed enough for transferring it into clinical use. Several alternative approaches with slower frame rate but more practical for implementation were suggested (see review by Tanter and Fink [16]). A major contribution to the field was done in a series of works by Fink and co-authors who have suggested several techniques to accelerate image acquisition.

One notable method suggested for ultrafast imaging is the use of plane-wave transmission. As explained in Sect. 7.3.3, in a conventional B-scan imaging utilizing a phased array transducer, focusing is applied to the beam during transmission (using "beam forming"). The received echo signals are also post-processed using beam forming so as to retrospectively localize the echo signal along a narrow beam. This provides reasonably sharp A-lines with good spatial resolution. However, image formation is relatively slow and is defined by the number of lines collected to complete the scan (see Fig. 7.23). Thus, image acquisition can be accelerated by

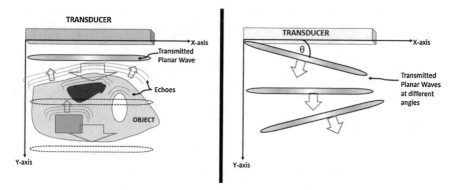

Fig. 7.38 (Left) Ultrafast imaging can be obtained by transmitting a single planar wave and retrospectively applying beam forming. (Right) In order to improve the spatial resolution, several plane waves are transmitted at different angles. The image is reconstructed by compounding all the images obtained from all transmissions

reducing the number of required transmissions. In planar wave imaging on the other hand, a single plane wave is transmitted, and beam forming is applied only to the collected echo signals, as shown schematically in Fig. 7.38. This allows very fast imaging since only one transmission is needed. For example, considering the fact that the speed of sound is about 1500 [m/s] in soft tissues, echo time to cover a 15 cm range is only 200 μsec (back and forth). This can easily yield a frame rate of 5000 images per second and even higher if more sophisticated transmission regimes are applied.

Although planar wave transmission yields superb temporal resolution, the spatial resolution and image quality are rather poor. In 2009 a method for overcoming this drawback was suggested by Montaldo et al. [17]. The method is termed coherent plane-wave compounding. The method allows ultrafast image acquisition but with a trade-off between frame rate and resolution. With this method, instead of using a single plane-wave transmission, electronic steering is applied, and several waves are sequentially transmitted with slightly different tilt angles (see Fig. 7.38). As a result, a set of images depicting acquisitions from different view angles is obtained. This naturally slows the acquisition rate in a proportional manner, but the spatial resolution is improved, speckle noise is reduced, and overall the image quality is improved.

References

1. Beyer RT, Letcher SV. Physical ultrasonic. New York: Academic Press; 1969.
2. Angelsen BAJ. Ultrasound imaging: waves, signals, and signal processing. Trondhein: Emantec AS; 2000.
3. Jensen JA. Estimation of blood velocities using ultrasound: a signal processing approach. New York: Cambridge University Press; 1996.
4. Kinsler LE, Frey P. Fundamentals of acoustics. New York: John Wiley & Sons; 1962.

5. Azhari H. Basics of biomedical ultrasound for engineers. Hoboken: Wiley–IEEE Press; 2010.
6. Gaitini D, Rothstein T, Gallimidi Z, Azhari H. Feasibility study of breast lesion detection using computerized contrast enhanced through-transmission ultrasonic imaging. J Ultrasound Med. 2013;32(5):825–33.
7. Katz-Hanani I, Rothstein T, Gaitini D, Gallimidi Z, Azhari H. Age related ultrasonic properties of breast tissue in-vivo. Ultrasound Med Biol. 2014;40(9):2265–71.
8. Greenleaf JF, Johnson SA, Lent AH. Measurement of spatial distribution of refractive index in tissues by ultrasonic computer assisted tomography. Ultrasound Med Biol. 1978;3(4):327–39.
9. Azhari H, Sazbon D. Volumetric imaging using spiral ultrasonic computed tomography. Radiology. 1999;212(1):270–5.
10. Ignee A, Atkinson NS, Schuessler G, Dietrich CF. Ultrasound contrast agents. Endosc Ultrasound. 2016;5(6):355–62.
11. Dietrich CF, Averkiou M, Nielsen MB, et al. How to perform contrast-enhanced ultrasound (CEUS). Ultrasound Int Open. 2017;3:E2–E15.
12. Ophir J, Cespedes I, Ponnekanti H, Yazdi Y, Li X. Elastography: a quantitative method for imaging the elasticity of biological tissues. Ultrason Imaging. 1991;13:111–34.
13. Sigrist RMS, Liau J, El Kaffas A, Chammas MC, Willmann JK. Ultrasound elastography: review of techniques and clinical applications. Theranostics. 2017;7(5):1303–29.
14. Nightingale K. Acoustic radiation force impulse (ARFI) imaging: a review. Curr Med Imaging Rev. 2011;7(4):328–39.
15. Bruneel C, Torguet R, Rouvaen KM, Bridoux E, Nongaillard B. Ultrafast echotomographic system using optical processing of ultrasonic signals. Appl Phys Lett. 1977;30(8):371–3.
16. Tanter M, Fink M. Ultrafast imaging in biomedical ultrasound. IEEE Trans Ultrason Ferroelectr Freq Control. 2014;61(1):102–19.
17. Montaldo G, Tanter M, Bercoff J, Benech N, Fink M. Coherent plane-wave compounding for very high frame rate ultrasonography and transient elastography. IEEE Trans Ultrason Ferroelectr Freq Control. 2009;56(3):489–506.

Exemplary Questions for Chap. 1

Part I: Verbal Questions. Kindly Answer Briefly Without Browsing the Text

1. What are the two basic information elements that an image has to provide?
2. What are the required stages for developing an imaging modality?
3. What is the definition of the "point spread function"?
4. What is the meaning of the term FWHM, and how is it related to the image properties?
5. What is the definition of "spatial resolution"?
6. What is the definition of a "voxel"?
7. What is the definition of "temporal resolution"?
8. When do we use physiological gating during image acquisition?
9. What features of an image are mostly provided by the low frequencies, and what features by the high frequencies?
10. What is MTF and how is it related to the PSF?
11. What is the definition of "contrast"?
12. What is the difference between image noise and image artifacts?
13. What is the definition of a "phantom"?
14. What is the meaning of the term CNR?
15. What is the meaning of the term AUC?
16. What are the three standard views used in radiology?

© Springer Nature Switzerland AG 2020
H. Azhari et al., *From Signals to Image*,
https://doi.org/10.1007/978-3-030-35326-1

Part II: Quantitative Questions

17. Using MATLAB® code or any other programing tools, build a Shepp-Logan numerical phantom. Use a 256 × 256 pixel matrix. The image should look like the one in the figure shown here (without the markings of course).

    A. Referring to the three encircled targets marked as **D**, estimate the baseline spatial resolution.
    B. Define three ROI in the targets **A**, **B**, and **C**. (If you have difficulty in selecting a circular ROI, you may use a square one, but make sure it contains most of the area within **B**.)
    C. Consider target **B** to be the clinically important region. Using Eq. 1.7, calculate the contrast when **A** is considered the background and then when the gray area between **B** and **C** is considered as the background.
    D. Apply a 2D-FT to the phantom (use *Log(Abs(f) + 1)* for display). Then, use low-pass filtering by including only the frequencies encompassed within radius *R* around the origin. Use incrementally increasing values of *R* (e.g., by five pixel increments), and calculate/estimate how it affects the spatial resolution and contrast calculated above.
    E. Now add zero mean and Gaussian white noise with variance of 0.01 to the image and recalculate the above terms.

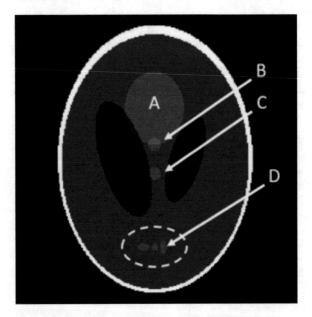

18. For Eq. 1.16 there is some theoretical basis for choosing *k* between 3 and 5, but an appropriate value for some imaging conditions can be estimated heuristically.

(a) From the chart below with simulated imaging data, is a k value of 3, 4, or 5 most appropriate for these conditions? (b) Sometimes, negative contrast lesions are called "cold" lesions and positive contrast lesions are called "hot" lesions. If not limited by spatial resolution and given similar photon count density, how does minimum detectable lesion size for a positive contrast lesion compare to a negative contrast lesion?

Table for question		Simulated imaging data	
Lesion number	Smallest observable lesion size (pixels)	Lesion photons per pixel	Background photons per pixel
1	3	49	55
2	4	93	103
3	5	194	216

19. In a paper by Goldberg-Zimring D, et al. (Automated detection and characterization of multiple sclerosis lesions in brain MR images. Magn Reson Imaging 1998;16(3):311–8), a method for automated detection of multiple sclerosis lesions in MRI using artificial neural networks was suggested. The following results were reported:

True positive	False negative	True negative	False positive
125	18	10,070	427

– What is the sensitivity and what is the specificity of this algorithm?

Solutions to Exemplary Questions for Chap. 1

Part I: Verbal Questions

– 1–16 All the answers can be found within the text.

Part II: Quantitative Questions

17. The results may vary according to several factors. Consult your instructor or TA regarding your results.

18. **Answer:**

(a) Although Eq. 1.16 has photon density n scaled as photons per centimeter and lesion size d in cm, the equation holds for any linear scale as long as nd^2 is unitless. If ε is the pixel size in cm, and n' and d' are background photons per pixel and lesion size in pixels, then we can write Eq. 1.16 as:

$$(n'/\varepsilon^2)(d'\varepsilon)^2 C^2 \geq k^2$$

$$n'd'^2 C^2 \geq k^2$$

which is the same equation with the photon count density and lesion size scaled in pixels. We need to assume that the image has on the average the background photon count density. These are all negative contrast lesions. For example, for the third lesion $C = (194{-}216)/216 = -0.1019$, or a contrast of 10.19%. Applying Eq. 1.16 at the limit where the equality holds, the calculation yields $k = 4.5, 3.9,$

© Springer Nature Switzerland AG 2020
H. Azhari et al., *From Signals to Image*,
https://doi.org/10.1007/978-3-030-35326-1

and 4 for lesions 1, 2, and 3, respectively. Picking an integer value of $k = 4$ would be reasonable for these imaging conditions.

(b) Equation 1.16 shows that the minimal detectable lesion size is inversely proportional to the magnitude of the contrast. The maximum magnitude of contrast that a cold lesion can have is $|C| = |(0 - B)/B|$ where B is the background photon count density and 0 shows that there are no detected photons in the region of the lesion. Since here, $C^2 = 1$, Eq. 1.16 gives the minimal size as:

$$d \geq k/\sqrt{n}$$

However, there is no mathematical limit to the magnitude of the contrast of a hot lesion, so there is no mathematical limit to the minimal size:

$$d \geq k/(C\sqrt{n})$$

where the contrast C can be quite large. Because of limits to their contrast, detection of cold lesions can be difficult. As a rough analogy, at twilight it can be much easier to see the first stars in the sky than to see a keyhole that is darker than its surroundings.

19. **Answer:**

Sensitivity $= 87.4\%$
Specificity $= 95.9\%$

Exemplary Questions for Chap. 2

1. Given the object shown at the upper part of the picture, which of the drawings depicted below describes its projection?

   A.
   B.
   C.
   None of the above

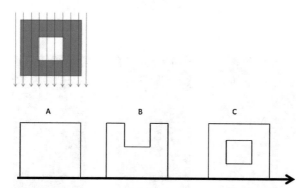

2. Given the following four binary images where white designates "1" and black designates "0," please plot schematically the corresponding projections for $\theta = 0$, i.e., onto the x-axis.

© Springer Nature Switzerland AG 2020
H. Azhari et al., *From Signals to Image*,
https://doi.org/10.1007/978-3-030-35326-1

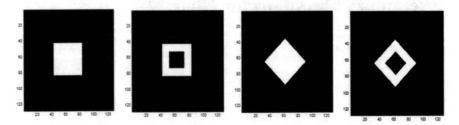

3. Given the object shown at the upper part of the picture shown below, which of the drawings depicted below describes its sinogram (Radon transform)?

A.
B.
C.
None of the above

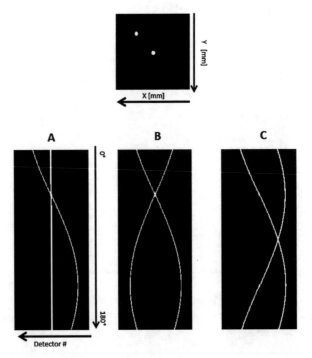

4. Given the object shown in this figure (upper left side) and its corresponding line in K-space at an angle of $\theta = 135°$, which of the objects shown below will have the same line in K-space at an angle of $\theta = 45°$?

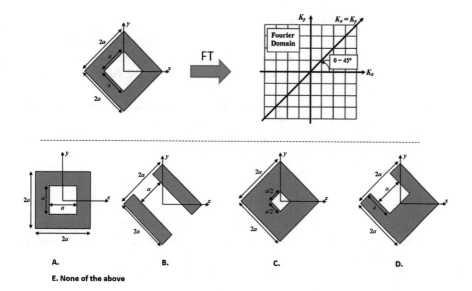

A.   B.   C.   D.

E. None of the above

5. The objects depicted in drawings A–D can be described in K-space. Please mark the incorrect statement.

   A. The values obtained in K-space for object B along the line shown in the drawing are the same as the values obtained along the K_x-axis for object D.

   B. The value at the central point of the K-space for object B is equal to that of object D.

   C. The value at the central point of the K-space for object A is greater than that of object C.

   D. Each object has at least one line through the origin in K-space which can be described by a single SINC function.

   E. None of the above.

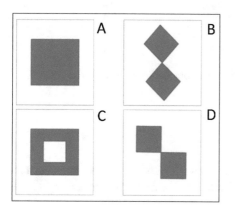

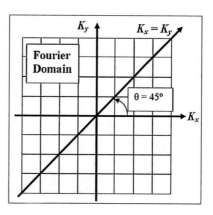

6. Which of the following statements is incorrect?

   A. For a rectangular object which is located at the center of the image so that its long axis is parallel to the y-axis, the horizontal width of the central lobe in K-space of the object is wider than its width along the vertical direction in K-space.

   B. In the Radon transform, the sum of all the values along each row (corresponding to the projection angle) is the same.

   C. Implementing FBP for image reconstruction without implementation of the K-filter will blur the boundaries in the image.

   D. For a single point located at the center of the image, the Radon transform will yield a straight line.

   E. One of the above statements is incorrect.

7. An object contains two symmetrically located hot inserts of identical tracer uptake level and density and an attenuating region A of thickness 1 cm with no tracer uptake and uniform attenuation μ.

   A SPECT acquisition is then performed, starting at $0°$ projection (horizontal in this case) by rotating the detector counterclockwise.

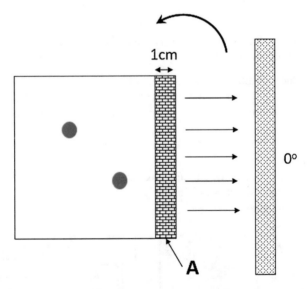

   – At what projection angle will the sum of the projection values be the lowest?

   A. $0°$
   B. $135°$
   C. $45°$
   D. $10°$

8. An object consisting of nine squares was scanned with CT. In order to reconstruct the attenuation coefficients, three horizontal projections and three vertical

projections with the marked values were acquired. If we use the difference ART method (one iteration) to reconstruct the image when each pixel represents a square, we will find that:

A. $A = B$
B. $A > B$
C. $A < B$
D. *Reconstruction* algorithm does not converge after a single iteration.

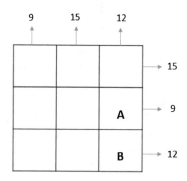

9. A facility consists of 16 identical square containers with the length of each side being a unit. The containers are arranged as described in the drawing. The dark containers were filled with a solid material with an attenuation coefficient μ. The containers marked with letters A, B, C, F, E, and D were filled with a homogeneous liquid solution of positron-emitting material, and the remainder of the containers (light squares) were filled with water. Both the water and the solution do not attenuate. The facility was inserted into a PET scanner, and the horizontal and vertical projections were measured only. (Note that the projections in PET stem from dual emissions at 180°; thus, the rays are attenuated similar to x-rays.)

– Which of the following statements is <u>incorrect</u>?

A. $P6 > P7$
B. $P4 > P3$
C. $P2$ has the highest value.
D. $P8$ has the highest value after $P1$
E. Not a single pair of projections has an equal value.

P1 P2 P3 P4

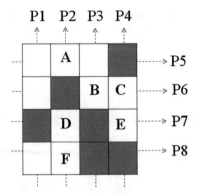

10. For an image consisting of four pixels, the projections were drawn. What will be the ratio of the signal reproduced in the square marked in gray by the first iteration using ART of the difference and the value obtained by the first iteration using ART of multiplications (the reconstruction is performed clockwise as described by the arrows, i.e., begin with the upper left projection) (Clarification: The value of the multiplications ART is in the denominator):

A. 0.9
B. 10/9
C. 9/5
D. 1
E. None of the above

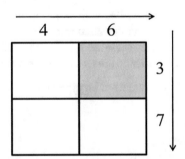

11. An object is given, with a 5 × 5 matrix, filled with zeros, except for two pixels, with values of 0.8, as shown in the figure below.

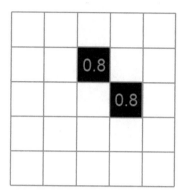

- Four projections are acquired at 0°, 45°, 90°, and 135°. Assume that the projections are calculated only at the center of the pixel and that the BP occurs only through central portion of each image pixel.
- A filter that enhances the edges is also given by $F = [0, -0.4, 1, -0.4, 0]$.

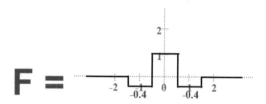

a. Use the given filter to perform FBP and compare the result to the original image.
b. Perform **nonfilter** back projection and compare the results to the filtered reconstruction.

12. Use MATLAB® or any other programing tool to create the Shepp-Logan numerical phantom. Then write a code for obtaining the Radon transform of the phantom for a given set of projection angles. Then write or use a library code for the inverse Radon transform. See how artifacts occur when the number of projections is reduced too much. Also see the effects of the K-filter and other available filters.

Solutions to Exemplary Questions for Chap. 2

1. Answer: **B**
2. Answer:

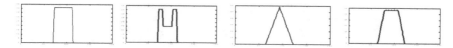

3. Answer: **A**

   A note to remember is that a single point at the center will produce a straight line in the sonogram, as shown in **A**. Also note that in this image the sinogram lines intersect at $45°$.

4. Answer: **D**

   Implementing the slice theorem, it follows that identical projections will have the same K-line along the same angle. And the projection of this shape at $45°$ is the same as that of the reference object.

5. Answer: **D**

   Although this statement is valid for the other three objects, it is not applicable to object C.

6. Answer: **E**

   All other statements are correct.

7. Answer: **D**

   Note that SPECT yields single-sided emission projections. At a $10°$ angle, the emitted rays from both sources will be attenuated by $e^{-\mu \cdot 1/\sin(10)}$, which is smaller than the value for $0°$, i.e., $e^{-\mu \cdot 1}$. At $45°$ one source will not be attenuated at all, and at $135°$ no attenuation will occur.

8. Answer: **C**

   We will get the following pixel values: $A = 3, \quad B = 4$.

© Springer Nature Switzerland AG 2020
H. Azhari et al., *From Signals to Image*,
https://doi.org/10.1007/978-3-030-35326-1

9. Answer: **E**

   Since $P4 = P7$

10. Answer: **B**

    Since the ration is $2/\frac{9}{5}$

11. Answer:

    First let us acquire the data of each the projection vectors: p_1, p_2, p_3, and p_4, respectively.

    Note that the spatial resolution dictates that:

    1 pixel for $0°$ and $90°$ projections and $1/\sqrt{2}$ pixel for $45°$ and $135°$ projections.

    Hence, the projections vectors are:

    $$p_1 = [0, 0, 0.8, 0.8, 0]; p_2 = \sqrt{2} \cdot [0, 0.8, 0, 0.8, 0];$$
    $$p_3 = [0, 0.8, 0.8, 0, 0]; p_4 = \sqrt{2} \cdot [0, 1.6, 0, 0, 0]$$

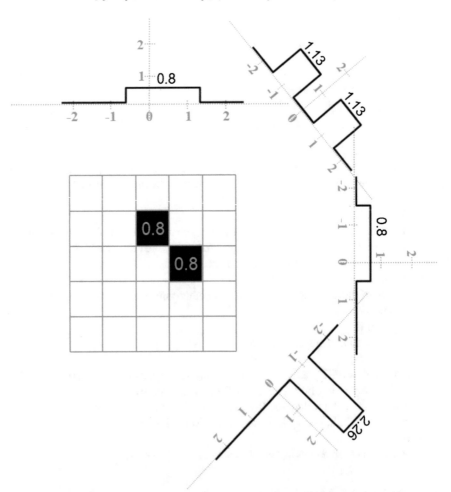

Next, each of the projections should be convoluted with the filter F, yielding the q vector: $q_i = p_i * F$

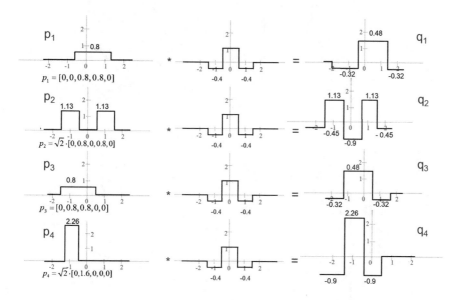

$q_1 = [0, -0.32, 0.48, 0.48, -0.32]$; $q_2 = [-0.45, 1.13, -0.9, 1.13, -0.45]$
$q_3 = [-0.32, 0.48, 0.48, -0.32, 0]$; $q_4 = [-0.9, 2.26, -0.9, 0, 0]$

Kindly remember that:

- For a given projection angle, a filtered projection q_i gives the same contributions to *all* pixels (x, y) in the image along the projection line, of the resulting matrix Q_i.
- The value of each pixel will be determined by the sum of *all* the projection contributions (over all projection angles).
- The final image is formed after normalization by the total number of projections.

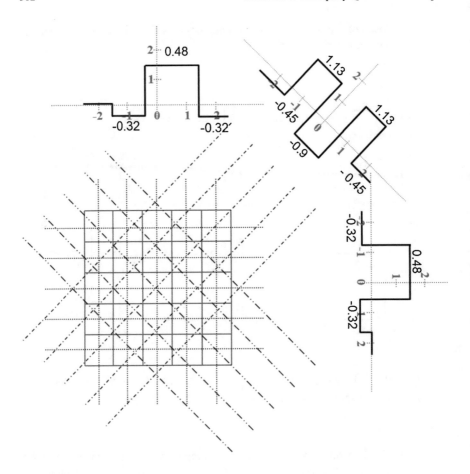

The contribution of the first filtered projection is Q_1:

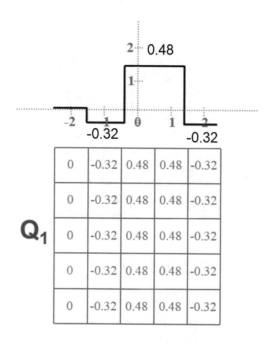

0	-0.32	0.48	0.48	-0.32
0	-0.32	0.48	0.48	-0.32
0	-0.32	0.48	0.48	-0.32
0	-0.32	0.48	0.48	-0.32
0	-0.32	0.48	0.48	-0.32

Q_1

The contribution of the second filtered projection is Q_2:

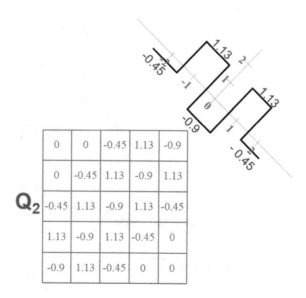

0	0	-0.45	1.13	-0.9
0	-0.45	1.13	-0.9	1.13
-0.45	1.13	-0.9	1.13	-0.45
1.13	-0.9	1.13	-0.45	0
-0.9	1.13	-0.45	0	0

Q_2

The contribution of the third filtered projection is Q_3:

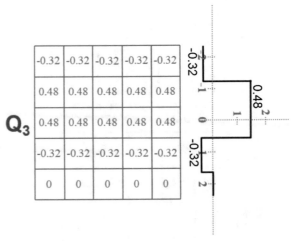

The contribution of the fourth filtered projection is Q_4:

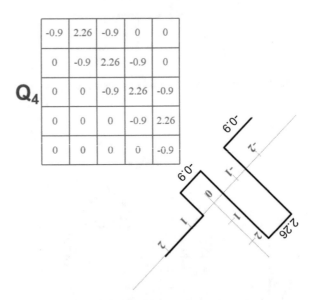

The final matrix FBP is the averaged matrices:

$$FBP = \frac{Q_1 + Q_2 + Q_3 + Q_4}{4}$$

-0.30	0.4	-0.29	0.32	-0.38
0.12	-0.29	1.08	-0.21	0.32
0.007	0.32	-0.21	1.09	-0.29
0.20	-0.38	0.32	-0.29	0.40
-0.22	0.20	0.007	0.12	-0.30

Note that only the pixels with the original higher values have values relatively
close to 1, while the rest have pixels that are less than 0.4. Naturally, more
projections will generate a more accurate result.

When rounding the values of matrix FBP, one obtains the exact original image.

When performing nonfiltered reconstruction, the same process is taking place,
only by using the original (unfiltered) projections:

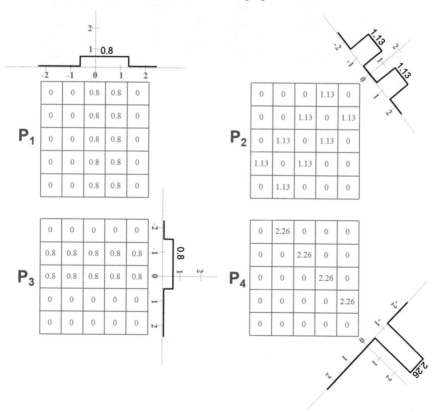

The final matrix BP is the averaged matrices:

$$BP = \frac{P_1 + P_2 + P_3 + P_4}{4}$$

0	0.57	0.2	0.48	0
0.2	0.2	1.25	0.4	0.48
0.2	0.48	0.4	1.25	0.2
0.29	0	0.48	0.48	0.57
0	0.29	0.2	0.2	0

12. Answer:

Note that as the number of projections increases, the reconstructed object in the image, becomes much more similar to the original phantom, and the "projection-line" artifact is reduced. With proper K-filter, the small details of the phantom are seen much more clearly in the final image, as expected.

Exemplary Questions for Chap. 3

Part I: Verbal Questions. Kindly Answer Briefly Without Browsing the Text

(i) Why is a very long x-ray exposure time undesired?

(ii) Which parameter influences most the image noise?

(iii) Which parameter influences most the image contrast?

(iv) What will be the features of an x-ray image acquired with a very short exposure time?

(v) Why can mammography be done using lower current and lower voltage, compared to chest x-ray?

(vi) Why is it crucial that during iodine-based angiography, the patient will stay still?

(vii) Why can iodine, barium, and gold be used as contrast agents?

(viii) How is it possible to generate a coronal view in CT, even though the scan itself generates only axial slices?

(ix) Why, during a chest x-ray, is it important that the patient will be as close as possible to the detection metal plate?

(x) What is the ALARA principle?

Part II: Quantitative Questions

1. For a combination of $mAs_0 = 250$ & $kV_0 = 100$, find a matching combination from the given options that would maintain approximately the same detector dose, but would decrease the patient dose. (a) $mAs_0 = 130$ & $kV_0 = 120$, (b) $mAs_0 = 500$ & $kV_0 = 80$, (c) $mAs_0 = 210$ & $kV_0 = 100$. What other image ramifications would you expect from that chosen combination?

© Springer Nature Switzerland AG 2020
H. Azhari et al., *From Signals to Image*,
https://doi.org/10.1007/978-3-030-35326-1

2. When considering the gray levels of illustrations, A–D, which of the four options might represent correctly a CT image of a phantom filled with water and three different inserts, simulating tissues (adipose tissue, soft tissue, and bone tissue)?

A)
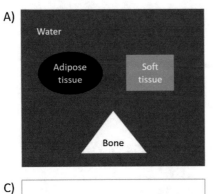

B)
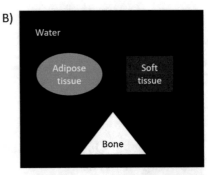

C)
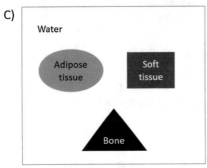

D)
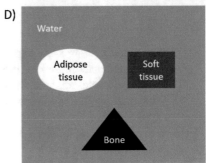

3. Calculate the DLP (dose length product) for a CT scan of ten slices, each with slice thickness of 1 mm and a pitch of 1.11, where the linear dose profiles for both central and peripheral measurements are $D_p(z) = a \cdot z + b$; $D_c(z) = c \cdot z + d$;.

4. When comparing these two scan combinations: (a) 100 mAs and 150 kV_p and (b) 150 mAs and 100 kV_p, mark the correct statement.

   (a) In respect to option (b), option a would have approximately 50% more patient dose, but 15% less detector dose.
   (b) In respect to option (b), option a would have less contrast and less noise.
   (c) Both options, (a) and (b), would have approximately the same patient dose, but option a would have better contrast.
   (d) Both options, (a) and (b), would have approximately the same detector dose, but option a would have better SNR.
   (e) Both options, (a) and (b), would result in the same image.

5. A patient undergoes a mammography scan, emitting a polychromatic spectrum of 30 kV_p and an average energy of about 18 keV. Mark the correct statement

regarding the radiation spectrum that reaches the detectors (after the beam passed through the patient).

(a) The total radiation intensity will be higher than the original intensity, the spectrum peak will be higher than 30 keV, and the average energy will be higher than 18 keV.
(b) The total radiation intensity will be higher than the original intensity, but the spectrum peak will be lower than 30 keV, and its average energy will be lower than 18 keV.
(c) The total radiation intensity will be lower than the original intensity; the spectrum peak will remain the same, i.e., $30 \, kV_p$; and the average energy will be higher than 18 keV.
(d) The total radiation intensity will be lower than the original intensity, the spectrum peak will be lower than 30 keV, and the average energy will be lower than 18 keV.
(e) The total radiation intensity will be lower than the original intensity, the spectrum peak will be lower than 30 keV, but the average energy will remain the same, i.e., 18 keV.

6. A 10 cm × 10 cm phantom is filled with a background medium, with CT value of 0HU, and contains several square (2 cm × 2 cm) inserts. The HU value of the inserts is known, and also, that two of them are filled with air.

(a) Sort in ascending order the signal that each of the five detectors measured.

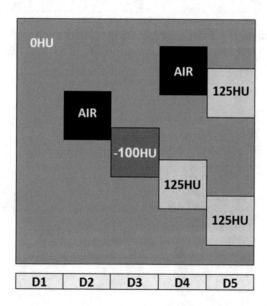

(b) How would your order be different if the phantom looked like that? (Compare to previous answer.)

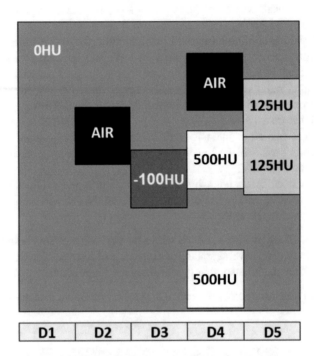

7. A woman with a rounded tumor is having a chest x-ray with two positioning configurations (A) and (B) and same exposure parameters (tube current, tube voltage, and exposure time).

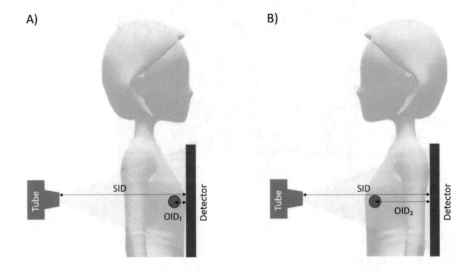

It is known that the SID is 100 cm and the focal spot size is 1.5 mm. In the first configuration, OID_1 is 3 cm, and in the second configuration, OID_2 is 12 cm. Mark the correct statement.

a. In configuration A, both the magnification and penumbra are larger compared to configuration B.
b. In configuration A, both the magnification and penumbra are smaller compared to configuration B.
c. As long as the x-ray exposure parameters are the same (tube current, tube voltage, and exposure time), the positioning configuration has no impact on the final image.
d. In configuration A, the magnification is larger, while the penumbra is smaller, compared to configuration B.
e. In configuration A, the magnification is smaller, while the penumbra is larger, compared to configuration B.

8. A phantom was scanned with CT twice, generating images A and B, presented with the same window level. The technician significantly changed only one parameter between the scans, either the voltage or the tube current, while the other remained exactly the same.

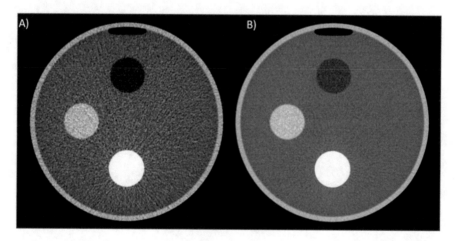

Based on the images, mark the correct statement.

(a) It is impossible to estimate, based only on the images, if they were taken with the same voltage and different current or taken with the same current and different voltage.
(b) Image (A) presents more contrast and more noise than image (B), so it is likely that image (A) was taken with lower voltage than image (B), but the same tube current was used for both images.

(c) Both images present the same contrast level, but image (B) presents less noise than image (A), so it is likely that the same tube current was used for both of them, but image (A) was scanned with higher tube voltage.

(d) Both images present the same noise level, but image (B) presents more contrast than image (A), so it is likely that the same tube voltage was used, but image (B) was scanned with higher tube current.

Solutions to Exemplary Questions for Chap. 3

Part I: Verbal Questions

   (i) X-ray is an ionizing radiation that is harmful to the body.
  (ii) Tube current
 (iii) Tube voltage (x-ray spectrum)
  (iv) The image would look whiter, due to insufficient radiation, giving the impression of high attenuation.
   (v) The breast is a much smaller organ and is less attenuating.
  (vi) Since one of the most common techniques to "extract" the blood vessels is to "subtract" the new image from the original, non-iodine image, resulting only in the iodine-filled vessels
 (vii) All of those materials are biocompatible, with high attenuation.
(viii) Since CT acquires a volume of data, which is actually a 3D matrix of voxels that can be "sliced" into slices along any direction one desire, including coronal, sagittal, or even oblique.
  (ix) If the patient would be standing far from the metal plate, the blurriness of the image, i.e., the penumbra, would be increased, according to Eq. (3.10). This might result in the loss of significant small findings in the image and would impair the image quality.
   (x) It states that the patients should be exposed to the minimal radiation dose that would still enable a good diagnostic of the image.

© Springer Nature Switzerland AG 2020
H. Azhari et al., *From Signals to Image*,
https://doi.org/10.1007/978-3-030-35326-1

Part II: Quantitative Questions

1. According to the 15% rule of thumb, when reducing the current by half (from 250 to 130) and increasing the voltage by 15% (from 100 to 120), the detector dose would remain about the same, but the patient dose will be reduced significantly. Another factor that should be taken into account is the fact that the image contrast would be lessen with the new combination. In any case, the ALARA concept should be the main guideline when choosing the optimal setting for scanning.

2. The gray shades of objects in a CT image are determined based on their HU values. Higher HU values are presented with brighter shades, while lower HU values appear darker in the image. It is possible to see that according to Eq. (3.12), higher attenuation yields a higher HU value; hence, the order of the gray shades from dark to bright would follow the order of attenuation coefficient. As was mentioned in the section "X-Ray Interaction with Body Tissues," adipose tissues have a lower attenuation than water, which has a lower attenuation than soft tissues, while the most attenuating body parts are the bones. From here it is easy to deduce that the order of gray shades from dark to bright would be adipose tissues, water, soft tissue, and bone. Illustration (A) follows that order of shades.

3.
$$\text{CTDI}_{100} = \frac{1}{nT} \int_{-50mm}^{50mm} D(z)dz \Rightarrow$$

$$\text{CTDI}_{100}^{\text{peripheral}} = \frac{1}{10} \int_{-50mm}^{50mm} (a \cdot z + b)dz = 10 \cdot b$$

$$\text{CTDI}_{100}^{\text{centeral}} = \frac{1}{10} \int_{-50mm}^{50mm} (c \cdot z + d)dz = 10 \cdot d$$

$$\text{CTDI}_w = \frac{2}{3}\text{CTDI}_{100}^{\text{peripheral}} + \frac{1}{3}\text{CTDI}_{100}^{\text{centeral}} \Rightarrow$$

$$\text{CTDI}_w = \frac{2}{3}10 \cdot b + \frac{1}{3}10 \cdot d = \frac{10}{3} \cdot (2b + d) \Rightarrow$$

$$\text{CTDI}_{\text{vol}} = \frac{\text{CTDI}_w}{\text{pitch}} = \frac{\frac{10}{3} \cdot (2b + d)}{10/9} = 3 \cdot (2b + d) \Rightarrow$$

$$\text{DLP} = nT \cdot \text{CTDI}_{\text{vol}} = 10 \cdot 3 \cdot (2b + d)$$

4. First let us approximate the patient dose and the detector dose for each option: It can be estimated that patient dose is proportional to the term $\text{mAs} \cdot \text{kV}^2 : \frac{\text{PD}_a}{\text{PD}_b} = \frac{100 \cdot 150 \cdot 150}{100 \cdot 100 \cdot 150} \Rightarrow \text{PD}_a \simeq 1.5\text{PD}_b$. For the detector dose, with the help of the 15% rule of thumb, it is easier to first find a combination with similar voltage: $\text{DD}_b(\text{mAs},$ $\text{kV}_p) \propto (150 \ \& \ 100) \simeq (75 \ \& \ 115) \simeq (38 \ \& \ 130) \simeq (19 \ \& \ 150)$.

Now that the voltage of both combinations is similar, the ratio of the detector dose is linear with the tube current; therefore,

$$\left. \begin{array}{l} DD_b\left(mAs, kV_p\right) \propto (19\&150) \\ DD_a\left(mAs, kV_p\right) \propto (100\&150) \end{array} \right\} \Rightarrow DD_a \simeq 5 \cdot DD_b$$

Now it is easy to rule out all the answers but the second one, since neither of the doses are similar, which guarantees different image quality outcome. The second answer is correct, as the voltage in option (a) is higher than the voltage in option (b), which means that the contrast in option a would be lessen. On the other hand, the dose is much higher in option (a), which would result less noise in the outcome image.

5. The total beam intensity that would reach the detectors will be smaller than the original intensity, since some of the energy that was emitted from the tube was lost, due to the attenuation by the body itself. On the other hand, the peak of the spectrum will remain unchanged, since most of the lost energy is of the lower energy photons. This is the reason for the beam hardening effect that causes an increase in the average energy of the spectrum. Therefore, the correct answer will be the third answer.

6. (a) From the fact that the background medium has a 0 HU, it has to be water, as this is one of the definitions for CT values. Let's call the attenuation coefficient of water μ_w. We can also assume that the original intensity is I_0. With the help of Eq. (3.5), one can find the signal that each of the five detectors measured:

   For the first detector, the signal is attenuated only due to the 10 cm of water; hence, $I_1 = I_0 \cdot e^{-10 \cdot \mu_w}$.

   For the second detector, the signal is only attenuated due to 8 cm of water, as air is not attenuating at all; therefore, $I_2 = I_0 \cdot e^{-8 \cdot \mu_w}$.

   For the inserts of 125 HU and -100 HU, it is possible to find the terms for the attenuation coefficient according to Eq. (3.13).

   For the insert with -100 HU:

$$HU = \frac{\mu_{insert} - \mu_{water}}{\mu_{water}} \cdot 1000 \Rightarrow -100 = \frac{\mu_{insert} - \mu_w}{\mu_w} \cdot 1000 \Leftrightarrow \mu_{insert} = 0.9 \cdot \mu_w$$

Remembering that the total attenuation is the sum of the attenuation of each object along the way, regardless of their order, thus the attenuation of the signal of the third detector would be comprised of the attenuation of 8 cm of water and 2 cm of insert:

$$I_3 = I_0 \cdot e^{(-8 \cdot \mu_w) + (-2 \cdot \mu_{insert})} = I_0 \cdot e^{-8 \cdot \mu_w} \cdot e^{-2 \cdot 0.9 \cdot \mu_w} = I_0 \cdot e^{-9.8 \cdot \mu_w}$$

For the insert with 125 HU:

$$HU = \frac{\mu_{\text{tissue}} - \mu_{\text{water}}}{\mu_{\text{water}}} \cdot 1000 \Rightarrow 125 = \frac{\mu_{\text{tissue}} - \mu_w}{\mu_w} \cdot 1000 \Leftrightarrow \mu_{\text{tissue}} = 1.125 \cdot \mu_w.$$

Consequently, the signal of the fourth and fifth detectors, respectively, will be

$$I_4 = I_0 \cdot e^{-6 \cdot \mu_w} \cdot e^{-2 \cdot \mu_{\text{insert}}} = I_0 \cdot e^{-6 \cdot \mu_w} \cdot e^{-2 \cdot 1.125 \cdot \mu_w} = I_0 \cdot e^{-8.25 \cdot \mu_w}$$
$$I_5 = I_0 \cdot e^{-6 \cdot \mu_w} \cdot e^{-4 \cdot \mu_{\text{insert}}} = I_0 \cdot e^{-6 \cdot \mu_w} \cdot e^{-4 \cdot 1.125 \cdot \mu_w} = I_0 \cdot e^{-10.5 \cdot \mu_w}$$

Summarize all of the results:

$$I_1 = I_0 \cdot e^{-10 \cdot \mu_w}; I_2 = I_0 \cdot e^{-8 \cdot \mu_w}; I_3 = I_0 \cdot e^{-9.8 \cdot \mu_w}; I_4 = I_0 \cdot e^{-8.25 \cdot \mu_w}; I_5$$
$$= I_0 \cdot e^{-10.5 \cdot \mu_w}$$

Now it is easy to sort the signal in ascending order: $I_5 < I_1 < I_3 < I_4 < I_2$.

(b) Remember the basic formula for calculating the HU values:

$$HU = \frac{\mu_{\text{tissue}} - \mu_{\text{water}}}{\mu_{\text{water}}} \cdot 1000 \Rightarrow 500 = \frac{\mu_{\text{tissue}} - \mu_w}{\mu_w} \cdot 1000 \Leftrightarrow \mu_{\text{tissue}} = 1.5 \cdot \mu_w.$$ Consequently, the

$$I_4 = I_0 \cdot e^{-4 \cdot \mu_w} \cdot e^{-4 \cdot \mu_{\text{insert}}} = I_0 \cdot e^{-4 \cdot \mu_w} \cdot e^{-4 \cdot 1.5 \cdot \mu_w} = I_0 \cdot e^{-10 \cdot \mu_w}$$

Summarize all of the results:

$$I_1 = I_0 \cdot e^{-10 \cdot \mu_w}; I_2 = I_0 \cdot e^{-8 \cdot \mu_w}; I_3 = I_0 \cdot e^{-9.8 \cdot \mu_w}; I_4 = I_0 \cdot e^{-10 \cdot \mu_w}; I_5$$
$$= I_0 \cdot e^{-10.5 \cdot \mu_w}$$

Now it is easy to sort the signal in ascending order: $I_5 < I_1 = I_4 < I_3 < I_2$.

Note that in this case, the signal for both detectors #1 and #4 is equal; therefore, if this was only a single projection scan, like in projectional radiography, there would be no way of telling the different objects along the beam path. In computed tomography, due to the many directions of projections, the object is clearly demonstrated, with all of its inner structures.

7. The two main factors that should be considered are the magnification and penumbra (blurriness). The magnification is calculated according to Eq. (3.9) and the penumbra according to Eq. (3.10).
   For the first configuration, the magnification factor and penumbra are

$$M_1 = \frac{SID}{SOD} = \frac{100}{100 - 3} = 1.03; w_1 = W \cdot (M_1 - 1) = 1.5 \cdot 0.03 = 0.045$$

For the second configuration, the magnification factor and penumbra are $M_2 = \frac{\text{SID}}{\text{SOD}} = \frac{100}{100-12} = 1.13; w_2 = W \cdot (M_2 - 1) = 1.5 \cdot 0.13 = 0.195$

The correct statement is #2.

Note that since $\text{SOD} \geq \text{SID} \Rightarrow M \geq 1$. This means that for the same focal spot size, larger magnification results larger penumbra and more blurriness around the edges.

8. The correct answer can be found by elimination. Remember that changing the contrast is possible by changing the voltage, not by changing the tube current. In addition, the total amount of radiation can be the same between two combination of voltage and current, only when changing both. Changing only one of them forces a change in the total dose, hence a change in the noise level of the resulted image.

For this reason, it is possible to eliminate the first answer, as the images do not present the same noise level and contrast.

For the case described in the third answer: Both images present the same contrast level, but image (B) presents less noise – a good estimation would be that the voltage was the same (same contrast), but the tube current for (B) was higher, resulting less noise. This is in contrast to what the rest of the third statement, so it is eliminated.

For the case described in the fourth answer: Both images present the same noise level, but image (B) presents more contrast – a good estimation would be that the voltage was NOT the same and that image B was taken with lower voltage to achieve better contrast. To maintain similar noise level, image B would also have to be taken with much higher current. This is in contrast to the rest of the fourth statement, so it is eliminated as well.

A good description of the images is the second answer, and it is the correct statement.

Exemplary Questions for Chap. 4

1. A phantom with an elliptical shape in the transverse plane is centered in the FOV and imaged on an Anger camera. Two identical, small sources are placed such that one is at the center of the phantom and one has an anterior offset from the center of $\frac{1}{4}$ the minor axis A (see the diagram below). Given that the activity of each source is I_0, and that the absorption coefficient of the emitted gamma rays is μ in water, what is the ratio R of the count rate S_M at detector position M with respect to the rate S_N at position N? In this case, what is the minimum value the ratio can have? What is the ratio $R(t)$ at time t, if the source at the center is not identical, but has a half-life τ_1, the other source has a half-life of τ_2, and I_0 refers just to the initial activities, which are the same? In the second case, what is the approximate value of $R(t)$ after many half-lives?

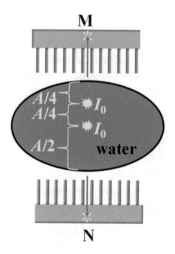

© Springer Nature Switzerland AG 2020
H. Azhari et al., *From Signals to Image*,
https://doi.org/10.1007/978-3-030-35326-1

2. If a source with activity A has a half-life τ, what fraction of its activity remains after half of a half-life?
3. Given a radioisotope with a half-life of τ, what is the approximate percent decrease in activity A per unit time?
4. Given a phantom of water and constant thickness L in the anterior/posterior direction (up/down) with an attenuation coefficient μ, show that the geometric mean of the acquired counts of two anterior/posterior detectors is independent of position, for point sources not aligned in the anterior/posterior direction, that is to say, that the relative intensity in an image composed of the geometric mean of the two detectors is independent of the up/down position of these non-aligned point sources.

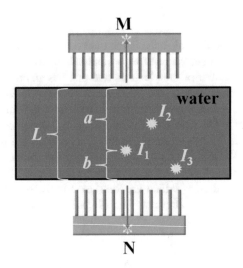

5. For a single pinhole collimated gamma camera, how does the detector sensitivity ε vary with the distance r of a point source place directly in front of the pinhole of diameter d?

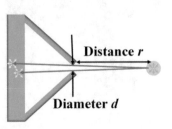

6. A ^{99}Mo/^{99m}Tc generator has been eluted so that there is virtually no ^{99m}Tc left in the generator. Show that the activity of ^{99m}Tc in the generator will maximize at about 1 day after elution.

7. Examine Table 4.1 and conjecture a reason that dual-isotope SPECT studies can be performed (e.g., acquiring separate ^{99m}Tc and ^{123}I radiotracer images at the same time) but that dual-isotope PET studies are not trivial.

8. For a given geometry, tungsten collimators generally provide superior imaging characteristics than lead collimators. Give a possible reason for this.

9. An engineer calculates that for optimal image quality, a ^{99m}Tc radiotracer SPECT scan should be performed on a dual-headed camera having projections every 2° with 60 seconds per projection, giving full sinograms. Is this reasonable for a phantom acquisition? For a patient acquisition?

10. A point source of ^{99m}Tc gives a radiation field of φ mSv/h at a distance L from the source. After three half-lives have passed, what is the field at half the distance $L/2$?

Solutions to Exemplary Questions for Chap. 4

1. **Answer:**

Attenuating each source for the path length in water ($A/2$ and $A/4$ for the first and second sources) into detector M with some sensitivity ε, the count rate can be written:

$$S_M = \varepsilon\left(I_0 e^{-\mu A/2} + I_0 e^{-\mu A/4}\right)$$

Attenuation in air is negligible for γ-rays, for these distances. In the other direction, the first source is attenuated as at detector M, but the second source is further away at $3A/4$ from the posterior surface:

$$S_N = \varepsilon\left(I_0 e^{-\mu A/2} + I_0 e^{-\mu A3/4}\right)$$

The ratio $R = S_M/S_N$ becomes

$$R = \frac{e^{-\mu A/2} + e^{-\mu A/4}}{e^{-\mu A/2} + e^{-\mu A3/4}}$$

$$R = \frac{1 + e^{+\mu A/4}}{1 + e^{-\mu A/4}}$$

Both the distance A and the attenuation coefficient μ are always positive, so $e^{\mu A/4}$ is always greater than $e^{-\mu A/4}$ and the numerator for R is always greater than the denominator, giving a minimum possible value of 1 for $A = 0$. If the sources have different half-lives, one can write the count rates as

© Springer Nature Switzerland AG 2020
H. Azhari et al., *From Signals to Image*,
https://doi.org/10.1007/978-3-030-35326-1

$$S_M = \varepsilon\left(I_0 e^{-\mu A/2} e^{-\lambda_1 t} + I_0 e^{-\mu A/4} e^{-\lambda_2 t}\right)$$

$$S_N = \varepsilon\left(I_0 e^{-\mu A/2} e^{-\lambda_1 t} + I_0 e^{-\mu A 3/4} e^{-\lambda_2 t}\right)$$

where $\lambda_1 = \ln(2)/\tau_1$ and $\lambda_2 = \ln(2)/\tau_2$, as in Eq. 4.2. The ratio $R(t)$ is now

$$R(t) = \frac{1 + e^{+\mu A/4} e^{-(\lambda_2 - \lambda_1)t}}{1 + e^{-\mu A/4} e^{-(\lambda_2 - \lambda_1)t}}$$

After a long time, if $\lambda_1 < \lambda_2$ (source at the center decays more slowly), the second terms of the numerator and denominator both approach 0, since $e^{-(\lambda_2 - \lambda_1)t}$ approaches 0, so $R \approx 1$. This makes sense, since if the offset source is comparatively "dead," only the center source predominates as a signal and the geometry is symmetrical. If the center source decays more quickly, the factor $e^{-(\lambda_2 - \lambda_1)t}$ becomes very large, since $\lambda_2 < \lambda_1$. In that case the left-hand term 1 in the numerator and denominator is comparatively small and can be ignored, so $R \approx e^{\mu A/2}$. This is like the case in which only the offset source exists.

2. **Answer:**
   One could use Eq. 4.2, noting that $A = \lambda N$, or one could simply reason that if f is the fraction of activity remaining after ½ of the time span, then f^2 is the fraction after the full time span. Since the full duration is a half-life $f^2 = ½$, so $f = 1/\sqrt{2}$. Therefore, about 71% of the activity remains after half of a half-life.

3. **Answer:**
   For a unit time duration of Δt, if it is relatively short, we can approximate

$$\frac{\Delta A}{\Delta t} \cong \frac{dA}{dt}$$

$$\frac{\Delta A}{\Delta t} \cong \frac{d}{dt}\left(A_0 e^{-\lambda t}\right)$$

where $\lambda = \ln(2)/\tau$ and the initial activity is A_0.

$$\frac{\Delta A}{\Delta t} \cong -\lambda A_0 e^{-\lambda t}$$

$$\frac{\Delta A}{\Delta t} \cong -\lambda A$$

$$100 \times \frac{\Delta A}{A} \cong -100 \times \lambda \Delta t$$

The left-hand side of the above equation is the percent decrease in activity, so for a unit time of $\Delta t = 1$, the percent decrease in activity per unit time is $100 \times \lambda$, or about $69.3/\tau$. For example, the half-life of ^{68}Ga is 67.6 minutes, so it decays at a rate of about 1% per minute. This does not mean that 100% is decayed in 100 minutes: the approximations above are valid for incremental changes over durations that are small compared to the half-life.

4. **Answer:**

If S_{M1} and S_{N1} are the counts due to a point of activity I_1 to the anterior and posterior detector, respectively, the geometric mean D is given by

$$D = \sqrt{S_{M1} \times S_{N1}}$$

$$D = \sqrt{(\varepsilon I_{M1}) \times (\varepsilon I_{N1})}$$

where ε is the detector sensitivity (relatively equal between similar detectors), and I_{M1} is the attenuated photon ray path from the I_1 source into the anterior detector M, and I_{N1} is the same into the posterior detector N. If a is the distance from the I_1 source through the water into detector N, and b is the distance from the same source through the water to detector M, the above can be written:

$$D = \varepsilon \sqrt{(I_1 e^{-\mu a}) \times (I_1 e^{-\mu b})}$$

$$D = \varepsilon I_1 \sqrt{e^{-\mu(a+b)}}$$

$$D = \varepsilon I_1 e^{-\mu L/2}$$

This holds true for all the other point sources (e.g., I_2 and I_3), regardless of the relative sizes of a and b, that is to say the placement of the point sources, as long as they are not aligned on the up/down axis here. Therefore, the relative intensities of the point sources in such an image made from the geometric mean of two opposing planar images all have the exact same scaling factor due to photon attenuation and detector efficiency. The relative intensities in such an image are true to the relative intensities of the sources. It is easy to show that this is not true for an arithmetic average of the two images.

5. **Answer:**
   Assuming little septal penetration, the sensitivity will be proportional to the fraction of emitted photons that enter the detector through the pinhole compared to the total number of photons that leave the source. This will be the ratio of the solid angle of the pinhole with respect to the point source to the total solid angle of the emitted radiation. Equivalently, this is the fraction of the photons that pass through the area of the pinhole compared to the total number of photons that have radiated out from the source to the imagined surface of a sphere at radius r.

   $$\varepsilon \propto \frac{\text{Area of pinhole}}{\text{Area of sphere at radius } r}$$

   $$\varepsilon \propto \frac{\pi(d/2)^2}{4\pi r^2}$$

   $$\varepsilon = k\frac{d^2}{r^2}$$

   where k is some constant. Detector sensitivity varies inversely as the square of the distance r from the source to the pinhole. Note that doubling the pinhole diameter quadruples the camera sensitivity in this model. However, a larger hole degrades resolution as is evident from the diagram: photons from a point origin are blurred across the detector face (in blue) because of the acceptance angle the diameter d presents. More exact models of pinhole sensitivity and resolution account for septal penetration.

6. **Answer:**
   Setting the derivative of Eq. 4.8 to zero enables one to find, for example, the time of the first maximum depicted in Fig. 4.13. If the activity of ^{99m}Tc is A_2 and for ^{99}Mo it's A_1, then Eq. 4.8 gives

   $$A_2(t) = kA_1(0)\frac{\lambda_2}{\lambda_2 - \lambda_1}\left(e^{-\lambda_1 t} - e^{-\lambda_2 t}\right)$$

   $$\frac{d}{dt}A_2(t) = kA_1(0)\frac{\lambda_2}{\lambda_2 - \lambda_1}\left(-\lambda_1 e^{-\lambda_1 t} + \lambda_2 e^{-\lambda_2 t}\right)$$

   Setting the derivative to zero, so that time t is now at the maximum:

   $$\lambda_2 e^{-\lambda_2 t} = \lambda_1 e^{-\lambda_1 t}$$

$$t = \frac{\tau_1}{(\tau_1/\tau_2) - 1} \cdot \frac{\ln(\tau_1/\tau_2)}{\ln(2)}$$

where τ refers to the half-life of each isotope and the decay rate constant $\lambda = \ln(2)/\tau$. The half-life of ^{99}Mo is 65.94 hours and the half-life of ^{99m}Tc is 6.01 hours, so t is about 23 hours.

7. **Answer:**

The imaging photons for PET radiotracers are all 511 keV. Therefore, they cannot be distinguished by setting separate energy windows in the acquisition spectrum. By contrast, for example, ^{99m}Tc has an energy peak at 141 keV and ^{123}I has a peak at 159 keV. Given sufficient energy resolution, two separate energy windows can be set, and the two images can be acquired simultaneously with SPECT. It is conceivable to separate two PET radiotracers simultaneous by using other characteristics, such as different dynamic biological uptake, but this is not trivial to do.

8. **Answer:**

The density of tungsten is 19.3 g/mL, whereas lead has a density of 11.3 g/mL. The greater density, and therefore greater electron density, of tungsten reduces septal penetration by the γ-rays and will provide better spatial resolution for a given geometry. However, tungsten is significantly more expensive than lead.

9. **Answer:**

A full sinogram has $360°$, so this means that each head needs to do $180°$ every $2°$. This is 90 stops at 60 s per stop, giving a 90-minute total acquisition time. For a phantom, the ^{99m}Tc has lost only about 15% of its activity, so this is reasonable. However, most patients cannot lie still for 90 minutes, so this is unreasonable. Practically, most patient scans are designed to be acquired in less than 45 minutes.

10. **Answer:**

From Eq. 4.2, we have that the amount of ^{99m}Tc decreases by a factor $(\frac{1}{2})^3$, for three half-lives of decay. From Eq. 4.5 we have that the field will increase by a factor $(d_1/d_2)^2$ where d_2 is the new distance and d_1 is the original distance. So the new field φ_2 will be

$$\varphi_2 = \varphi \left(\frac{1}{2}\right)^3 \left(\frac{L}{L/2}\right)^2$$

So the new field is half the value of the original field: $\varphi/2$ mSv/h.

Exemplary Questions for Chap. 5

1. A vial with 8.3 GBq of ^{18}F-FDG is used to supply an activity of about 370 MBq, about once every 20 minutes, for each patient to be scanned on PET. How many patients can be supplied from this vial?

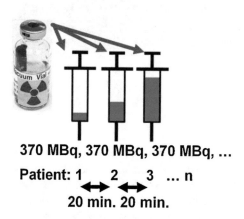

370 MBq, 370 MBq, 370 MBq, ...

Patient: 1   2   3   ... n

20 min. 20 min.

2. Calculate the PET projection $P_{\theta=0}(x=t)$ of a centered circular phantom with a homogeneous radiotracer density ρ MBq/cm^2, radius a, in the transaxial plane onto the x-axis. The attenuation coefficient of the phantom material is μ cm^{-1}.
3. With sufficient energy resolution, two energy peaks can be distinguished in dual isotope studies in some SPECT studies. For example, dual isotope studies using I^{123}-MIBG to image enervation and Tc^{99m}-mibi to image cardiac perfusion have been performed simultaneously. Why cannot this method be used for a PET dual isotope image using ^{18}F-DOPA (fluorodopa) and ^{82}Rb (rubidium-82; see Table 4.1)?

© Springer Nature Switzerland AG 2020
H. Azhari et al., *From Signals to Image*,
https://doi.org/10.1007/978-3-030-35326-1

4. A patient is scanned with ^{68}Ga-PSMA (prostate-specific membrane antigen) PET for 30 minutes. Because of a technical problem, the scan must be repeated immediately after. Approximately what duration must the second scan be to obtain a similar number of counts as the first scan?

5. Approximately, what temporal resolution of a TOF scanner would be required to obviate the need for an image reconstruction algorithm for clinical PET?

6. The phantom schematic depicts an elliptical phantom filled with water, having an attenuation coefficient μ of about 0.1 cm^{-1}. The breadth of the minor axis is 20 cm and for the major axis it's 40 cm. There is a positron source at the very center. Approximately what is the ratio of the count rate R_A for detector pair A for an LOR across the minor axis compared to the count rate R_B for detector pair B for an LOR across the major axis?

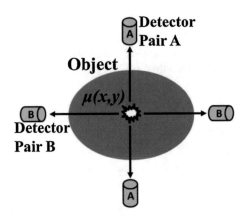

7. Scintillating crystals comprise a major cost in PET scanner design. One of the first whole-body PET scanners used bismuth germanate (BGO) crystals with a 55-cm-diameter bore (d_1) and 15 cm axial FOV (h_1). A newer design uses BGO crystals with a 74-cm-diameter bore (d_2) and 25 cm axial FOV (h_2). The crystal depth is the same in both designs (30 mm along the radial direction). Approximately how many times more crystal volume does the second design have compared to the first?

8. A phantom study is run to emulate a 10 mCi injection of ^{18}F-FDG to a 70 kg patient with a 60 minute wait time before scanning. What should the average voxel value be for the radiotracer concentration of the phantom in Bq/mL, at scan time?

9. A PET image $f(x,y)$ is defined by four unit squares (1 × 1) which have a magnitude of 1 (radiotracer concentration of 1) in their interiors. The four are centered at $(-1, -1)$, $(-1, 1)$, $(1, -1)$, and $(1, 1)$ respectively.

   (a) Find the projection $P_{\theta = 0}(t = x)$ onto the x-axis.

(b) Using the Fourier slice theorem, find a function describing the lateral axis of the 2D Fourier transform of this image.

Use the following definitions:

RECT($|x| < B$) = 1 if $|x| < B$ and RECT($|x| < B$) = 0 if $|x| \geq B$.

and $\mathfrak{F}[\text{RECT}(|x| < B)] = 2\sin(BK_x)/K_x$, where K_x is the Fourier index in the lateral direction and $\mathfrak{F}$ means the Fourier transform. This is using the non-unitary, angular frequency form of the Fourier transform. Other forms will give equivalent results.

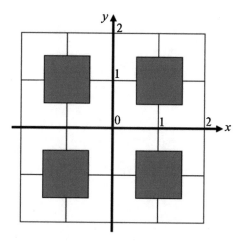

10. Use Fig. 5.1, Table 5.1, and Sect. 5.1.3 to estimate what percent of the positron energy is lost before annihilation occurs, for ^{18}F. It is useful to know that the momentum of the photon p_γ is related to the energy of a photon E_γ by the relation $p_\gamma = E_\gamma/c$, where c is the speed of light in a vacuum.

Solutions to Exemplary Questions for Chap. 5

1. **Answer:**

   If all the patients were injected at the same time, the answer is simply $8300/370 = {\sim}22$ patients. Because of the 20-minute wait, the ^{18}F is decaying, so the concentration of the radiotracer is diminishing, and each patient will require a larger volume for injection to get the same activity (see the figure).

   If time $t_0 = 0$ when the vial is at 8.3 GBq, the first patient can receive the activity (A_1) immediately, and it is equal to the patient injected activity 370 MBq, A_p. For the second patient, after a delay t_d of 20 minutes, the dose must be A_p, so at time t_0, this portion must be larger, back-corrected by a 20-minute decay factor. The same is true for patient 3, but the delay is $2 \times t_d$, and so on. The individual dose for each patient at time t_0 can be written as

$$A_1 = A_p$$

$$A_2 = A_p/e^{-\lambda t_d}$$

$$A_3 = A_p/e^{-\lambda(2t_d)}$$

$$A_i = A_p/e^{-\lambda(i-1)t_d}$$

where i refers to the ith patient and λ is the decay rate constant and the decay factor is $\exp(-\lambda t_d)$ for each 20-minute period, based on Eq. 4.1. Therefore, the sum A_0 needed at time t_0 for n patients is

© Springer Nature Switzerland AG 2020
H. Azhari et al., *From Signals to Image*,
https://doi.org/10.1007/978-3-030-35326-1

$$A_0 = \sum_{i=1}^{n} A_p / e^{-\lambda(i-1)t_d}$$

$$A_0 = A_p \sum_{i=1}^{n} e^{+\lambda(i-1)t_d}$$

This is the sum of a geometric series commonly solved as

$$A_0 = A_p \left(1 + e^{\lambda t_d} + e^{2\lambda t_d} + \ldots + e^{\lambda(n-2)t_d} + e^{\lambda(n-1)t_d} \right)$$

$$A_0 e^{\lambda t_d} = A_p \left(e^{\lambda t_d} + e^{2\lambda t_d} + \ldots + e^{\lambda(n-2)t_d} + e^{\lambda(n-1)t_d} + e^{\lambda n t_d} \right)$$

Taking the difference of the last two lines:

$$A_0 \left(e^{\lambda t_d} - 1 \right) = A_p \left(e^{\lambda n t_d} - 1 \right)$$

$$n = \frac{\tau}{t_d \ln(2)} \ln \left[1 + \frac{A_0}{A_p} \left(2^{t_d/\tau} - 1 \right) \right]$$

where τ is the half-life of F^{18} (109.8 minutes from Table 4.1) and $\lambda = \ln(2)/\tau$. This gives $n = 11.01$, so there is enough in the vial for 11 patients. L'Hôpital's rule can be used to show that the limit for n as t_d approaches 0 is about 22 patients, as initially calculated.

2. **Answer:**

Without attenuation, the projection will be scaled according to ρL, where L is the length of the line integral of each ray path of interest through the phantom. Since, in PET, the attenuation is dependent on the total length of the ray path through the attenuation material, because two photons at 180° must be detected, the attenuation factor is $\exp(-\mu L)$ for each ray path. From the image below, on the support $|x| \leq a$, projection onto the x-axis is parallel to the y-axis and the limits of the line integral will be from $y = -\sqrt{a^2 - x^2}$ to $y = +\sqrt{a^2 - x^2}$. So $P_0(x)$ is just $P_0(x) = \int \rho \exp(-\mu L) dy$ with ρ a constant over the line integral that covers length L. Therefore, the projection is

$$P_{\theta=0}(x) = 2\rho \left(\sqrt{a^2 - x^2} \right) \exp \left(-2\mu \sqrt{a^2 - x^2} \right) \text{ for } |x| \leq a \text{ and 0 elsewhere.}$$

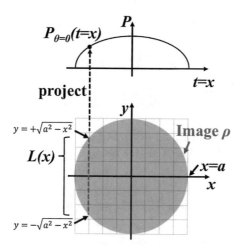

3. **Answer:**

It is conceivable to distinguish the ^{99m}Tc image and the ^{123}I image by using energy windows centered on their respective energy peak values (141 keV and 159 keV, respectively) and generating two separate emission data sets. However, in PET, the energy peak is always at 511 keV, the rest mass of the positron and the electron that have been annihilated. Consequently, energy discrimination alone cannot be used to discriminate two images in PET simultaneously.

4. **Answer:**

A first estimate would simply decay-correct the Ga^{68} activity for the ½ hour delay. The decay factor is

$$f = (1/2)^{t/\tau}$$

where $t = 30$ minutes here and τ is a half-life of 67.6 minutes for Ga^{68} (Table 4.1). This gives $f = 0.7352$ and the second scan should be about 40.8 minutes. However, this does not account for the decay that occurs during both scans. Since the number of events counted per second is proportional to the decays per second (activity), it's the integration of the activity A that is proportional to the total number of events n for the duration t_1 of the whole scan:

$$n = k \int_0^{t_1} A \, dt$$

where k is a proportionality constant accounting for the branching ratio (percent of decays giving rise to positrons) and the efficiency (events are lost to scatter, attenuation, and detector efficiency). For the first scan:

$$n = kA_0 \int_0^{t_1} \exp(-\lambda t)dt$$

$$n = \frac{kA_0}{-\lambda}[\exp(-\lambda t)]_0^{t_1}$$

$$n = \frac{kA_0}{\lambda}[1 - \exp(-\lambda t_1)]$$

where A_0 is the activity at the start of the first scan and λ is the decay constant and $\lambda = \ln(2)/\tau$. For a second scan of duration t_2, and initial activity of fA_0, this total number of events is

$$n = \frac{kfA_0}{\lambda}[1 - \exp(-\lambda t_2)]$$

Equating the two totals and solving for the second duration give

$$t_2 = \frac{-\tau}{\ln(2)} \ln\left(2 - 2^{t_1/\tau}\right)$$

This gives a slightly longer time of 43.6 minutes to account for decay during the scan. Note what happens if the first scan is longer than one half-life (e.g., for a C^{11} scan): there is no duration which is long enough since more than half of the activity is gone at the end of the first scan.

5. **Answer:**
   In Sect. 5.1.2 the spatial resolution of a clinical scanner is listed as good as 4 mm, and in Sect. 5.3 it is noted that detector crystal size can be as small as 3 mm. It is reasonable to take 2 mm as the best spatial resolution conceivable for a clinical scanner. Applying Eq. 5.2 (using meters and seconds) where our uncertainty in position is Δr:

$$\Delta r = c\Delta t/2$$

$$2 \times 10^{-3} = 3 \times 10^8 \Delta t/2$$

$$\Delta t = 1.\dot{3} \times 10^{-11}$$

That is, with a temporal resolution of 13 ps, the PET annihilation events could theoretically be plotted directly into 2 mm $\times$ 2 mm $\times$ 2 mm voxels in 3D space, without image reconstruction. This temporal resolution is an order of magnitude faster than current clinical PET detectors.

6. **Answer:**

From Eq. 5.3 we can write

$$R_A/R_B = e^{-\mu L_1}/e^{-\mu L_2}$$

where L_1 is the minor axis and L_2 is the major axis. Substitution gives the ratio R_A/R_B as e^2, which is about 7.4. These dimensions approximate a slender adult torso, suggesting that the anterior/posterior LORs receive substantially more counts than the lateral LORs.

7. **Answer:**

The scintillating crystals of the detector modules define a surface of a cylinder in the gantry of a clinical PET camera. Comparing the detector surface area in each case approximates the amount of BGO needed:

$$V_2/V_1 = A_2/A_1$$

where the 1 and 2 refer to the first and second cases with V and A referring to the volume of the BGO and the surface area presented by the detectors, respectively. So:

$$V_2/V_1 = (\pi d_2 h_2)/(\pi d_1 h_1)$$

$$V_2/V_1 = \left(\frac{d_2}{d_1}\right)\left(\frac{h_2}{h_1}\right)$$

The amount of BGO needed increases approximately linearly with the bore diameter and its axial length. The newer design requires about 2.24 times as much BGO.

8. **Answer:**

As in Eq. 5.7 we can use mass as a volume surrogate. The "patient" has a volume V of approximately 70,000 mL. From Table 4.1, the half-life τ of ^{18}F is 109.8 minutes, and from Sect. 4.1.1, an activity A_0 of 10 mCi is 370 MBq. Equation 4.2 accounts for decay. The concentration C at scan time after a wait of t is then

$$C = A_0 \times \left(\frac{1}{2}\right)^{t/\tau}/V$$

$$C = 370 \times 10^6 \times \left(\frac{1}{2}\right)^{60/109.8} /70000$$

The expected voxel value is about 3619 Bq/mL.

9. **Answer:**

(a) Because projection is an additive process, the principle of superposition applies. Visually, the projection is simple (see diagram); it is a rectangular function of height 2 with a "bite" taken out of it. The total width goes from $-3/2$ to $+3/2$, and the "bite" ranges from $-\frac{1}{2}$ to $\frac{1}{2}$. So the projection $P_{\theta = 0}(t = x)$ onto the x-axis can be written as

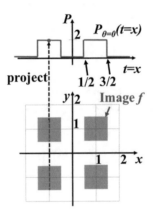

$$P_0(x) = 2[RECT(|x| < 3/2) - RECT(|x| < 1/2)]$$

(b) According to the Fourier slice theorem, the 1D Fourier transform of this projection onto the x-axis gives the K_x-axis of the 2D Fourier transform of the 2D image $f(x,y)$. Applying the Fourier transform to the projection gives

$$\mathfrak{F}[P_0(x)] = 2\left[\frac{2\sin\left(\frac{3}{2}K_x\right)}{K_x} - \frac{2\sin\left(\frac{1}{2}K_x\right)}{K_x}\right]$$

$$\mathfrak{F}[P_0(x)] = \frac{4}{K_x}[\sin(3K_x/2) - \sin(K_x/2)]$$

If $\mathfrak{F}[f(x,y)] = F(K_x, K_y)$, then by the Fourier slice theorem, the lateral axis of this 2D Fourier transform is

$$F\left(K_x, K_y = 0\right) = \frac{4}{K_x}\left[\sin\left(3K_x/2\right) - \sin\left(K_x/2\right)\right]$$

An equivalent answer can be made with one term. Another identity for the non-unitary, angular frequency form of the Fourier transform is

$$\mathfrak{F}\left[\frac{1}{2}\left[\delta(x - A) + \delta(x + A)\right]\right] = \cos\left(AK_x\right)$$

where δ is the sampling function such that $f(A) = \int_{-\infty}^{\infty} f(x)\delta(x)dx$. There is also the common result that convolution between two functions in real space translates to multiplication of their Fourier transforms in Fourier space. If g and h are two functions such that P is their convolution $P = g*h$, then the Fourier transform of P is given by

$$\mathfrak{F}[P] = \mathfrak{F}[g]\,\mathfrak{F}[h].$$

By inspection the projection $P_0(x)$ can be written as a convolution:

$$P_0(x) = 2[\delta(x - 1) + \delta(x + 1)] * RECT\left(|x| < \frac{1}{2}\right)$$

$$\mathfrak{F}[P_0(x)] = \mathfrak{F}\left[4\left[\frac{1}{2}[\delta(x - 1) + \delta(x + 1)]\right] * RECT\left(|x| < \frac{1}{2}\right)\right]$$

$$\mathfrak{F}[P_0(x)] = 4\cos\left(K_x\right) \cdot \frac{2\sin\left(K_x/2\right)}{K_x}$$

$$\mathfrak{F}[P_0(x)] = \frac{8}{K_x}\cos\left(K_x\right)\sin\left(K_x/2\right)$$

There are a number of methods to show that this is equivalent to the previous result.

10. **Answer:**

Assume that the electron in the annihilation does not contribute much to the momentum of the photons. The positron has some momentum p_β in the direction of travel, called x in the diagram. The component of the momentum of the photons p_γ the x-direction should equal the momentum of the positron right before annihilation. The diagram has been re-oriented from Fig. 5.1, but the difference in the directions of the photons remains 180°, less a total deviation of θ due to momentum. The momentum of the positron is along x, and before annihilation, it is

$$P_\beta = mv$$

where m is the rest mass of the positron and v is its speed.

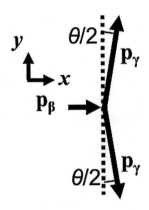

$$P_\beta = \sqrt{m^2 v^2}$$

$$P_\beta = \sqrt{2mE_\beta}$$

where E_β is the classical kinetic energy ($E_\beta = mv^2/2$) of the positron before annihilation. The component of the two annihilation photons in the x-direction $[2p_\gamma]_x$ is

$$[2p_\gamma]_x = 2(E_\gamma/c)\sin(\theta/2)$$

The largest value of θ is $0.25°$, so the small angle approximation is valid:

$$[2p_\gamma]_x = 2(E_\gamma/c)(\theta/2)$$

$$[2p_\gamma]_x = E_\gamma \theta/c$$

if θ is converted to radians. This must be equal to the x-component of the momentum before annihilation p_β, so

$$\sqrt{2mE_\beta} = E_\gamma \theta/c$$

$$E_\beta = E_\gamma^2 \theta^2 / (2mc^2)$$

where mc^2 is the energy of each annihilation photon ($E_\gamma = 511$ keV), so

$$E_\beta = E_\gamma \theta^2 / 2$$

Substituting $\theta = 0.25\pi/180$ radians gives an upper value for the energy of the positron before annihilation of 5 eV. Section 5.1.2 notes that the average positron energy is about 40% of the maximums listed in Table 5.1. Positrons emitted from ^{18}F are emitted with an average energy of about 0.25 MeV, so virtually 100% of this kinetic energy is lost before annihilation. It is easy to show that a positron with 5 eV of kinetic energy is not travelling at relativistic velocities, so using the classical kinetic energy for E_β is reasonable.

Exemplary Questions for Chap. 6

Part I: Verbal Questions. Kindly Answer Without Browsing the Text

1. What are the three types of susceptibility for a matter positioned within a magnetic field?
2. What are the three physical properties that affect the magnitude of M_0 and how?
3. What relaxation times are used in MRI?
4. What are the main sources of contrast in MRI?
5. What is the advantage of spin echo?
6. What are the advantages and disadvantages of EPI?
7. What are the advantages and disadvantages of TOF flow imaging?

Part II: Quantitative Questions

8. During an MRI scan, an RF pulse defined by $B_1 = A_0 \sin (\pi \cdot t/\tau)$ was applied. Given that the transmission duration was τ, the magnetization flip angle will be:

    A. $2A_0 \cdot \gamma$
    B. 0
    C. $A_0 \gamma 2\tau/\pi$
    D. $A_0 \gamma \tau/\pi$
    E. None of the above

9. During an MRI scan we would like to flip the magnetization vector so that $M_z = -M_0$. For that aim an RF pulse which generates a magnetic field $B_1(t) = \sin (A \cdot \pi t)$, where $A = 1[1/\text{sec}]$, is transmitted. What should be the transmission duration τ in order to obtain the required flipping?

© Springer Nature Switzerland AG 2020
H. Azhari et al., *From Signals to Image*,
https://doi.org/10.1007/978-3-030-35326-1

A. $\tau = \arcsin\left(1 + \frac{\pi^2}{\gamma}\right)$

B. $\tau = \arccos\left(1 + \frac{\pi^2}{\gamma}\right)$

C. $\tau = \frac{1}{\pi} \cdot \arcsin\left(1 - \frac{\pi^2}{\gamma}\right)$

D. $\tau = \frac{1}{\pi} \cdot \arccos\left(1 - \frac{\pi^2}{\gamma}\right)$

E. None of the above

10. During an MRI scan, using a scanner with $B_0 = 1$ [Tesla], we wanted to select a slice by transmitting an RF pulse which angular frequency is given by $(1 + \alpha) \cdot \omega_0$. Due to a the technical fault, the transmitted RF pulse had a frequency of $(1 - \alpha) \cdot \omega_0$. Given that the slice selection gradient is G_0[Tesla/meter] along the Z direction, what is the error (in [m]) in the slice selection position?

A. None of the answers below

B. $\frac{(1+2\alpha)\cdot\omega_0}{\gamma\cdot G_0}$

C. $\frac{\alpha\cdot\omega_0}{\gamma\cdot G_0}$

D. $\frac{2\alpha\cdot\omega_0}{\gamma\cdot G_0}$

E. $\frac{\omega_0}{\gamma\cdot G_0}$

11. We would like to scan objects by MRI, using the K-space trajectory depicted in the figure shown below. Given that the required spatial resolution is Δ along all directions and that the analytic equations describing the trajectory are given by (where t is given in msec)

$$\begin{cases} K_x = K_{x,\max} \cdot \sin\left(\frac{\pi}{18} \cdot t\right) \\ K_y = K_{y,\max} \cdot \sin\left(\frac{\pi}{9} \cdot t\right) \end{cases}$$

kindly answer the following:

A. What is the value of $K_{x,\max}$?
B. What is the time required for reaching $K_{x,\max}$ for the first time?
C. What are the gradient fields G_X, G_Y needed to perform this scan?
D. At what time points will the absolute value of $|G_y|$ be maximal?

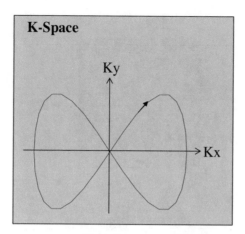

12. In an MRI scan, it is required to obtain an image with rectangular-shaped pixels, for which the horizontal spatial resolution is Δ_x and the vertical one is $\Delta_y = 2\Delta_x$. The scanning trajectory in K-space is described schematically in the upper figure below. It consists of two stages: During the first stage, the sampling pointer in K-space is moved to point P. The trajectory during the second part is given by the following pair of equations:

$$\left\{ \begin{array}{l} K_x = K_x(p) + B \cdot t \\ K_y = K_y(p) \cdot [1 - \sin(\Omega t)] \end{array} \right\}$$

Kindly answer the following:

A. What are the coordinates of point P in K-space, i.e., $K_x(p)$; $K_y(p)$?
B. If during the first stage trapezoidal gradients are applied along the two directions, what should be the value of t_0 for the gradient field $G_y(t)$?
C. What should be the gradient fields during the second stage?
D. What is the time τ required for the second stage?
E. What is the value of the constant B?
F. Assuming now that τ is known, and that the initial flip angle was $30°$, and given that $T_1 \gg T_2$, what is the value of the magnetization M_{xy} (in terms of M_0)?

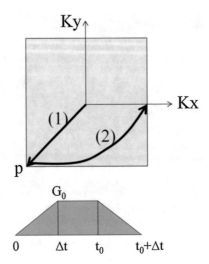

Solutions to Exemplary Questions for Chap. 6

Part I: Verbal Questions

1. Diamagnetic, paramagnetic, and ferromagnetic
2. The magnetic field, the proton density, and the temperature. The first two increase the magnitude of M_0 when increased, while the temperature decreases it (a reciprocal ratio).
3. Spin-lattice relaxation characterized by T_1. Spin-spin relaxation characterized by T_2. Effective spin-spin relaxation marked as T_2^* which is shorter than T_2 due to the magnetic field inhomogeneity (this property is machine dependent). And T_{1rho} which occurs during the spin-lock sequence
4. Proton density, the above-listed relaxation times, and phase in flow imaging and thermal imaging. In addition, the time to echo TE and time to repeat TR may also play an important role in the image contrast.
5. Spin echo allows us to overcome the effects stemming from the magnetic field inhomogeneity. Thus, we can measure and image the actual T_2, rather than T_2^*, which changes from one scanner to another.
6. Echo planar imaging (EPI) offers a very rapid scan time, as it covers the entire K-space in one TR (or a few if interleaving is applied). However, it requires strong gradient fields with high slew rate, and the SNR is reduced due to the long readout time.
7. TOF flow imaging allows us to image blood vessels without using any contrast-enhancing material. However, it can depict only flow which passes through the slice and cannot image flow which occurs within the slice. In addition, it does not show the flow direction.

© Springer Nature Switzerland AG 2020
H. Azhari et al., *From Signals to Image*,
https://doi.org/10.1007/978-3-030-35326-1

Part II: Quantitative Questions

8. Answer: **C**
   The flip angle is calculated by

$$\alpha = \gamma \int_0^\tau B_1 dt = \gamma \int_0^\tau A_0 \sin(\pi \cdot t/\tau) dt = A_0 \gamma 2\tau/\pi$$

9. Answer: **D**
   The requirement for $M_z = -M_0$ implies that $\alpha = \pi$. Thus,

$$\pi = \gamma \int_0^\tau B_1 dt = \gamma \int_0^\tau \sin(1 \cdot \pi \cdot t) dt$$

$$\Rightarrow \frac{\pi^2}{\gamma} = [1 - \cos(\pi \cdot \tau)]$$

$$\Rightarrow \tau = \frac{1}{\pi} \cdot \arccos\left(1 - \frac{\pi^2}{\gamma}\right)$$

10. Answer: **C**
    The planned relation was

$$(1 + \alpha) \cdot \omega_0 = \gamma \cdot B_0 + \gamma G_0 \cdot Z_1$$

$$\Rightarrow Z_1 = \frac{\alpha \cdot \omega_0}{\gamma G_0}$$

And the actual position was

$$(1 - \alpha) \cdot \omega_0 = \gamma \cdot B_0 + \gamma G_0 \cdot Z_1$$

$$\Rightarrow Z_1 = \frac{-\alpha \cdot \omega_0}{\gamma G_0}$$

Thus, the error is given by

$$\frac{2\alpha \cdot \omega_0}{\gamma \cdot G_0}$$

11. Answers:

A. $K_{x,\max} = \frac{\pi}{\Delta}$

B. To reach the maximal spatial frequency along K_x, the following equation should apply:

$$K_x(t) = K_{x,\max} \cdot \sin\left(\frac{\pi}{18} \cdot t\right) = K_{x,\max}$$
$$\Rightarrow \sin\left(\frac{\pi}{18} \cdot t\right) = 1$$
$$\Rightarrow \frac{\pi}{18} \cdot t = \frac{\pi}{2}$$
$$\Rightarrow t = 9[\text{msec}]$$

C. As recalled the spatial frequency relates to the gradient by

$$K_x = \gamma \int\limits_0^t G_x dt$$

It follows therefore that

$$\frac{1}{\gamma}\frac{dK_i}{dt} = G_i$$

$$\Rightarrow \begin{cases} G_x = \dfrac{1}{18\gamma} \cdot \dfrac{\pi^2}{\Delta} \cdot \cos\left(\dfrac{\pi}{18}t\right) \\ G_y = \dfrac{1}{9\gamma} \cdot \dfrac{\pi^2}{\Delta} \cdot \cos\left(\dfrac{\pi}{9}t\right) \end{cases}$$

D. From the previous equation, it follows that

$$\text{for } G_{y,\,max} \quad \Rightarrow \cos\left(\frac{\pi}{9}t\right) = 1$$

$$\Rightarrow \frac{\pi}{9}t = n \cdot \pi \quad n = 0, 1, 2, 3. \ldots$$

$$\Rightarrow t = n \cdot 9[\text{msec}]$$

12. Answers:

A. The point P is located at the lowest point on the negative (left) side of K-space. Therefore, its coordinates are

$$\begin{cases} K_{x,\,max} = -\dfrac{\pi}{\Delta_X} \\[3mm] K_{y,\,max} = -\dfrac{\pi}{\Delta_Y} = -\dfrac{\pi}{2\Delta_X} \end{cases}$$

B. As recalled,

$$K_x(t) = \gamma \int\limits_0^t G_x(t)dt$$

To move the pointer in K-space, we need to travel a distance which equals $|K_{y,\,max}|$. In this case, the integral is simply given by the area of the trapezoidal shape

$$K_{y,\,max} = \frac{\pi}{2\Delta_X} = 2 \cdot \left(\frac{1}{2}\gamma G_0 \cdot \Delta t\right) + \gamma G_0 \cdot (t_0 - \Delta t)$$

$$\Rightarrow t_0 = \frac{\pi}{\gamma G_0 \cdot 2\Delta x}$$

C. Applying the relation given in the previous question, we obtain

$$G_i = \frac{1}{\gamma}\frac{dK_i}{dt}$$

$$\Rightarrow \begin{cases} G_x = \dfrac{1}{\gamma} \cdot B \\[3mm] G_y = \dfrac{1}{\gamma} \cdot \dfrac{\pi \cdot \Omega}{2\Delta x} \cdot \cos\left(\Omega t\right) \end{cases}$$

D. The following relation should apply:

$$K_y = -\frac{\pi}{2\Delta x}[1 - \sin(\Omega \cdot \tau)] = 0$$
$$\Rightarrow \sin(\Omega \cdot \tau) = 1$$
$$\Rightarrow \Omega \cdot \tau = \frac{\pi}{2}$$
$$\Rightarrow \tau = \frac{\pi}{\Omega \cdot 2}$$

E. The following relation should apply:

$$K_x(\tau) = -\frac{\pi}{\Delta x} + B \cdot \tau = \frac{\pi}{\Delta x}$$
$$\Rightarrow B \cdot \tau = \frac{2\pi}{\Delta x}$$
$$\Rightarrow B = \frac{2\pi}{\tau \cdot \Delta x}$$

F. $M_{xy} = \frac{M_0}{2} e^{-\frac{1}{T_2}(t_0 + \Delta t + \tau)}$

Exemplary Questions for Chap. 7

1. Why do we smear gel on the patient's skin when conducting an ultrasonic scan?

   A. To prevent irritation caused by the acoustic vibrations
   B. To prevent heating of the skin stemming from the acoustic energy
   C. To prevent trapping of air between the skin and the transducer
   D. To amplify the echoes amplitude
   E. All of the above

2. Which of the following statements is <u>incorrect</u>:

   A. The velocity of the ultrasound waves depends on the matter density.
   B. Sound waves do not propagate in vacuum.
   C. The TGC in ultrasound imaging of the same object should have larger gain values as the frequency increases.
   D. The amplitude of the wave passing from one medium to another may be higher than the wave amplitude of the impinging wave.
   E. One of the above answers is incorrect.

3. During an experiment in the imaging lab, a B-scan of a plastic tube dipped in water and containing air was conducted. The image shown on the left (Fram#1) was obtained (the transmission is from top to bottom). Then, water was injected into the tube and the image shown on the right (Frame#2) was obtained. As you can see, a double bright strip now appears from below. How can we explain the difference between the two images obtained for the same tube?

   A. The flow of water in the pipe produces an artifact in the picture.
   B. The rapid water flow in the pipe caused a large Doppler shift, and this caused an aliasing folding effect which duplicated the upper wall image.
   C. The acoustic impedance for air is so high that all the waves are reflected from the top wall. Thus, one cannot see the bottom wall of the tube on the left picture.

© Springer Nature Switzerland AG 2020
H. Azhari et al., *From Signals to Image*,
https://doi.org/10.1007/978-3-030-35326-1

D. The reflection coefficient between the air and the pipe is so high that no echoes were obtained from the bottom pipe wall in the left picture.

E. None of the above.

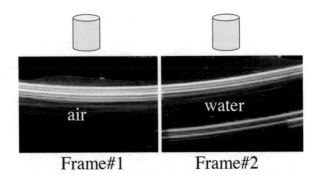

Frame#1   Frame#2

4. The amplitude of an ultrasound beam was decreased after passing a certain distance S within the tissue to 64% of its initial value. Assuming that the attenuation coefficient is linearly dependent on the frequency, what will be the relative amplitude value reaching a point located at a quarter of the previous distance (i.e., $S/4$), if the transmission frequency is doubled?

A. 16%

B. 40%

C. 80%

D. 64%

E. None of the above

5. Given the image of an MRI signal in K-space (left image) obtained from a blood vessel surrounded by soft tissues and also given the speed of sound in the medium $-c$, the speed of blood flow $-v$, and the frequency of the transmitted ultrasound wave $-f_0$, what would be the frequency shift that this blood vessel will induce in the Doppler mode when an ultrasound wave is transmitted at it with an angle of 45° relative to the x-axis (right image)?

A. 0

B. $\frac{\sqrt{2}f_0v}{c}$

C. $\frac{f_0v}{c}$

D. $\frac{2f_0v}{c}$

E. None of the above

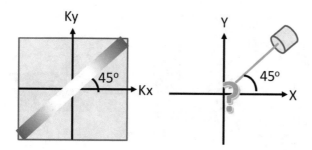

6. When imaging an embryo in the uterus using the B-scan mode, it is preferable to use:

   A. High frequency to achieve deep beam penetration, but that will reduce the spatial resolution.

   B. Low frequency to achieve high spatial resolution, but that will reduce the beam penetration.

   C. High frequency to achieve good spatial resolution, but that will reduce the beam penetration.

   D. Low frequency to achieve deep beam penetration, but that will reduce the spatial resolution.

   E. There is no connection between frequency and imaging quality.

   F. None of the above.

7. When using M-mode ultrasonic imaging of the heart's left ventricle, a series of echoes were detected. Given that the attenuation coefficient μ for all tissues is the same, and that the transmission amplitude is A_0, the distances are as depicted in the drawing. Also given that the acoustic impedance for the myocardium is Z_1, for the blood it is Z_2, and for all other external tissues it is Z_0, what is the theoretical value for the second echo amplitude detected by the receiver?

   A. $A_0 e^{-2\mu(s_1+s_2)} \left(\frac{2Z_1}{Z_0+Z_1} \right) \left(\frac{Z_2-Z_1}{Z_2+Z_1} \right)$

   B. $A_0 e^{-2\mu(s_1+s_2)} \left(\frac{4Z_0 \cdot Z_1}{(Z_0+Z_1)^2} \right) \left(\frac{Z_2-Z_1}{Z_2+Z_1} \right)$

   C. $A_0 e^{-\mu(s_1+s_2)} \left(\frac{2Z_1}{Z_0+Z_1} \right) \left(\frac{Z_2-Z_1}{Z_2+Z_1} \right)$

   D. $A_0 e^{-\mu(s_1+s_2)} \left(\frac{4Z_0 \cdot Z_1}{(Z_0+Z_1)^2} \right) \left(\frac{Z_2-Z_1}{Z_2+Z_1} \right)$

   E. None of the above

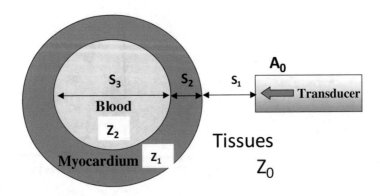

8. We would like to scan the abdomen of an obese person using B-scan. However, we have only two transducers: one transmitting at a frequency of 3 MHz and the other and at 5 MHz. Which of the following statements is correct in this context?

   A. You should always choose a 5 MHz transducer because its resolution is better.
   B. You should always choose a 3 MHz transducer because its resolution is better.
   C. It is usually better to choose a 3 MHz transducer because the waves will penetrate deeper.
   D. It is usually better to choose a 5 MHz transducer because the waves will penetrate deeper.
   E. It does not matter in this case, since the frequency is important only when using only the Doppler mode.

9. An ultrasound signal is sent from transducer "a" to receiver "b" through a cylindrical ultrasound phantom full of water in which the speed of sound is C_2. A plastic rectangular block of thickness L is placed in the phantom as shown in the diagram. The speed of sound in plastic is C_1 and is faster than the speed of sound in water $C_1 > C_2$. How much sooner does a signal arrive when travelling from transducer "a" to receiver "b" than from transducer "c" to receiver "d"?

   A. $L\left(\frac{1}{C_2} - \frac{1}{C_1}\right)$
   B. $2D/C_2$
   C. $1/(C_1 - C_2)$
   D. $(D - L)\left(\frac{C_1 - C_2}{C_1 C_2}\right)$
   E. None of the above

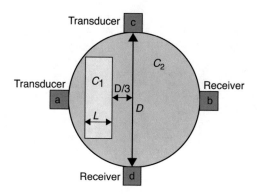

Solutions to Exemplary Questions for Chap. 7

1. Answer: **C**

   In order to ensure good acoustic coupling, we must prevent the trapping of air between the skin and the transducer.

2. Answer: **E**

   Because all the other statements are correct

3. Answer: **D**

   The acoustic impedance of air is very low. Thus, the reflection coefficient between the pipe wall and the air is very high and all the echoes are obtained only from the top wall.

4. Answer: **C**

   From the initial transmission values, it is concluded that

   $$A_1 = A_0 \cdot e^{-\alpha_1 f \cdot S} = 0.64 A_0$$
   $$\Rightarrow e^{-\alpha_1 f \cdot S} = 0.64$$

   By doubling the frequency and changing the distance, we obtain

   $$A_2 = A_0 \cdot e^{-\alpha_1 \cdot 2f \frac{S}{4}} = 0.64 A_0$$
   $$\Rightarrow A_0 e^{-\alpha_1 f \frac{S}{2}} = \sqrt{0.64} \cdot A_0 = 0.8 A_0$$

5. Answer: **A**

   Whatever is narrow in K-space is wide in the object domain and vice versa (see Chap. 1). Therefore, it can be concluded that the blood vessel is actually orthogonal to the beam and the Doppler shift will be zero.

© Springer Nature Switzerland AG 2020
H. Azhari et al., *From Signals to Image*,
https://doi.org/10.1007/978-3-030-35326-1

6. Answer: **C**

Since the embryo is very small, the important factor in this case is the spatial resolution. In order to overcome the high attenuation and poor beam penetration, an intravaginal transducer may be used.

7. Answer: **B**

$$A = A_0 e^{-\mu(s_1)} \left(\frac{2 \cdot Z_1}{(Z_0 + Z_1)} \right) e^{-\mu(s_2)} \left(\frac{Z_2 - Z_1}{Z_2 + Z_1} \right) e^{-\mu(s_2)} \left(\frac{2Z_0}{(Z_0 + Z_1)} \right) e^{-\mu(s_1)}$$

8. Answer: **C**

In this case the penetration is more important; hence, the lower frequency of 3MHz is preferred because the waves will penetrate deeper into the body.

9. Answer: **A**

$$t_1 = \frac{D}{C_2}, t_2 = \frac{D - L}{C_2} + \frac{L}{C_1}$$

The difference is therefore

$$L \left(\frac{1}{C_2} - \frac{1}{C_1} \right)$$

Medical Imaging: Final Exam A

Name: _____ ID No: _____

Duration: 3 hours

You may use only **two** A4-sized pages (both sides each) of equations and your own notes.

Any use of calculators or other electronic equipment is forbidden!

Good Luck

Part A [40%]

1. **[20%]** A kidney was scanned along a certain line (A–B) by three different imaging modalities. The width of the kidney at the site being scanned is L, and in the middle (see illustration), there is a stone with diameter $D = L/4$. Two detectors suitable for the reception of all types of radiation and having area $= S$ are located as shown in the picture. For simplicity, we would also assume that the kidney attenuation coefficient is uniform and is the same for the SPECT and PET radiation and its value is μ. Similarly, the attenuation of the stone is μ_0. A radioactive material is now injected into the body and its concentration during the "wash-in" period is given by $C(t) = C_0[1 - e^{(-t/\tau)}]$ for $0 \leq t \leq 4\tau$. It can be also assumed that the material is spread throughout the kidney evenly (except, of course, at the stone).

   A. **[3%]** The first scan was performed in PET configuration. The positron emission per unit volume was $P_0 \cdot C(t)$. Calculate the intensity that will be recorded as a function of time along the scanned line.

   B. **[6%]** What would be the total radiation that will be recorded during the entire "wash-in" time (4τ seconds)?

© Springer Nature Switzerland AG 2020
H. Azhari et al., *From Signals to Image*,
https://doi.org/10.1007/978-3-030-35326-1

C. **[6%]** A second scan was performed in the SPECT configuration by the right detector (B). Calculate the intensity that will be recorded as a function of time at that position if the gamma radiation emission per unit volume is given by $G_0 \cdot C(t)$.

D. **[5%]** Now we switch to x-ray imaging when the left element (A) transmits with radiation intensity I_0 and the right element (B) is used for reception. If a new contrast agent is injected into the kidney, for which the attenuation coefficient is given by $\mu(t) = \mu_0 \cdot C(t)$, what will be the intensity that will be recorded as a function of time along that axis, when is it maximal during the wash-in time, and what is its value?

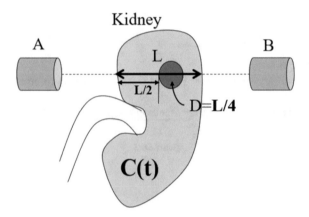

2. **[20%]** We want to carry out an MRI scan along a spiral trajectory given by the equation $\overline{K} = t \cdot A \cdot e^{-j \cdot \Omega \cdot t}$ (where the real part relates to the K_x direction and the imaginary part to the K_y direction). The required resolution is Δ along both directions. The scanned object is a container of liquid with relaxation times $T_1 \gg T_2$ (i.e., the effect of the spin-lattice relaxation is negligible during the scan).

A. **[1%]** What is the value of Ky_{max}?

B. **[5%]** If we want to flip the magnetization to a 30° angle by transmitting a pulse given by $B_1 = B \cdot \sin(\pi t / \tau)$, what should be the transmission time t_1?

C. **[5%]** If we want to finish the trajectory scan described in the picture at time $t = T_2$, what should be the value of Ω?

D. **[5%]** Now given that $\Omega = 10\pi/T_2$, write the explicit expression for each of the gradients $G_x(t)$, $G_y(t)$.

E. **[4%]** If at the end of the scan we want to return to the axes origin point by using a trapezoid gradient as shown in the lower figure, what should be the value of t_0, for the G_x application?

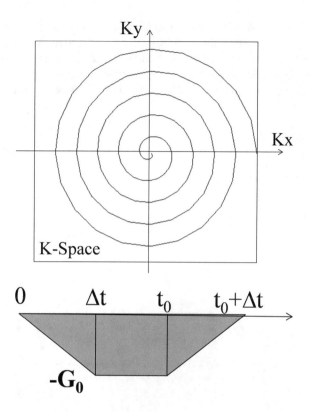

Part B [60%]

Please Mark the Correct Answer

1. [5%] Two materials were scanned using MRI. Both materials have equal relaxation times T_1, T_2, but the proton density of the first material is <u>twice as large</u> as that of the second material. The magnetization vector is flipped in the first material by $30°$ and in the second material by $90°$, and the signals are sampled at time TE. Which of the following statements is **correct**?

A. The signal received from the first material at time TE will be stronger.
B. The signal received from the second material at time TE will be stronger.
C. Both signals will have the same intensity.
D. The strength of the signal depends on the size of $B1$ and therefore one cannot tell.
E. None of the above.

2. [5%] An ultrasound wave with an amplitude A_0 is transmitted from the right transducer to the left transducer through water and a tissue sample of thickness L with another layer of water behind it as described in the figure. The impedance for water is Z_0 and for the tissue is Z_1, the attenuation coefficient of the tissue is α, and the attenuation in the water is negligible. What is the amplitude of the received wave?

A. $A_0 \cdot \dfrac{4Z_1^2}{(Z_0+Z_1)^2} \cdot e^{-\alpha L}$

B. $A_0 \cdot \dfrac{4Z_0Z_1}{(Z_0+Z_1)^2} e^{-\alpha L}$

C. $A_0 \cdot \dfrac{4Z_0^2}{(Z_0+Z_1)^2} \cdot e^{-\alpha L}$

D. $A_0 \cdot e^{-\alpha L}$

E. None of the above

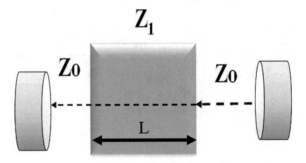

3. [5%] Data projections of the 3×3 pixel image are given as described. Reconstruct the image by ART of multiplications and then by ART of differences in the order indicated by the arrows. (The reconstruction is performed clockwise as described by the arrows, i.e., starting with the upper left projection.) Which statement **is incorrect**?

A. In both methods we get convergence.
B. In both methods we get the same values at the bottom row.
C. In both methods we get the same values at the top row.
D. The obtained middle pixel value is different for the two methods.
E. For both methods the sum of all pixels $= 6$.

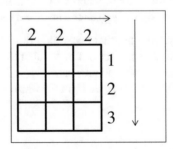

4. [5%] A phantom is scanned by an x-ray device with an average radiation energy of 70 keV. The radiation after the rays pass through the phantom is measured. Which of the following statements is **correct**?

A. The radiation intensity will be higher than the intensity before entering the phantom and its average energy will be higher than 70 keV.

B. The radiation intensity will be lower than the intensity before entering the phantom but its average energy will be higher than 70 keV.

C. The radiation intensity will be lower than the intensity before entering the phantom but its average energy will be equal to 70 keV.

D. The radiation intensity will be lower than the intensity before entering the phantom and its average energy will be less than 70 keV.

E. None of the above.

5. [5%] In an MRI scanner with a field of B_0, we want to select a slice of thickness T located at a distance Z_1 from the center of the field. The slice selection gradient is G_0 [Tesla/meter] – along the Z direction, what is the frequency band (**in Hertz**) that we need to transmit?

A. $\gamma G_0 T$

B. $\frac{1}{2\pi} \cdot \gamma G_0 (Z_1 - T)$

C. $\frac{1}{2\pi} \cdot \gamma G_0 (Z_1 + T)$

D. $\frac{1}{2\pi} \cdot \gamma G_0 T$

E. None of the above

6. [5%] An apparatus consists of two square containers with identical <u>internal</u> dimensions. The upper container has a wall thickness of T and the lower has a wall thickness of $2T$ as shown in the figure. The containers were filled with a positron-emitting material with an initial intensity I_0 and half-life τ. The upper container was filled 2τ minutes after its production and the bottom was filled τ minutes after its production. The walls are made of a solid material with an attenuation coefficient μ. The apparatus was inserted into a PET scanner and the horizontal projections A and B were measured. What would be the A/B ratio?

A. 1

B. $\frac{1}{4} e^{2\mu T}$

C. $\frac{1}{2} e^{2\mu T}$

D. $\frac{1}{2} e^{-2\mu T}$

E. None of the above

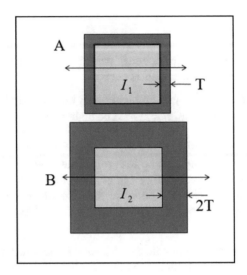

7. **[5%]** For the development of a SPECT scanner, an experiment of an ex vivo animal brain was performed as described in the figure. Two point-emission sources were installed (marked by the stars). They emitted gamma radiation with identical intensity and their locations were as indicated in the picture. The thickness of each side of the brain is T and the attenuation coefficient is uniform and its value is μ. In the right hemisphere, there is a bleeding area of diameter D that does not attenuate the rays. What is the ratio between the signal measured by detector A and the signal measured by detector B (i.e., A/B)?

A. 1.

B. $\dfrac{1+e^{-\mu\frac{T}{2}}}{e^{-\mu(T-D)}+e^{-\mu\frac{3}{2}T}}$

C. $\dfrac{1+e^{-\mu\frac{T}{2}}}{e^{+\mu D}\left[e^{-\mu\frac{T}{2}}+e^{-\mu T}\right]}$

D. $\dfrac{e^{-\mu\frac{T}{2}}}{e^{-\mu(T-D)}}$

E. None of the above

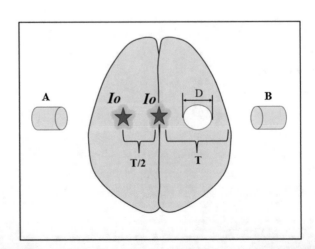

8. [5%] A CT scanner can, due to certain constraints, acquire only 31 projections per scan. How many detectors must be used to sample each projection so that we shall not obtain sampling artifacts in reconstructing an image with a uniform resolution?

A. Approximately 31
B. Approximately 62
C. Approximately 20
D. Approximately 40
E. None of the above

9. [5%] Water density <u>decreases</u> when cooled in the temperatures that range from 4 °C to 0 °C. For water cooled from 4° to 2°, the effect on ultrasound will be:

A. The speed of sound will increase and the acoustic impedance will increase.
B. The speed of sound will increase and the acoustic impedance will decrease.
C. The speed of sound will decrease and the acoustic impedance will decrease.
D. The speed of sound will decrease and the acoustic impedance will increase.
E. None of the above.

10. [5%] For the images described in drawings **a–d**, each black pixel $= 1$ and each white pixel $= 0$. Please mark the **correct** sentence.

A. Each of the four images has a line in the K plane described by only one SINC function.
B. The value at the axes origin in K plane is different for each image.
C. The function obtained in the K plane along the K_y-axis is equal for images **a** and **b**.
D. The function obtained in the K plane along the K_x-axis is equal for images **a** and **d**.
E. None of the above.

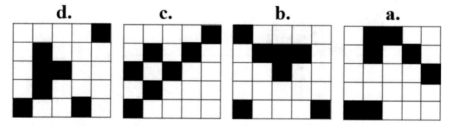

d.   c.   b.   a.

11. [5%] Which of the following statements is correct (please mark only one correct sentence)?

A. In a PET scan, the photon energy emitted from the source depends on the magnitude of the current set in the imaging device.
B. In a PET scan, the photon energy obtained by using the ^{11}C source is lower than that emitted from source ^{18}F.
C. In a PET scan, the wavelengths of the photons emitted depend on the body being scanned.
D. In a PET scan, the emitted photons energy will be reduced in time according to the half-life of the source.
E. None of the above.

12. [5%] A one-dimensional gamma camera is made up of only two detectors. The width of each detector is 3 inches. When a gamma ray hits any X_0 point on the surface of the instrument's mica, a light spot is formed on its surface. The spot is approximated by a trapezoidal function with a base width of 4 inches and its height is I_0 as shown in the picture (units are in inches). Given that the weighting matrix is $\mathbf{W}' = [0.5/I_0 \ 1.5/I_0]$, calculate the **error** (in inches) obtained when evaluating the position of a ray hitting at $X_0 = 2$.

A. 0.25
B. 0
C. 0.5
D. 1
E. None of the above

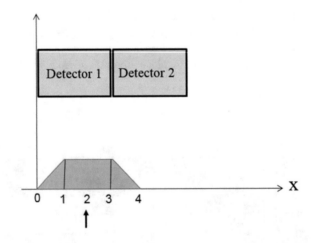

Solutions to Exemplary Exam A

Part 1: First Question

A. The positron emission as function of time:

$$P(t) = P_0 \cdot C_0 \cdot \left(1 - e^{-\frac{t}{\tau}}\right) \cdot S \cdot \left(\frac{3L}{4}\right) \cdot e^{-\frac{3L}{4}\cdot\mu} \cdot e^{-\frac{L}{4}\cdot\mu_0}$$

B. The total radiation that will be recorded during the entire "wash-in" time (4τ seconds):

Let us define all the time-independent factors as $A \triangleq P_0 \cdot C_0 \cdot S \cdot \left(\frac{3L}{4}\right) \cdot e^{-\frac{3L}{4}\cdot\mu} \cdot e^{-\frac{L}{4}\cdot\mu_0}$.

The total recorded radiation is thus given by

$$A \cdot \int_0^{4\tau} \left(1 - e^{-\frac{t}{\tau}}\right) dt = A \cdot \tau \cdot \left(3 + e^{-4}\right)$$

C. The SPECT intensity that will be recorded by the right detector (B) as a function of time:

The contribution made by the right part is $G_{\text{Right}} = G_0 \cdot C(t) \cdot S \cdot$

$$\left(\int_0^{\frac{L}{4}} e^{-\mu\xi}\right) d\xi = G_0 \cdot C(t) \cdot S \cdot \frac{1}{\mu} \cdot \left(1 - e^{-\mu\frac{L}{4}}\right).$$

© Springer Nature Switzerland AG 2020
H. Azhari et al., *From Signals to Image*,
https://doi.org/10.1007/978-3-030-35326-1

Similarly, the contribution made by the left part is $G_{\text{Left}} = G_0 \cdot C(t) \cdot S \cdot \frac{1}{\mu} \cdot$
$\left(1 - e^{-\mu\frac{L}{2}}\right)$.

Accounting for the stone attenuation and the attenuation made by the right part, the overall signal is given by

$$G(t) = G_0 \cdot S \cdot \frac{1}{\mu} \cdot C_0\left[1 - e^{(-t/\tau)}\right] \cdot \left[\left(1 - e^{-\mu\frac{L}{4}}\right) + \left(1 - e^{-\mu\frac{L}{2}}\right) \cdot e^{-\mu_0\frac{L}{4}} \cdot e^{-\mu\frac{L}{4}}\right]$$

D. The x-ray radiation intensity as a function of time is given by

$$I(t) = I_0 \cdot e^{-\mu_0\frac{L}{4}} \cdot e^{-\mu_0\frac{3L}{4}\cdot\left[1 - e^{(-t/\tau)}\right]} \cdot e^{-\mu\frac{3L}{4}}$$

The maximal value during the wash-in is at time $t = 0$ and is given by $I_{\text{Max}} = I_0 \cdot e^{-\mu_0\frac{L}{4}} \cdot e^{-\mu\frac{3L}{4}}$.

Part 1: Second Question

A. The value of $K_{y_{\text{max}}}$ is $\frac{\pi}{\Delta}$.
B. In order to flip the magnetization to a 30° angle by transmitting a pulse given by the sine function, the following relation should be fulfilled:

$$\alpha = \frac{\pi}{6} = \gamma \int_0^{t_1} B \cdot \sin(\pi t/\tau) dt$$

$$\Rightarrow \frac{\pi}{6} = \gamma B \frac{\tau}{\pi}[1 - \cos(\pi t_1/\tau)]$$

The corresponding transmission time can be extracted from this equation to yield

$$t_1 = \frac{\tau}{\pi} \cdot \cos^{-1}\left[1 - \frac{\pi^2}{6 \cdot \gamma \cdot B \cdot \tau}\right]$$

C. The value of Ω:
At the end of the trajectory, the coordinates in K-space are given by $\left\{K_x = \frac{\pi}{\Delta}; K_y = 0\right\}$.

The horizontal coordinate is given by $K_x = \text{Re al}\{\overline{K}(t)\} = t \cdot A \cdot \cos(\Omega t)$ which is equal at time $t = T_2$ to $T_2 \cdot A \cdot \cos(\Omega T_2) = \frac{\pi}{\Delta}$.

It follows therefore that $\Omega = \frac{1}{T_2} \cdot \cos^{-1}\left(\frac{\pi}{A \cdot \Delta \cdot T_2}\right)$.

However, from the other condition, it also follows that
$$\begin{cases} K_y = 0 = T_2 \cdot \sin(\Omega \cdot T_2) \\ \Rightarrow \Omega = \left(\frac{n \cdot \pi}{\Delta \cdot T_2}\right) \end{cases}$$

Thus, both conditions should be fulfilled simultaneously.

D. Given that $\Omega = 10\pi/T_2$, write the explicit expression for each of the gradients:
In order to find the explicit expression for each of the gradients, let us start first by writing the K-space values:

$$\begin{cases} K_x = \text{Re al}\{\overline{K}(t)\} = t \cdot A \cdot \cos\left(\frac{10\pi}{T_2}t\right) \\ K_y = \text{Im}\{\overline{K}(t)\} = -t \cdot A \cdot \sin\left(\frac{10\pi}{T_2}t\right) \end{cases}$$

As recalled the gradient is interconnected to K-space by the following relations:

$$K_i = \gamma \int G_i(t)dt$$
$$\Rightarrow G_i = \frac{1}{\gamma}\frac{\partial K_i}{\partial t}$$

The solution is thus given by

$$\begin{cases} G_x = \frac{A}{\gamma} \cdot \left[\cos\left(\frac{10\pi}{T_2}t\right) - t\frac{10\pi}{T_2}\sin\left(\frac{10\pi}{T_2}t\right)\right] \\ G_y = \frac{A}{\gamma} \cdot \left[-\sin\left(\frac{10\pi}{T_2}t\right) - t\frac{10\pi}{T_2}\cos\left(\frac{10\pi}{T_2}t\right)\right] \end{cases}$$

E. For the trapezoid gradient:
As recalled, $K_x = \gamma \int G_x(t)dt$, and therefore,

$$K_x = \frac{\pi}{\Delta} = \gamma \cdot \Delta t \cdot G_0 + \gamma \cdot (t_0 - \Delta t) \cdot G_0 = \gamma \cdot t_0 \cdot G_0$$

And the required time is given by

$$t_0 = \frac{\pi}{\gamma \cdot \Delta \cdot G_0}$$

Part 2: Multiple-Choice Questions

1. (C) Both signals will have the same intensity.
2. (B) $A_0 \cdot \frac{4Z_0 Z_1}{(Z_0 + Z_1)^2} e^{-\alpha L}$
3. (D) The obtained middle pixel value is different for the two methods.
4. (B) The radiation intensity will be lower than the intensity before entering the phantom but its average energy will be higher than 70 keV.
5. (D) $\frac{1}{2\pi} \cdot \gamma G_0 T$
6. (C) $\frac{1}{2} e^{2\mu T}$
7. (C) $\dfrac{1 + e^{-\mu \frac{T}{2}}}{e^{+\mu D} \left[e^{-\mu \frac{T}{2}} + e^{-\mu T} \right]}$
8. (C) Approximately 20
9. (B) The speed of sound will increase and the acoustic impedance will decrease.
10. (D) The function obtained in the K plane along the K_x-axis is equal for images a and d.
11. (E) None of the above.
12. (B) 0

Medical Imaging: Final Exam B

Duration: 3 hours

You may use only **two** A4-sized pages (both sides each) of equations and your own notes.

Any use of calculators or other electronic equipment is forbidden!

Good Luck

Part A [40%]

1. **[20%]** A patient with three tumors in his leg, and attached to his bone a metal plate is implanted (see figure). For simplicity, it can be assumed that the problem is two-dimensional. The tumors are identical and have a square shape with a size T, and with distance S between them (see also "magnifying glass"), and with an attenuation factor μ_T. The shin is approximately a rectangle with a width of d and an attenuation coefficient μ_L. The bone and metal plate are rectangular with widths B and M and have attenuation coefficients μ_B and μ_P, respectively.

We would like to evaluate the expected signals that will be obtained from the patient's leg by several methods for the scanning field of:

$$0 \le y \le 3S + 2T$$

A. **[5%]** The first scan was performed using x-rays, where the right element (R) transmits with radiation intensity I_0 and the left element (L) is used for reception. What will be the intensity $I(y)$ that will be recorded as a function of location y along the vertical axis for the above-described scanning field?

© Springer Nature Switzerland AG 2020
H. Azhari et al., *From Signals to Image*,
https://doi.org/10.1007/978-3-030-35326-1

B. **[7%]** Next, the patient was sent to a SPECT scan after being injected with radioactive material <u>specific to the tumors</u> that emits gamma radiation with the intensity G_0per unit volume. Assuming that the two detectors are identical and have a unit area, what is the projection $P(y)$ that will be obtained by the L detector for that scanned field?

C. **[5%]** Finally, the patient was sent to a PET scan. This time he was injected with positrons emitting intensity P_0per unit volume. What will be the projection $P(y)$ obtained for the above-described scanning field?

D. **[3%]** If the concentration of the contrast material in the tumors changes in time according to $c(t) = c_0 \cdot t \cdot e^{-t \cdot \lambda}$, what is the optimal time to perform a PET scan?

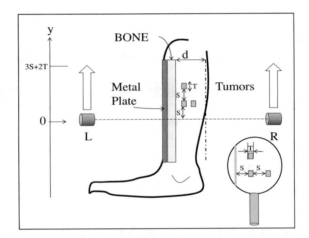

2. **[20%]** We want to perform an MRI scan along the trajectory schematically described in the figure. The required resolution is Δx along the X direction, as well as $\Delta y = 2\Delta x$ along the Y direction. The scanned object is a container of liquid with relaxation times $T_1 \gg T_2$ (i.e., the effect of the spin-lattice relaxation is negligible during the scan).

A. **[3%]** First, we want to bring the sampling point to the upper right corner. It is given that the gradient in the X direction is activated for a time period of τ and changes by $G_X = G_0 \cdot \sin(\pi t/\tau)$. What is the value of τ in order to reach the right side of the K plane?

B. **[3%]** If we assume that τ is known and that the gradient in Y is also activated for a time period of τ and changes by $G_y = G_0 \cdot t$, what is the value of G_0 in order to reach the upper right corner?

C. **[7%]** Once we have reached the upper right corner, we want to move to the upper left corner by a trajectory given by $K_y(t) = A \cdot K_x^2(t)$ (see figure). For this purpose, the gradient $G_X = -G_0$ is applied, meaning that it is a negative

RECT with a height of G_0. What should be the corresponding value of the gradient $G_y(t)$?

D. [4%] If we want to flip the magnetization to an angle of 45° by transmitting a pulse given by $B_1 = B(1 - e^{-t/\tau})$ for a period of τ, what should be the value of τ?

E. [3%] After flipping the magnetization by an initial angle of 45°, what is the M_{xy} magnetization value at the **end** of the trajectory (hint: it is easier to calculate the time for K_X)?

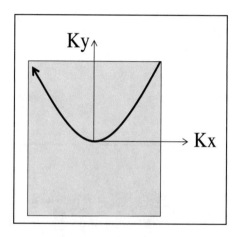

Part B [60%]

Please Mark the Correct Answer

1. [5%] During an MRI scan, we want to flip the magnetization vector to obtain $M_{xy} = \frac{\sqrt{2}}{2} M_0$. In order to flip it, an RF pulse is transmitted and creates a magnetic field $B_1(t) = Ct^2$, $C = const$. How long should the pulse be transmitted in order to obtain the desired flip?

A. $\sqrt{\frac{\pi}{4\gamma C}}$

B. $\frac{\pi}{8\gamma C}$

C. $\sqrt{\frac{\pi}{8\gamma C}}$

D. $\sqrt[3]{\frac{3\pi}{4\gamma C}}$

E. None of the above

2. [5%] An ultrasound wave with an amplitude A_0 is transmitted through a muscle with thickness L and a kidney. At a distance D within the kidney, there is a stone as shown in the figure. The impedance of the muscle is Z_0 and Z_1 for the kidney, and the reflection coefficient of the stone is R. The attenuation coefficient of the muscle and the kidney is α, and there is no energy loss for the reception of the transmitter. What is the amplitude of the wave reflected from the stone?

A. $R \cdot A_0 \cdot \frac{4Z_1^2}{(Z_0+Z_1)^2} \cdot e^{-\alpha(L+D)}$

B. $R \cdot A_0 \cdot \frac{4Z_0Z_1}{(Z_0+Z_1)^2} e^{-2\alpha(L+D)}$

C. $R \cdot A_0 \cdot \frac{4Z_0Z_1}{(Z_0+Z_1)^2} e^{-\alpha(L+D)}$

D. $R \cdot A_0 \cdot e^{-\alpha(L+D)}$

E. None of the above

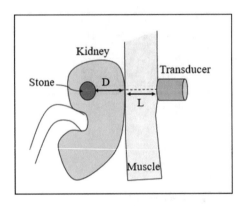

3. [5%] An object consists of nine pixels with the attenuation coefficients values μ as described. If we calculate the projections and apply reconstruction using the ART method of **multiplication** in one iteration according to the trajectory described by the arrows (first columns and then rows), which of the following statements is **incorrect** for the obtained reconstruction?

A. Only five pixels will be exactly reconstructed.
B. The reconstructed pixel at the lower right corner is six times larger than the one at the top left corner (pixels 9 and 1 in the figure).
C. The middle row will be fully reconstructed.
D. The reconstructed pixel at the lower right corner is three times larger than the one at the top right corner (pixels 9 and 3 in the figure).

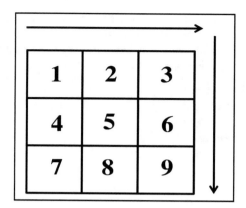

4. [5%] A hand was imaged twice using an x-ray device. The first time a regular image was obtained and the second time a metal plate was placed above the hand (see illustration). The plate attenuates the radiation significantly. Which of the following statements is **correct**?

A. The first image will be brighter than the second one. But the bone-tissue contrast will improve in the second image.

B. The first image will be darker than the second one. But the bone-tissue contrast will improve in the second image.

C. The first image will be brighter than the second one. But the bone-tissue contrast will be worse in the second image.

D. The first image will be darker than the second one. But the bone-tissue contrast will be worse in the second image.

E. None of the above.

5. [5%] In an MRI scanner with a field of B_0, a slice selection gradient is activated along the Z direction with a value of $G_Z = (|B_0|/N)$[Tesla/meter]. If we transmit a field B_1 with ω frequency, what will be the distance of the selected slice Z_1 from the center of the field?

A. $\left(\frac{\omega}{|B_0|\gamma} - 1\right) \cdot N$

B. $\left(1 - \frac{N \cdot \omega}{|B_0|\gamma}\right)$

C. $\left(\frac{\omega}{|B_0|\gamma}\right) \cdot N$

D. $\left(1 - \frac{|B_0|\gamma}{\omega}\right) \cdot N$

E. None of the above

6. [5%] An assembly consists of nine identical square containers arranged as described in the figure. The dark containers were filled with a solid material with an attenuation coefficient μ. The containers marked with letters A, B, C, and D were filled with a homogeneous liquid solution of a positron-emitting material, while the remaining tanks (the blank squares) were filled with water (the water do

not attenuate). The assembly was inserted into a PET scanner and only the horizontal and vertical projections were measured.

If we use ART of multiplication (one iteration clockwise) to reconstruct the image when each pixel represents a container, we will find that:

A. Pixel B is the strongest
B. Pixel C is the strongest
C. Pixel A is the strongest
D. Pixel D is the strongest
E. All of the pixels above are the same
F. None of the above

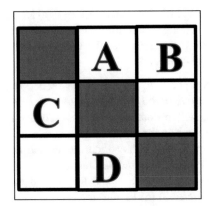

7. [5%] For the development of a SPECT scanner, an experiment of an ex vivo animal brain was performed as described in the figure. Three emitting point sources were installed (marked by the stars). They emitted gamma radiation with initial intensities and locations as indicated in the figure. The thickness of each side of the brain is T and the attenuation coefficient is uniform and its value is μ. It is known that the half-life of the material in the center is 2τ minutes, whereas for the material on both sides, it is only τ minutes. What is the signal intensity expected to be measured at time 2τ in detector A?

A. $I_0\left(4e^{-\mu\frac{T}{2}} + 2e^{-\mu T} + 4e^{-\mu 2T}\right)$

B. $I_0\left(e^{-\mu\frac{T}{2}} + e^{-\mu T} + e^{-\mu 2T}\right)$

C. $I_0\left(2e^{-\mu\frac{T}{2}} + e^{-\mu T} + 2e^{-\mu 2T}\right)$

D. None of the above

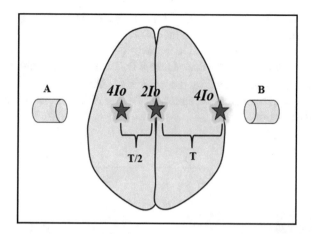

8. [5%] In an x-ray CT scanner, the detectors are arranged on a straight mount. The size of each detector is D. For radiation safety reasons, only 100 projections can be collected in this scanner. When applying reconstruction using filtered back projection, in order to obtain an image without artifacts (such as streaks), the size of the imaging window (reconstruction circle) is limited to:

A. There is no relation between the size of the imaging window and the appearance of artifacts
B. $\pi \cdot 100 \cdot d/2$
C. $200 \cdot d/\pi$
D. $100 \cdot d/\pi$
E. None of the above

9. [5%] Which of the following statements is **incorrect**?

A. In ultrasound, the amplitude of the through-transmitted wave from one medium to another can be bigger than that of the impinging wave.
B. In ultrasound imaging, the TGC of the same object should have larger values as the frequency increases.
C. The velocity of the ultrasound waves is not wavelength dependent.
D. Sound waves do not progress in vacuum.
E. Changing the density of the material will change both the speed of sound and the acoustic impedance.
F. One of the above answers is incorrect.

10. [5%] For the objects described in Figs. **A–D**, the length of each rectangular side and the diameter of the circle equal $2a$. The side of the hole in Fig. C is in the length of a. The projection on the X-axis was measured.
Please mark the **incorrect** sentence.

A. The ratio of the middle value in the projection of C divided by that of B is $1/\sqrt{2}$.

B. The ratio of the middle value in the projection of B divided by that of D is the same as the ratio of B to that of A.

C. The ratio of the middle value in the projection of B divided by that of C is double the ratio of D to that of A and D. The ratio of the middle value in the projection of A divided by that of B is $1/\sqrt{2}$.

D. In K-space, the values along the K_x line for object A can be described by a SINC function.

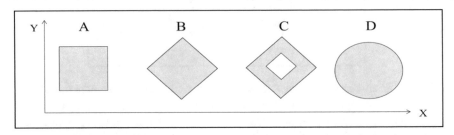

11. [5%] Two PET scans were carried out on the same patient (at different times), one with a substance containing O^{15} and the other with a substance containing F^{18}. (The number indicates the atomic number of the material.) The substances' half-lives are different! Which of the following statements is true?

A. The wavelength received at the first scan will be longer than the second scan according to the ratio 18/15.

B. The wavelength received on the first scan will be shorter than the second scan according to the ratio 15/18.

C. The wavelength of the material with the shorter half-life will be shorter.

D. The wavelength of the material with the shorter half-life will be longer.

E. The wavelengths received in both scans will be equal.

12. [5%] An Anger camera consists of three $2a$-wide PMT tubes. Their distance from the crystal is a. (Assume that the problem is one-dimensional.) It is given that the signals from tubes 1 and 3 are I_1 and the center signal is $I_2 = 6I_1$. If the weight function for determining the position (X) is $W = [1, \beta, 1] \cdot \left(\frac{a}{I_1}\right)$, what is the value of β if given that the error in the position estimation of the described event is 0 (please simplify the expression as much as possible)?

A. 1

B. 1/3

C. 1/2

D. 1/6

E. None of the above

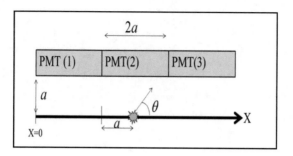

Solutions to Exemplary Exam B

Part 1: First Question

A. For the x-ray scan

Let us divide the scan along axis Y into three distinct regions:

Region (i) – The first region relates to locations where there are no tumors and corresponds to $0 \le y \le S; S + T \le y \le 2S + T; 2(S + T) \le y \le 3S + 2T$

In this region, the x-rays will be attenuated only by the tissues and the metal plate. Thus, the intensity will be

$$I(y) = I_0 \cdot e^{-M \cdot \mu_P} \cdot e^{-B \cdot \mu_B} \cdot e^{-d \cdot \mu_L}$$

Region (ii) – The second region relates to locations where there are two tumors and corresponds to

$$S \le y \le S + T$$

In this region, the x-rays will be attenuated also by the tumors; thus, the intensity will be

$$I(y) = I_0 \cdot e^{-M \cdot \mu_P} \cdot e^{-B \cdot \mu_B} \cdot e^{-(d-2T) \cdot \mu_L} \cdot e^{-2T \cdot \mu_T}$$

Region (iii) – The third region relates to locations where there is only one tumor and corresponds to

$$2S + T \leq y \leq 2(S+T)$$

In this region, the x-rays will be attenuated also by the single tumor; thus, the intensity will be

$$I(y) = I_0 \cdot e^{-M \cdot \mu_P} \cdot e^{-B \cdot \mu_B} \cdot e^{-(d-T) \cdot \mu_L} \cdot e^{-T \cdot \mu_T}$$

B. For the SPECT scan

Region (i) – There will be no emitted radiation along the horizontal line in this region; thus, $P(y) = 0$.

Region (ii) – In this region each tumor will emit radiation but the rays will sense different attenuation:

$$P(y) = G_0 \cdot e^{-M \cdot \mu_P} \cdot e^{-B \cdot \mu_B} \cdot e^{-S \cdot \mu_L} \left[\int_0^T e^{-\mu_T \xi} d\xi \right] + G_0 \cdot e^{-M \cdot \mu_P} \cdot e^{-B \cdot \mu_B} \cdot e^{-2S \cdot \mu_L}$$

$$\cdot e^{-T \cdot \mu_T} \left[\int_0^T e^{-\mu_T \xi} d\xi \right]$$

After integration this becomes

$$P(y) = G_0 \cdot e^{-M \cdot \mu_P} \cdot e^{-B \cdot \mu_B} \cdot e^{-S \cdot \mu_L} \cdot \frac{1}{\mu_T} \left(1 - e^{-T \cdot \mu_T}\right) + G_0 \cdot e^{-M \cdot \mu_P} \cdot e^{-B \cdot \mu_B} \cdot e^{-2S \cdot \mu_L}$$

$$\cdot e^{-T \cdot \mu_T} \frac{1}{\mu_T} \left(1 - e^{-T \cdot \mu_T}\right)$$

which after simplification becomes

$$P(y) = G_0 \cdot e^{-M \cdot \mu_P} \cdot e^{-B \cdot \mu_B} \cdot e^{-S \cdot \mu_L} \cdot \frac{1}{\mu_T}$$

$$\times \left[\left(1 - e^{-T \cdot \mu_T}\right) + e^{-S \cdot \mu_L} \cdot e^{-T \cdot \mu_T} \cdot \left(1 - e^{-T \cdot \mu_T}\right) \right]$$

Region (iii) – For the third region:

$$P(y) = G_0 \cdot e^{-M \cdot \mu_P} \cdot e^{-B \cdot \mu_B} \cdot e^{-S \cdot \mu_L} \cdot \frac{1}{\mu_T} \left(1 - e^{-T \cdot \mu_T}\right)$$

C. For the PET scan

Region (i) – There will be no emitted radiation along the horizontal line in this region; thus, $P(y) = 0$.

Region (ii) – In this region the two tumors will emit radiation; thus:

$$P(y) = P_0 \cdot 2T \cdot e^{-M \cdot \mu_P} \cdot e^{-B \cdot \mu_B} \cdot e^{-(d-2T) \cdot \mu_L} \cdot e^{-2T \cdot \mu_T}$$

Region (iii) – For the third region, only one tumor is accounted for:

$$P(y) = P_0 \cdot T \cdot e^{-M \cdot \mu_P} \cdot e^{-B \cdot \mu_B} \cdot e^{-(d-T) \cdot \mu_L} \cdot e^{-T \cdot \mu_T}$$

D. The optimal time to perform a PET scan

– The maximal signal will be when the concentration of the radiopharmaceutical will be maximal. In order to find that time point, we should first calculate the temporal derivative of the wash-in-wash-out curve, i.e.:

$$\frac{dc(t)}{dt} = c_0 \cdot e^{-t \cdot \lambda} - c_0 \cdot t \cdot \lambda \cdot e^{-t \cdot \lambda}$$

– Equating the derivative to zero and solving for t yield

$$t = \frac{1}{\lambda}$$

Part 1: Second Question

A. Moving the sampling point to the upper right corner (solving for transmission time):

At the upper right corner of the K-space, the coordinates are $\left[K_{x_{max}}, K_{y_{max}}\right]$. And the value of $K_{x_{max}}$ is $\frac{\pi}{\Delta x}$. It follows then that the following relation should hold:

$$K_{x_{max}} = \frac{\pi}{\Delta x} = \gamma \int_0^\tau G_0 \cdot \sin\left(\frac{\pi t}{\tau}\right) dt = \frac{\gamma \cdot G_0 \cdot 2\tau}{\pi}$$

– And the transmission time is therefore

$$\tau = \frac{\pi^2}{\Delta x \cdot \gamma \cdot G_0 \cdot 2}$$

B. Moving the sampling point to the upper right corner (solving for the gradient): Similar to the previous question, the following relation should hold:

$$K_{y_{max}} = \frac{\pi}{2\Delta x} = \gamma \int_0^{\tau} G_0 \cdot t \cdot dt = \frac{\gamma \cdot G_0 \cdot \tau^2}{2}$$

- And therefore the gradient value is given by

$$G_0 = \frac{\pi}{\gamma \cdot \Delta x \cdot \tau^2}$$

C. Moving the sampling point to the upper left corner: The K_x coordinate is given by

$$K_x = -\gamma G_0 \cdot t$$

- Thus, the K_y coordinate is given by

$$K_y(t) = A \cdot K_x(t)^2 = A \cdot (\gamma G_0 \cdot t)^2$$

- As recalled the gradient is interconnected to K-space by the following relations:

$$K_i = \gamma \int G_i(t)dt$$
$$\Rightarrow G_i = \frac{1}{\gamma} \frac{\partial K_i}{\partial t}$$

- Therefore,

$$G_y(t) = 2A \cdot \gamma \cdot t(G_0)^2$$

D. Calculating the transmission time for the required flip angle: As recalled,

$$\alpha = \gamma \int_0^\tau B_1(t)\,dt$$

$$\Rightarrow \frac{\pi}{4} = \gamma \int_0^\tau B \cdot \left(1 - e^{-t/\tau}\right) dt$$

– Solving for τ yields

$$\tau = \frac{\pi}{4\gamma \cdot B} \cdot e^1$$

Part 2: Multiple-Choice Questions

1. (D) $\sqrt[3]{\frac{3\pi}{4\gamma C}}$
2. (B) $R \cdot A_0 \cdot \frac{4Z_0 Z_1}{(Z_0 + Z_1)^2} e^{-2\alpha(L+D)}$
3. (D) The reconstructed pixel at the lower right corner is three times larger than the one at the top right corner (pixels 9 and 3 in the figure).
4. (D) The first image will be darker than the second one. But the bone-tissue contrast will be worse in the second image.
5. (A) $\left(\frac{\omega}{|B_0|\gamma} - 1\right) \cdot N$
6. (C) Pixel A is the strongest.
7. (B) $I_0 \left(e^{-\mu\frac{T}{2}} + e^{-\mu T} + e^{-\mu 2T}\right)$
8. (C) $200 \cdot d/\pi$
9. (E) One of the above answers is incorrect.
10. (A) The ratio of the middle value in the projection of A divided by that of B is $1/\sqrt{2}$.
11. (E) The wavelengths received in both scans will be equal.
12. (D) 1/6

Index

A

Acoustic impedance, 326–329, 332, 338, 360, 433, 435, 439, 447, 452, 459

Adiabatic-pulse, 270, 271

ALARA, 149, 151, 387, 394

Algebraic reconstruction tomography (ART)
projection differences, 58–63
projection ratio, 63, 64, 83

A-line, 341–344, 346–350, 362

Alpha (α), 82, 83, 91, 92, 103, 161, 162, 165–167, 169, 174, 325, 326, 340, 341, 345, 352, 424, 430, 439, 444, 450, 452, 456, 467

Alpha-particles (α-particles), 146, 161, 165

A-mode, 4, 341–347

Annihilation photons, 162, 163, 165–167, 217, 219, 223, 224, 227–229, 231, 233, 236, 239, 240, 244, 247, 420, 421

Area under the curve (AUC), 41, 42, 365

Artifacts, 4, 12, 33–36, 58, 61, 76, 77, 112, 126, 128, 129, 136–138, 180, 181, 192, 198–201, 313, 365, 377, 386, 433, 447, 459

Attenuation, 4, 65, 79, 80, 95, 103–106, 109, 110, 112, 113, 115–117, 120–122, 124, 130, 133, 134, 141, 144, 147, 152, 154–156, 168, 192, 198–203, 220, 221, 231, 234–240, 247, 325–327, 340, 345, 352, 353, 374, 375, 379, 393–395, 400, 403, 405, 409, 410, 414, 415, 434, 435, 440–442, 444, 446, 450, 453, 456–458, 464

Attenuation correction, 198–203, 231, 234–240

Axial, 42, 78, 120, 125–127, 131, 134–139, 143, 148, 185, 189, 190, 194, 195, 197, 198, 213, 223–226, 228, 231, 240, 247, 249, 282, 294, 305, 337, 339, 387, 410, 417

Axial resolution, 337, 339

B

B_0, 260, 262–265, 267–269, 273–276, 279, 280, 282, 302, 424, 430, 445, 457, 467

B_1, 267–273, 276, 283, 285, 286, 293, 296, 306, 423, 430, 442, 443, 455, 457, 467

Back projection, 5, 57–58, 60–63, 74, 75, 85, 87, 199, 200

Barium, 1, 31, 121, 122, 149, 387

Bayesian, 47, 90–92, 205, 236, 248

Beam forming, 336, 362, 363

Beam hardening, 99, 395

Becquerel (Bq), 3, 163, 164, 206, 231, 241, 410, 418

Bessel, 333

Beta-particles (β-particles), 146, 162, 163, 165, 166, 175

Blank scans, 231, 244

Bloch equations, 276–277, 286, 297, 299, 304

Blur, 102, 110, 203, 219, 226, 227, 239, 248, 374

Bremsstrahlung, 100–101, 161, 162, 165, 166, 175–176

B-scan, 4, 347–351, 354, 355, 357–360, 362, 433, 435, 436

Bulk modulus, 325

© Springer Nature Switzerland AG 2020
H. Azhari et al., *From Signals to Image*,
https://doi.org/10.1007/978-3-030-35326-1

C

Cameras, Anger, 3, 180–183, 185, 187–189,
   193, 203, 204, 207, 208, 223,
   399, 460
Characteristic radiation, 100–102, 108, 114
Coincidence detection, 3, 162, 217, 219, 220,
   223, 226–229, 237
Collimation, 110, 180–181, 212, 213, 217, 220,
   225–228
Color flow mapping (CFM), 357–359
Colormap, 142
Compressibility, 325
Compton scatter, 101, 103, 104, 106, 155, 165,
   181, 182, 184, 198, 203
Computed radiography (CR), 107–110,
   118, 119
Cone-beam, 118, 129
Contrast
   agents, 95, 113, 119–123, 139, 140, 142,
      154–156, 387, 442
   Michelson, 32
   Weber, 31, 32
Contrast enhancing materials (CEM), 314–315,
   317, 329, 360
Contrast to noise ratio (CNR), 36–38,
   115, 365
Coronal, 42, 51, 134, 143, 206, 246, 248, 282,
   387, 393
CTDI$_{vol}$, 148, 394
Curie (Ci), 163
Cyclotron, 3, 178, 179, 218

D

Decibel (dB), 36, 326, 360
Delta function, 16, 55, 67, 284, 341
Density, 38–40, 91, 95, 99, 105, 111, 116, 117,
   128, 153–156, 180, 182, 198, 205, 218,
   239, 241, 262, 263, 277, 278, 286, 297,
   304, 305, 324, 327, 367, 369, 370, 374,
   407, 409, 427, 433, 443, 447, 459
Detectability, 37–41, 115, 236
Diamagnetic, 259, 427
Digital radiography (DR), 107–110, 114,
   118, 119
Digital subtraction angiography
   (DSA), 121, 122
Directivity function, 333, 337
DLP, 148, 388, 394
Doppler, 4, 7, 11, 12, 322, 330–331, 354–359,
   361, 433, 434, 436, 439
   imaging, 7, 354–359

shift, 11, 330–331, 356, 357, 433, 439
Dose, 47, 92, 99, 100, 114–119, 121, 131, 139,
   145–149, 151, 152, 163, 167, 168, 174,
   190, 196, 205, 209, 210, 242, 243, 387,
   388, 393–395, 397, 413
Dual energy, 108, 153–155
Dynamic range, 12–16, 38, 107, 110, 119, 141,
   195, 206, 343, 360

E

Echo, 4, 5, 7, 11, 279, 293, 301–311, 328, 330,
   337–347, 349, 354, 355, 359, 361–363,
   423, 427, 433–435, 439
Echo planar imaging (EPI), 5, 307–310, 423,
   427
Effective field, 269–271
Elastography, 4, 323, 324, 361–362
Electrocardiogram (ECG), 23, 24, 138, 139,
   191, 348
Electron capture (EC), 161–163, 174, 218
Electronvolts, 96
Emission imaging, 160, 163, 170, 171, 220
Energy window, 184, 202–204, 221, 225, 239,
   407, 415
Expectation-maximization (EM), 47, 86, 88,
   89, 91, 92, 235, 236

F

False negative, 40, 367
False positive, 39, 367
Fan beam, 78–80, 126, 128, 135, 351, 353
Faraday, 254, 271
Far field, 334
Fast spin-echo, 306, 307
Ferromagnetic, 259, 427
Field of view (FOV), 18, 50, 131, 133, 135,
   149, 194, 197, 206, 207, 224–229, 231,
   232, 240, 244, 247, 249, 291, 292,
   348–350, 354, 399, 410
Filtered back-projection (FBP), 47, 55, 66,
   69–76, 79, 82, 92, 152, 197, 200, 201,
   204, 205, 225, 228, 235, 238, 374, 377,
   384, 385, 459
Fission, 177
Flip angle, 268, 270, 272, 274, 275, 281, 293,
   294, 296–298, 306, 311, 423,
   425, 430, 466
Fluid-attenuated inversion recovery (FLAIR),
   298, 300, 301

Fluorodeoxyglucose (^{18}F-FDG), 210, 218, 219, 221, 228, 239, 241, 242, 244–247, 409, 410
Focal point, 98, 110–113, 129, 136, 192–194, 334–337, 361, 362, 391, 397
Fourier, 5, 23, 25, 26, 29, 46, 49, 67–70, 74, 205, 228, 236, 284, 285, 287–290, 296, 312, 356, 411, 418, 419
Free induction decay (FID), 273, 310, 311
Full width at half maximum (FWHM), 18, 20, 39, 203, 204, 206, 208, 240, 365
Functional MRI (fMRI), 5, 37, 254

G
Gadolinium, 31, 200, 314, 315
Gamma, 3–4, 7, 35, 96, 163, 167, 169, 172–175, 178, 180, 184–190, 193, 194, 200, 203, 211, 212, 214, 217, 220, 223, 235, 245, 399, 400, 442, 446, 448, 454, 458
    cameras, vii, 3–4, 7, 172, 173, 180, 184, 188–190, 194, 211, 212, 214, 217, 220, 223, 235, 245, 400, 448
    constant, 169
    rays, 35, 96, 163, 165, 167, 175, 178, 180, 185–187, 194, 203, 399, 442, 446, 448, 454, 458
Gating, 23, 24, 138, 139, 190–192, 247, 365
Gauss, 256, 257
Generator, 177, 178, 218, 256, 271, 401
Gold standard, 33, 39, 40
Gradient
    echo, 293–295, 297, 300, 303, 304, 306, 307, 309, 313
    field, 279–282
Gray level, 12–17, 25, 29–32, 34, 35, 37, 50, 58, 62, 106, 124, 141, 142, 144, 195, 277, 279, 342–344, 346, 347, 388

H
Half-life, 161, 164, 165, 167, 172–174, 177, 196, 218, 222, 243, 399, 400, 404, 405, 407, 414–417, 445, 448, 458, 460
Half-value layer (HVL), 165–167
Hilbert transform, 342, 343
Hounsfield units (HU), 124, 133, 134, 140–142, 201, 202, 236, 389, 394–396
Hybrid, 4, 6, 192–193, 201, 206–207, 221, 233–235, 243–244, 246, 247
Hydrogen, 7, 102, 121, 255, 259–263

I
Intensity, 12, 22, 26, 27, 34, 38, 41, 50, 97–99, 107, 110, 112, 113, 132, 136, 137, 142, 144, 147, 150–152, 155, 165, 166, 171, 173, 174, 186, 193, 196–200, 222, 232, 237, 242, 243, 317, 325–327, 330, 361, 362, 389, 395, 400, 441–443, 445, 446, 450, 452–454, 458, 463, 464
Inversion recovery, 298–301
Iodine, 3, 31, 120–122, 139, 140, 143, 144, 154, 155, 161, 163, 169, 174, 176, 387, 393
Isotopes, 160, 161, 163, 165, 167, 168, 172–175, 200, 207, 218, 221, 401, 407, 409
Iterative reconstruction, 5, 11, 66, 89, 91, 152, 197–199, 203–205, 236, 238, 240, 248

K
Kilovolts peak (kVp), 100, 113, 114, 131, 147, 152, 153, 388, 389, 394
K-space, 26–29, 67–69, 75–77, 285–297, 303, 307, 309, 310, 312, 313, 372–374, 424, 425, 427, 431, 432, 434, 439, 450, 451, 460, 465, 466

L
Larmor, 265, 266, 268, 269, 272, 280, 282–287, 296, 302, 303
Lateral resolution, 334–338
Least squares, 46, 83–86, 186
Likelihood, 46, 86–88, 91, 92, 148, 182, 205
Line of response (LOR), 220, 225–227, 229–231, 236–238, 240, 410
Line spread function (LSF), 16–18

M
Magnetic resonance angiography (MRA), 313–318
Magnetic resonance imaging (MRI), vii, 5, 7, 13, 15, 20, 22, 31, 32, 35–37, 48, 92, 196, 233, 247, 248, 253–318, 367, 423–425, 434, 442, 443, 445, 454, 455, 457
Magnetic resonance spectroscopy (MRS), 255
Magnetization, 258–266, 268–279, 285, 286, 298–303, 310–311, 316, 423, 425, 433, 442, 443, 450, 455
Main lobe, 333, 334, 337

Maximal intensity projection (MIP), 142, 144, 222, 246, 248, 315, 317
Maximum likelihood expectation maximization (MLEM), 86–90
M-mode, 345–348, 356, 435
Modulation transfer function (MTF), 29, 30, 365

N
Near field, 333, 334
Neutrons, 161, 163, 166, 177, 179
Noise
  Gaussian, 34, 35, 85
  speckle, 35, 363
  white, 34, 35, 366
Noise equivalent count rate (NECR), 232–233, 245
Nuclear magnetic resonance (NMR), 5, 254–256, 265

O
Ordered subsets EM (OSEM), 47, 89–90, 197, 235, 236, 238

P
Paramagnetic, 259, 314, 427
Penumbra, 111, 112, 391, 393, 396, 397
Phantoms, 19, 20, 33, 34, 36, 75, 76, 88, 132, 147, 186, 189, 196, 204, 206–209, 231, 232, 241, 244, 245, 346, 365, 366, 377, 386, 388–391, 399–401, 407, 409, 410, 414, 436, 445, 452
  Shepp-Logan numerical phantom, 33, 34, 36, 75, 76, 88, 366, 377
Phase contrast, 6, 317–318
Phased array, 4, 335–337, 349, 350, 362
Photoelectric effect, 101, 102, 104, 106, 108, 109, 155, 165, 181, 182, 198
Photoelectron, 102, 108, 183
Photomultiplier tubes (PMTs), 180, 182–188, 194, 195, 213, 217, 224, 226, 233, 240, 247, 248, 460
Photon counting, 38–40, 74, 84, 86, 89, 154–156, 230, 233, 367, 369, 370
Photon pair, 98, 116, 124, 219, 220, 224, 225, 227–229, 238, 249
Photo peak, 175
Piezoelectric, 11, 331–332, 336, 353
Pixel, 11, 12, 14, 16, 20–22, 26, 27, 30, 33, 35, 36, 39, 49–51, 57–60, 62–66, 69, 74–76,
78, 80, 81, 83, 85, 87, 92, 106–110, 115, 124, 126, 133–135, 141, 142, 144, 153, 154, 186, 188, 195–198, 200, 201, 203, 205–208, 243, 291, 292, 297, 343, 344, 346, 347, 356–359, 366, 367, 369, 375–377, 379–381, 385, 425, 444, 447, 452, 456, 458, 467
Point spread function (PSF), 16–18, 29, 205, 236, 240, 336–338, 365
Positron, vii, 3, 7, 48, 104, 160, 162, 163, 170, 173, 180, 217–249, 375, 410, 411, 415, 419–421, 441, 445, 449, 454, 457
Positron emission tomography (PET), 3–4, 7, 48, 63, 74, 86, 89, 92, 160–162, 166, 167, 170, 180, 187, 192, 193, 195–198, 203, 205, 210, 212, 217–249, 375, 401, 407, 409, 410, 414–417, 441, 445, 447, 448, 454, 458, 460, 464, 465
Power Doppler, 359
Precession, 264–266, 269, 270
Projection, 5, 36, 47–87, 89, 90, 95, 96, 106–124, 126, 129, 131, 134, 136, 139, 142, 144, 146–148, 152, 180, 186, 195, 197, 199–201, 203–205, 212–214, 222, 223, 226, 227, 235, 239, 240, 247, 274, 315, 351–354, 371, 374–377, 379–381, 383–386, 396, 401, 409, 410, 414, 418, 419, 444, 447, 454, 456, 458–460, 467
Proton density (PD), 262, 263, 277–279, 286, 297, 304, 305, 427, 443
Pulse height analysis, 184
Pulse sequence, 277, 293–312

R
Radiation, 6, 39, 47, 92, 95–102, 107, 108, 110, 111, 113–119, 121, 129, 131, 136, 138, 139, 143–149, 151, 152, 159, 160, 162, 163, 165–170, 172–177, 193, 203, 207, 209, 210, 214, 247, 361, 362, 389, 393, 397, 401, 406, 441, 442, 445, 446, 449, 450, 452–454, 457–459, 464, 465
Radiofrequency (RF), 254, 255, 267, 269, 270, 272–279, 281–286, 293, 309–313, 316, 342, 343, 423, 424, 455
Radionuclides, 161, 164, 209, 219, 236
Radon transform, 46, 52–57, 67, 69, 78, 79, 372, 374, 377
Random events, 226
Rayleigh scatter, 101–103, 106
Receiver operator curve (ROC), 39–42
Reflection, 7, 223, 328–330, 332, 337, 340, 344, 360, 434, 439, 456

Refraction, 328–330

Region of interests (ROIs), 30–32, 37, 40, 121, 141, 142, 194, 197, 206, 214, 242, 247, 249, 349, 354–360, 362, 366

Repetition time/time to repeat (TR), 293–298, 303–308, 310, 311, 427

Resonance frequency, 255, 265, 269, 271, 273, 276

Root mean square (RMS), 32, 66, 219

Rotating reference frame, 266–270, 276, 279, 287, 293, 302, 306, 317

Row action maximum likelihood algorithm (RAMLA), 92, 236

S

Safety, 143–152, 167–170, 257, 327, 459

Sagittal, 42, 134, 143, 222, 246, 282, 393

Scatter, 6, 103, 104, 106, 133, 182, 184, 198, 199, 201–204, 211, 214, 225–227, 231–233, 236, 239–240, 245, 338, 415

Scintillating crystal, 180–184, 186, 187, 194, 203, 211, 217, 223, 224, 233, 410, 417

Semiconductor, 187, 188

Sensitivity, 11, 12, 28, 41, 42, 107, 110, 112, 132, 146, 180–182, 190, 196, 197, 206, 211, 212, 214, 217, 223, 234, 245, 247, 249, 312, 367, 370, 400, 403, 405, 406

Septa, 9, 10, 180–182, 210, 224–226, 228, 233, 406

Shear modulus, 324

Side lobe, 333

Signal to noise ratio (SNR), 36–37, 122, 129, 150, 236, 263, 312, 322, 351, 388, 427

Slice selection, 282–285, 294, 303, 311, 312, 424, 445, 457

Silicon photomultipliers (SiPMs), 187, 233, 234

Single-photon emission computed tomography (SPECT), 3–4, 7, 9, 10, 46, 48, 51, 56, 63, 65, 74, 86, 89, 92, 159–214, 217, 222, 223, 226, 227, 233, 236, 239, 240, 243, 247, 374, 379, 401, 407, 409, 441, 442, 446, 449, 454, 458, 464

Sinogram, 52–57, 69, 70, 73–75, 78–80, 83, 90, 126, 130, 197–200, 226–228, 236, 238, 372, 379, 401, 407

Snell's law, 328

Spatial frequencies, 23–29, 74, 76, 197, 287, 288, 293, 431

Spatial resolution, 17–22, 38–40, 98, 109–111, 119, 129, 180–182, 188, 192, 201, 203–205, 208, 211–214, 219, 223, 240, 245, 247, 254, 291, 296, 354, 362, 363, 365, 366, 380, 407, 416, 424, 425, 435, 440

Specificity, 41, 42, 367, 370

Speed of sound, 324–325, 327–328, 330, 341, 344, 349, 352, 353, 358, 362, 363, 434, 436, 447, 452, 459

Spin echo, 248, 301–310, 423, 427

Spin-lattice relaxation, 273, 274, 278, 279, 298–300, 427, 442, 454

Spin-spin relaxation, 273–276, 279, 298, 299, 302, 427

Spiral CT, 78, 79

Standardized uptake value (SUV), 241–243, 245

Steady state, 310–311

Susceptibility, 145, 258–259, 423

T

T_1, 273–279, 297–301, 305, 310, 425, 427, 440, 442, 443, 454

T_2, 274–276, 279, 285, 297, 298, 302, 304, 305, 307, 310, 315, 425, 427, 433, 442, 443, 451, 454

TE, 278, 279, 293, 297, 298, 304, 306, 310, 427, 443

Temporal resolution, 22–24, 127, 190, 230, 231, 247, 363, 365, 410, 416

Tesla, 233, 256, 262, 275, 287, 424, 445, 457

Three-dimensional (3D), 7, 8, 21, 29, 45, 46, 49, 50, 57, 78, 124, 130, 137, 144, 162, 180, 189, 194, 217, 224, 225, 228–230, 235, 236, 270, 282, 283, 287, 289, 290, 312–313, 315, 317, 322, 350–351, 354, 357, 393, 416

Time gain compensation (TGC), 342, 344–347, 433, 459

Time of flight (TOF), 229–231, 315–316, 352

Time-of-flight acquisitions, 229–231

Tomographic slice theorem, 57, 66–69, 79, 200, 227

Tomography, vii, 2–4, 6, 45–92, 95–156, 170, 180, 189, 190, 195–197, 205, 213, 217–249, 352–354, 396

Torque, 263–265

Transducer, 4, 11, 324, 329–338, 340, 341, 344–355, 357, 359, 362, 433, 436, 439, 440, 444
Transmission imaging, 170, 220, 351–354
T_1 weighted, 298, 299, 305, 314
T_2 weighted, 305, 315
Two-dimensional (2D), 8, 21, 25–29, 49, 50, 56, 57, 63, 65, 67–70, 81, 82, 95, 115, 117, 125, 130, 135, 142, 187, 194, 198, 224–229, 233, 236, 286, 288, 290, 312, 315, 317, 322, 341, 345, 347–350, 357–359, 366, 411, 418, 436, 453

U
Ultrasonic computed tomography (UCT), 4, 351–354
Ultrasound, 4, 7, 11, 31, 48, 123, 321–363, 433, 434, 436, 444, 447, 456, 459

V
Voxel, 20–22, 78, 79, 82, 85, 86, 92, 124, 125, 134, 142, 198, 206, 222, 230, 231, 236, 240–242, 245, 260, 261, 263–265, 272–280, 286, 299, 302, 314, 315, 317, 365, 393, 410, 416, 418

W
Wave
  cylindrical, 324
  longitudinal, 323, 325, 361
  planar, 324, 326, 328, 363
  shear, 323, 324, 361, 362
  spherical, 324
Weight, 63–66, 79, 83–85, 92, 146, 160, 184, 186, 196, 241, 242, 259, 305, 460
Well counter calibration, 231, 232

X
X-rays, vii, 1–3, 6, 7, 12, 20, 29, 31, 35, 36, 38–41, 46–50, 53, 56, 57, 78–80, 86, 95–156, 159–163, 165–167, 170, 174, 189, 193, 194, 196, 232, 235, 247, 321, 325, 353, 375, 387, 390, 391, 393, 394, 442, 445, 450, 453, 457, 459, 463, 464
X-ray tube, 2, 79, 80, 96–100, 110, 115–118, 120, 123, 127–129, 135, 160, 167, 170

Y
Young's modulus, 324, 362